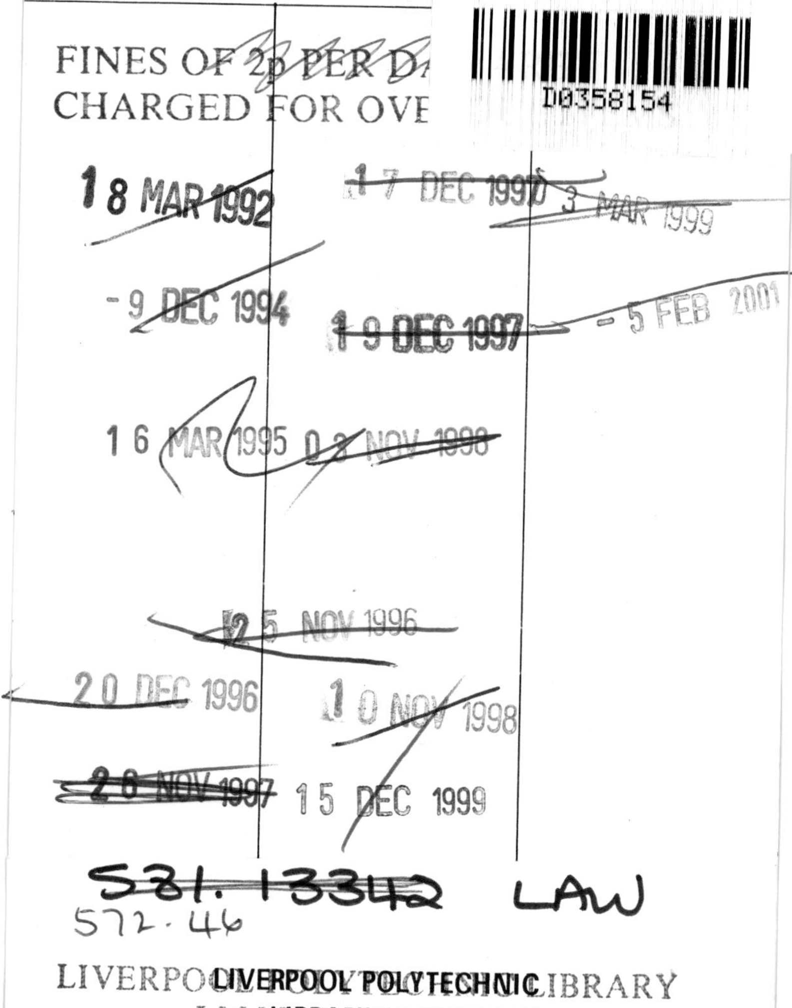

D.W. Lawlor

Photosynthesis: Metabolism, Control, and Physiology

Copublished in the United States with
John Wiley & Sons, Inc., New York

Longman Scientific & Technical,
Longman Group Limited
Longman House, Burnt Mill, Harlow
Essex CM20 2JE, England
Associated companies throughout the world

Copublished in the United States with
John Wiley & Sons, Inc., 605 Third Avenue, New York, NY 10158

First published 1987

British Library Cataloguing in Publication Data
Lawlor, D.W.
Photosynthesis: metabolism, control and physiology.
1. Photosynthesis
I. Title
581.1'3342 QK882

ISBN 0-582-44633-3

Library of Congress Cataloging-in-Publication Data
Lawlor, David, 1941-
Photosynthesis: metabolism, control, and physiology.
Includes bibliographies and index.
1. Photosynthesis. I. Title.
QK882.L36 1987 581.1'3342 85-15888
ISBN 0-470-206810 (USA Only)

Produced by Longman Singapore Publishers (Pte) Ltd.
Printed in Singapore.

Contents

Preface

This book provides a simplified description of the partial processes of photosynthesis at the molecular, organelle, cell and organ levels of organization in plants, which contribute to the complete process. It considers how photophysics and biochemistry determine the physiological characteristics of plants and production of plant dry matter. The text links the fundamentals of light capture by pigment molecules to the generation of high energy organic molecules and their consumption in carbon dioxide, nitrate and sulphate reduction. The mechanisms are related to the structure and function of the leaf and to control of energy and material fluxes. Photosynthesis in leaves is analysed as the resultant of light activation, biochemical demand, and the supply of CO_2. Photosynthetic processes are related to plant environment, sun and shade leaves for example, and C4 and CAM photosynthesis are analysed as ecophysiological variants of a basic process. Leaf photosynthesis is put into context as one, albeit the major, determinant of productivity of vegetation. Plant productivity is the result of the interaction of many sub-systems, all driven by the primary energy capture. Photosynthesis by the whole plant is to be seen as a series of balanced but dynamic interactions between individual molecular and physiological mechanisms.

The text is intended for undergraduate and graduate study in plant biology courses and for non-specialists in other disciplines who wish to understand photosynthetic mechanisms and control. The approach is qualitative, but there is some emphasis on quantitative aspects to encourage progress towards more rigorous analysis of the photosynthetic system, for example by modelling and systems analysis. References are mainly from the secondary or review literature, from which detailed arguments and the primary literature may be obtained. Several recent textbooks have considered more specialized aspects of photosynthesis and there are many excellent reviews; this book is intended to point the reader to them. I hope that this text will enable readers to appreciate the hierarchy of organization that enables plants to produce dry matter within the confines of the environment. With current exciting possibilities for genetically engineering plants for specific purposes, to give greater dry matter production or improved growth in particular environments for example, it is essential that the role of molecular events in the functioning of the whole system in relation to environment should be well understood.

I extend my thanks to all those who have helped in producing this book; to Professor Terry Mansfield who suggested that I should write it and Professor Dennis Baker who gave much valued editorial advice, suggestions and support throughout. Longman I thank for the opportunity to attempt this project and for their often tried patience and support. Dr Alfred Key's suggestions and constructive criticisms on a large part of the text were very valuable. Also Dr Stephen Gutteridge, Professor Peter Lea, Dr Keith Parkinson and Dr Roger Wallsgrove contributed ideas, information and comments on parts of the text. I have gained from their efforts but the omissions, distortions and factual mistakes are mine. Many people have given me much appreciated help, support and encouragement to explore the complexities of photosynthesis and plant functions, my parents and amongst others Professors A.J. Rutter, P.J. Kramer, C.P. Whittingham, H. Fock, C.B. Osmond and G. Farquhar.

In production of the manuscript Mrs Anita Webb and Janet Why and my wife gave excellent technical help with infinite patience over a long period; I am most grateful. The unstinting support of Gudrun, Kirsten and Kurt is beyond thanks; they deserve better reward.

D.W. Lawlor
Rothamsted Experimental Station
June 1985

Acknowledgements

We are grateful to the following for permission to reproduce copyright material:

Academic Press for our fig 9.3 adapted from figs 3, 4 (Hatch 1976b); Annual Reviews Inc. for our figs 3.9, 3.10 adapted from fig 1 (Myers 1971), our fig 6.3 adapted from fig 8 (Junge 1977), our fig 9.4 adapted from fig 1 (Osmond 1978); the author, W.L. Ogren, and the Biochemical Society, London, for our fig 7.5 from fig 2 (Ogren 1978); Elsevier Biomedical Press for our fig 3.4a from fig 1 (Kok 1961), our figs 4.2, 4.3 from Plates 1a and 1d (Coombs and Greenwood 1976), our fig 5.2b adapted from fig 1 (Sargeson and Critchley 1984), our fig 7.11 adapted from fig 4 (Heber 1975); Harper and Row for our fig 2.2 adapted from fig 3 (Gates 1962); Pergamon Press Ltd for our fig 3.6 adapted from fig 4 (van Ginkel and Kleinen Hammans 1980); Springer-Verlag for our fig 5.6 adapted from fig 2 (Hall and Rao 1977), our figs 11.2, 11.3, 11.6 adapted from fig 9.16 p 319 (Osmond, Björkman and Anderson 1980); Mr R.H. Turner and Rothamsted Experimental Station, Harpenden for our figs 4.1, 4.8a,b, 4.10 from unpublished electron micrographs; University Park Press for our table 9.2 from table 5 (Edwards, Huber, Ku *et al.* 1976); Verlag der Zeitschrift für Naturforschung for our fig 3.4b from fig 3 (Döring *et al.* 1969).

CHAPTER *1*

Introduction to the photosynthetic process

'Life is woven out of air by light' – I. Moleschott

Photosynthesis is the process by which living organisms convert the energy of light into the chemical energy of organic molecules. This process increases the total free energy available to organisms and directly or indirectly provides the energy for the whole of the living world. In a simplified form this may be written:

$$\text{low energy chemical state} + \text{light energy} \xrightarrow[\text{organisms}]{\text{photosynthetic}} \text{high energy chemical state} \qquad [\mathbf{1.1}]$$

Sunlight is the ultimate energy source for all biological processes. Without the input of light to change matter from a lower to a higher energy state life would not be possible. Energy is needed to rearrange electrons in molecules and to synthesize chemical bonds, but a complex process may not take place spontaneously and a mechanism is required; this book examines the mechanisms and how they function to capture energy and transduce it to form complex biochemical products from simple inorganic molecules.

According to the laws of thermodynamics, biological processes will tend to go from a high energy to a low energy condition losing energy in the process, unless energy is available to drive the reaction in the reverse direction. Living organisms are in an unstable thermodynamic state and require energy to keep chemical constituents in a highly ordered condition and to do work against the thermochemical energy gradient, in accumulating matter, such as ions or gases from the environment, or to grow, move, etc., all of which characterize the living state in contrast to the world of inanimate matter.

The movement of matter, chemical interconversion or changes in energy state cannot proceed with absolute efficiency, and involve the loss of some of the energy, usually as heat. Once a biological system has accumulated free energy it can convert it to different chemical forms or into physical energy or exchange it between organisms etc., but with time the useful energy will be lost and thermodynamic equilibrium (i.e. death) will be attained.

Without continuous supply of high energy 'food' living organisms cannot survive. All non-photosynthetic organisms, such as animals, fungi and bacteria are dependent upon preformed materials. Some organisms, principally bacteria, can utilize the energy of bonds in inorganic molecules as an energy source. However only those organisms able to use the supply of energy from the sun can increase the total free energy of living material and are independent of the limitations imposed by other energy sources. Photosynthesis by some bacteria, blue-green algae (also called cyanobacteria), algae and higher plants is achieved by a mechanism able to capture the fleeting energy of a light particle and make it available to biochemistry. Given an abundant supply of sunlight they can survive, grow and multiply using only inorganic forms of matter readily available in their environment.

Sunlight is the only form of energy which adds to the total energy supply of the earth and drives not only the weather but also the biological cycles. Solar energy dominates the earth although geochemical processes also contribute to the energy balance. The earth is bathed in a sea of energy in the form of electromagnetic waves, differing in wavelength and energy, derived from the thermonuclear reactions in the sun. Short wavelength radiation, such as X-rays, is highly energetic and may destroy complex molecules by ionization. Ultraviolet light, of wavelengths greater than X-rays but shorter than visible light, breaks bonds within organic molecules and destroys many biological tissues. Infra-red radiation is of longer wavelength than visible light and of low energy. It causes chemical bonds to stretch and vibrate but is not very active in biological processes; however it is important because it causes heating. The energy of visible light is sufficient to cause changes in the electronic states of the valency electrons of many molecules and can thus be used by living organisms to effect the transition from a low to a high energy state. Most molecules are stable in visible light (even if absorbing it) thus allowing the evolution of complex organic molecular 'living' systems, using light as their ultimate energy source.

Light was exploited early in evolution as a source of energy to drive biosynthetic processes. Considering the need for a continuous supply of energy, it is perhaps not surprising that sunlight provides the basis of life and extraterrestrial solar radiation provides more than 99.9 per cent of the energy used in the biosphere. The physical characteristics of light and of the molecules with which it reacts, are crucial to the process of capturing energy and will be considered in Chapter 2.

Concepts of photosynthesis

Photosynthesis in all chlorophyll-containing organisms (except photosynthetic bacteria) is the light driven synthesis of carbohydrates using carbon dioxide and water from the environment with the subsequent release of molecular oxygen. Energy is gained in the chemical bonds of the carbohydrates (see eqn 1.2):

$$CO_2 + H_2O + \begin{matrix}\text{light}\\\text{energy}\end{matrix} \xrightarrow[\text{containing plants}]{\text{chlorophyll-}} (CH_2O) + O_2 + \begin{matrix}\text{chemical}\\\text{energy}\end{matrix} \qquad [\mathbf{1.2}]$$

With the formation of glucose ($C_6H_{12}O_6$) the energy required is 2879 kJ mol^{-1}. This reaction requires energy in the form of 'high energy bonds' of the phosphorylated compound adenosine triphosphate (ATP) and reducing power as the reduced pyridine nucleotide, nicotinamide adenine dinucleotide phosphate (NADPH); these are synthesized by complex biochemical processes driven by light energy. The transformations of energy and material require many individual chemical steps (perhaps thousands), if the processes required to form the whole organism are counted, as they must be in a complex system.

ATP and reductant are also used to assimilate other inorganic compounds. Nitrate ions (NO_3^-) are reduced to ammonia, which is then consumed in the synthesis of amino acids:

$$NO_3^- + 9\,H^+ + 8e^- \xrightarrow[\text{enzymes (nitrate and nitrite reductase)}]{\text{light, chlorophyll,}} NH_3 + 3\,H_2O \qquad [1.3]$$

Photosynthetic bacteria and blue-green algae, but not higher algae or plants, assimilate atmospheric nitrogen to form ammonia:

$$N_2 + 6\,H^+ + 6e^- \xrightarrow[\text{enzyme (nitrogenase)}]{\text{light, chlorophyll,}} 2\,NH_3 \qquad [1.4]$$

Sulphate ions are also reduced before entering metabolism:

$$SO_4^{2-} + 9\,H^+ + 8e^- \xrightarrow[\text{enzymes}]{\text{light, chlorophyll,}} HS^- + 4\,H_2O \qquad [1.5]$$

Many algae, in the absence of oxygen, are able to produce hydrogen gas from water using light energy captured by chlorophyll; the enzyme hydrogenase catalyses the reaction.

$$H_2O \xrightarrow[\text{hydrogenase}]{\text{light, chlorophyll,}} H_2 + \tfrac{1}{2}\,O_2 \qquad [1.6]$$

These examples (considered in more detail later) show that photosynthesis is more than the assimilation of CO_2 with the production of oxygen but is a process with many possible products and capable of being used biologically in many ways.

Photosynthesis as an oxidation-reduction process

All the photosynthetic processes summarized in eqns [1.2]–[1.6] involve oxidation and reduction. Reduction is the transfer of an electron (e^-) or electron plus proton (H^+) from a donor (D) molecule to an acceptor (A); the donor is oxidized and the acceptor reduced. When electrons are donated to a compound which is electrically neutral the reduced compound becomes negatively charged and may accept a proton to restore electrical neutrality. In water, or dilute solution where the photosynthetic and other metabolic reactions take place, protons are freely available:

$$D + A \rightarrow D^+ + A^- + H^+ + e^- \rightarrow D + AH \qquad [1.7a]$$

Addition of e^- to an oxidized acceptor (A^+) reduces A but without H^+ transfer:

$$A^+ + e^- \rightarrow A \qquad [1.7b]$$

Biological reduction–oxidation (redox) reactions are usually catalysed by enzymes. Examples in plant metabolism are the reduction of oxaloacetic acid to malic acid by malate dehydrogenase:

$$COOH\,CO\,CH_2\,COOH + 2H \rightarrow COOH\,CH_2O\,CH_2\,COOH \qquad [1.8]$$

and reduction of 1,3-diphosphoglycerate to glyceraldehyde-3-phosphate by reaction with NADPH, catalysed by NADPH glyceraldehyde-3-phosphate dehydrogenase; this is an essential step in CO_2 fixation:

$$CH_2OP\,CHOH\,COOP + H \rightarrow CH_2OP\,CHOH\,CHO + P_i \qquad [1.9]$$

Reduction and oxidation reactions are of fundamental importance to our understanding of the mechanisms of photosynthesis. The primary reaction of photosynthesis, linking the physical energy of chlorphyll molecules excited by light with biochemical processes, is the transfer of electrons from a special form of chlorophyll to an acceptor molecule driven by the energy captured. The acceptor is reduced and the special form of chlorophyll oxidized. Electrons are then donated from different sources to the oxidized chlorophyll, reducing it and allowing the process to be repeated (Fig. 1.1). Discussion of the mechanism of the processes is the subject of much of this book.

Photosynthetic bacteria use electrons from many sources (e.g. hydrogen sulphide, H_2S), but not water, and light provides the energy:

$$H_2S \xrightarrow[\text{bacteriochlorophyll}]{\text{light,}} S + 2H^+ + 2e^- \qquad [1.10a]$$

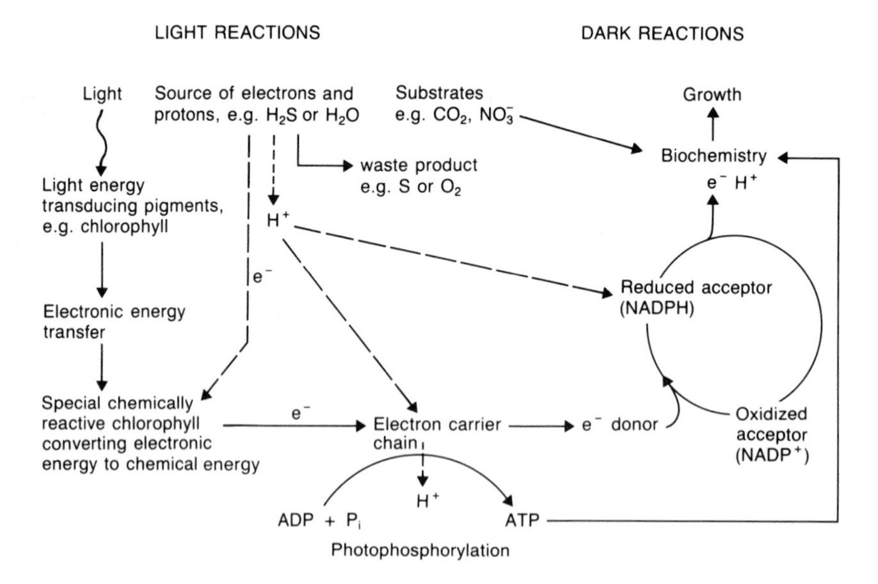

FIG. 1.1 Essential features of photosynthesis in all organisms.

Oxidation produces elemental sulphur. All other photosynthetic organisms use water as the source of reductant and light energy is needed to 'split' it:

$$2\,H_2O \xrightarrow[\text{chlorophyll}]{\text{light,}} O_2 + 4\,H^+ + 4\,e^- \qquad [\mathbf{1.10b}]$$

This is oxidation of water with the liberation of oxygen; H^+ and e^- can be used for chemical reactions. Protons also accumulate in the chloroplast and form a gradient of concentration, the energy of which is used to produce 'high energy bonds' of ATP (see p. 110).

The process of photosynthesis is divisible into the 'light reactions' where light energy generates NADPH and ATP, and 'dark reactions' where these compounds reduce a variety of inorganic and organic compounds in darkness.

Nature of the light-gathering process and generation of reductant and ATP

Equation 1 includes chlorophyll but ignores the many complex reactions of the oxidation–reduction processes, only some of which are directly dependent on light. Absorption of light by chlorophyll leads to ejection of an electron from a special form of chlorophyll; the electron is then captured in a chemical form as a reduced acceptor (a quinone molecule) in an electron transport chain. Electrons pass from a higher to lower energy state along a chain of electron acceptors and donors, which are alternatively reduced and oxidized in the process, until a stable reduced compound is formed. This is ferredoxin, which passes electrons to inorganic compounds, such as nitrate ions, or to secondary reductants, such as oxidized pyridine nucleotide ($NADP^+$). Energy transformations in living organisms take place at nearly isothermal conditions and involve rather small energy changes between each reaction or stage, which are rather easier to control than large 'jumps'. Also, the multiple steps contribute to achieving optimum rates of processes and efficiency. It is not possible to obtain both maximum rates and efficiency in a system; organisms may have evolved to maximize the energy output at rather lower efficiency.

Energy is required to carry out the biological catalytic reactions leading to the reorganization of the primary reactants into complex products and is provided by the hydrolysis of ATP, or other related phosphorylated adenylate compounds. ATP has three phosphate groups. When the bond joining the terminal group is hydrolysed it provides energy and releases phosphate, which may be donated to other compounds, activating them; phosphorylation is an essential step in many biochemical reactions.

$$ATP + H_2O \rightarrow ADP + P_i + \text{energy}\,(-31\ \text{kJ mol}^{-1}) \qquad [\mathbf{1.11}]$$

ATP is resynthesized by photophosphorylation. The movement of electrons (driven by light) along the chain of electron carriers also pumps H^+, producing a concentration gradient of H^+ which is coupled to ATP synthesis (Ch. 6).

The essential nature of the photosynthetic process is therefore a chemical oxidation–reduction reaction, catalysed by proteins, driven against the thermodynamic energy gradient by the energy of light captured by green plants. The light-transducing mechanism in the photosynthetic process generates reductant in the form of organic molecules. These pass the reductant to biochemical reactions, which are independent of the direct effects of light, although light is important in control processes. In addition, light energy drives the synthesis of ATP, which is also essential for biochemical reactions. The study of photosynthesis considers the atomic and molecular processes underlying capture of light energy and conservation, the relation between the production of energetically-favourable compounds and the assimilation of inorganic molecules. It considers how the plant's anatomy and physiology may be interpreted as a consequence of, and requirement for, the basic processes of photosynthesis. Ultimately the fundamental processes of photosynthesis are related to plant performance at an ecological level.

Inter-relation of photosynthesis and respiration

Non-photosynthetic organisms (or photosynthetic ones in darkness) respire preformed, complex compounds of high energy content, such as carbohydrates (CH_2O), to gain energy. In a series of oxidation–reduction reactions electrons pass from (CH_2O) to oxygen, producing water and oxidizing the (CH_2O) to CO_2. Anaerobic respiration (fermentation) without O_2 as final electron acceptor, is only 10 per cent as efficient as oxidative respiration and is not a major part of the biological energy cycle, although in early periods of earth's history it was the dominant respiratory process. Today organisms with such a metabolism are found in anaerobic environments such as stagnant water and the animal gut.

If the respiratory and photosynthetic processes are compared with the starting materials on the left-hand side of eqn 1.12 and the products on the right-hand side, the net result is energy conversion and closed cycles of carbon and water.

$$\text{Photosynthesis: } CO_2 + H_2O + \text{light energy} \xrightarrow[\text{catalysts}]{\text{biological}} (CH_2O) + O_2$$

$$\text{Respiration: } (CH_2O) + O_2 \xrightarrow[\text{catalysts}]{\text{biological}} CO_2 + H_2O + \text{heat energy} \qquad [1.12]$$

Net result of respiration and photosynthesis: light energy → heat energy

Thus photosynthesis and respiration work in opposition, forming a cyclic, closed system for matter. The physical energy of light is converted into chemical energy and ultimately heat.

Energy and electron transport

In photosynthesis, electrons pass from a donor to an acceptor, the two forming a redox pair or couple. The ability of electrons to transfer is determined by the energy required, called the redox potential. The redox potential is determined by comparison with a standard hydrogen electrode (see Chappell (1977) for a description of the system and theory) which has, by definition, a voltage ($E°$) of 0 volt under standard conditions and pH of 0. Biological reactions take place in solution close to pH7, so that the redox potential of biological redox substances is measured at pH7 and is called E'; E' is -0.42 V compared to $E°$. The relation between redox potential and concentration of oxidized and reduced substances [ox] and [red], is given by:

$$E' = E° + \frac{RT}{n\mathrm{F}} \ln \frac{[\mathrm{ox}]}{[\mathrm{red}]} \qquad [\mathbf{1.13}]$$

when n is the number of reducing equivalents (e^-) transferred, R is the gas constant, T the temperature and F the Faraday constant (F = 96 485 Coulomb mol^{-1}). Redox potentials depend on concentration and are given as a midpoint potential when the two forms are equal in concentration. Midpoint potentials of some redox systems relevant to photosynthesis, and discussed later, are; $ferredoxin_{red}/ferredoxin_{ox}$ which transfers 1 e^-, -0.43 V; $NADPH_{(red)}/NADP^+_{(ox)}$, which transfers 2 e, -0.320 V; $plastohydroquinone_{(red)}/plastoquinone_{(ox)}$, 2 e^- transported, $+0.110$V; chlorophyll reaction $centre_{red} \rightarrow rc_{ox}$, 1 e^- transferred, + 0.43V and $O_2/2H_2O$, $4e^-$ transported, + 0.82V.

Substances with more negative redox potential are energetically able to reduce (donate e^- to) others of more positive potential; it is impossible in the reverse direction unless energy and a mechanism are available. Hence ferredoxin is the e^- acceptor in photosynthesis and reaction centre chlorophyll can remove e^- from water. The redox potential expresses the possibility of transfer and defines the order of compounds in a redox or electron transport chain and the points at which energy must be added to the system for electrons to move or where energy may be extracted to do work, but says nothing about the mechanism. The maximum useful energy from a reaction is given by the difference in redox potential ΔE; the free energy of a reaction, ΔG, is related to redox potential by $\Delta E = -\Delta G/nF$.

Occurrence of photosynthesis

Photosynthesis is distinguished from other light-sensitive processes in animals and plants because it alone accumulates energy. It occurs in organisms as diverse as bacteria and trees (Table 1.1) and is similar in all, with light capture by a pigment and conversion of the energy to chemical form. Differences between organisms reflect their evolution, the greatest difference is between

Table 1.1 Main groups of photosynthetic organisms, their structure and photosynthetic characteristics

Prokaryotes	Eukaryotes
Single cell or little complexity	Mainly multicellular, complex intercellular interactions
Cell nucleus without membrane	Cell nucleus with membrane
Photosynthesis in vesicular membranes, not in discrete compartment	Photosynthesis in vesicular membranes in discrete compartment – chloroplast
Anoxygenic forms	Anoxygenic forms: none
Examples: Photosynthetic bacteria (purple sulphur – Rhodospirillaceae, green sulphur – Chromatiaceae)	
Process: Do not evolve O_2. Many obligate anaerobes. Source of reductant hydrogen sulphide, sulphur, thiosulphite, hydrogen, organic compounds, never water. Can reduce gaseous nitrogen	
Oxygenic forms	Oxygenic forms
Examples:	*Examples:*
Blue-green algae	Algae (green – Chlorophyceae, red – Rhodophyceae) Higher plants (Bryophyta, Angiospermae)
Processes: Evolve O_2. Source of reductant water. Reduce gaseous nitrogen	*Processes:* Evolve O_2. Source of reductant water. Do not reduce gaseous nitrogen

the photosynthetic bacteria, a very diverse group unable to oxidize water, and blue-green algae, algae and higher plants which oxidize water and evolve O_2.

The physical nature of light, but not its intensity or spectral distribution, has been a constant feature of the environment since the origin of the earth, and it is to be expected that the different pigment systems for light capture share many features. Chlorophylls of all organisms are very similar and related biosynthetically and light absorption and transport of electrons take place in membranes. In photosynthetic bacteria there is only a single light driven process, whereas in all other organisms there are two, one of which oxidizes water. Bacteria use a variety of substrates to provide reductant suggesting that many substrates were available at early stages of evolution but later the abundance of water, and perhaps shortage of other sources of reductant, made it the preferred source.

In photosynthesis the oxidized donor may accumulate, for example sulphur from H_2S in bacteria and O_2 from water in higher organisms. Van Niel established that the photosynthetic processes in all organisms conform to eqn 1.14:

$$\text{donor} - H_2 + \text{acceptor} \xrightarrow[\text{(bacterio)chlorophyll}]{\text{light,}} AH_2 + \text{donor} \qquad [1.14]$$

Electron transport coupled to ATP synthesis was an early feature of photosynthetic systems, requiring a closed membrane vesicle separating regions of high and low proton concentration which are needed for ATP generation. The gradient is from high proton concentration, $[H^+]$, outside to low $[H^+]$ inside the photosynthetic bacterial cell (light energy is used to pump H^+ out of the cell) and from high $[H^+]$ in the thylakoids to low $[H^+]$ in the cytosol of blue-green algae or in the chloroplast stroma of higher plants. Thylakoid membranes, although internal, are equivalent to the external membrane of bacterial cells. The space within the thylakoid is equivalent to the medium surrounding a photosynthetic bacterium. Thus the mechanisms driving ATP synthesis are comparable, despite the great structural differences between groups. Although the photosynthetic bacteria are of limited practical importance, they are valuable as a relatively simple system for analysis of photosynthesis.

Photosynthesis in relation to other plant functions

Photosynthesizing plants are autotrophs providing their metabolic requirements for energy and complex assimilates from light and inorganic materials from the environment. Higher plants produce carbohydrates (including storage polymers), lipids, amino acids, protein, pigments and hormones, from CO_2, nitrate or ammonia and inorganic salts. Photosynthetic bacteria and blue-green algae are also able to assimilate molecular nitrogen (N_2), and thus depend only on very simple starting materials. The ability of plants to synthesize many types of organic materials is remarkable and the key is the ability to trap light energy.

To capture the often diffuse light, and accumulate CO_2 from the low concentration in the aerial environment, terrestrial algae and higher plants have developed thalli and leaves which provide large surface areas for light capture and diffusion of gases. To obtain nutrients and water, absorbing organs such as roots have evolved. To extract nutrients from very dilute solutions and accumulate them at high concentrations is thermodynamically unstable and energetically unfavourable, so metabolic energy is required. In some photosynthetic bacteria, which have few specialized features, metabolism is linked directly to photosynthesis and growth occurs only in the light. However, such a direct connection between photosynthesis and metabolism is inflexible.

Higher plants, with their complex anatomy, specialization of function and with growth and reproduction spread over time and place, produce assimilates during illumination which are stored or translocated to roots, stems and reproductive organs and stored or used there. In darkness or poor illumination the stored assimilates are respired to generate energy and metabolites for protein and nucleic acid synthesis, membrane production and maintenance of ion balance (Fig. 1.2).

Photosynthesizing tissues respire and utilize assimilates. If photosynthesis and respiration were not controlled the products of photosynthesis would soon be consumed in a futile reduction/oxidation cycle (eqn 1.12). Elaborate biochemical mechanisms in the cell regulate photosynthesis and respiration and link them to growth of the cell and organism, to produce a stable system

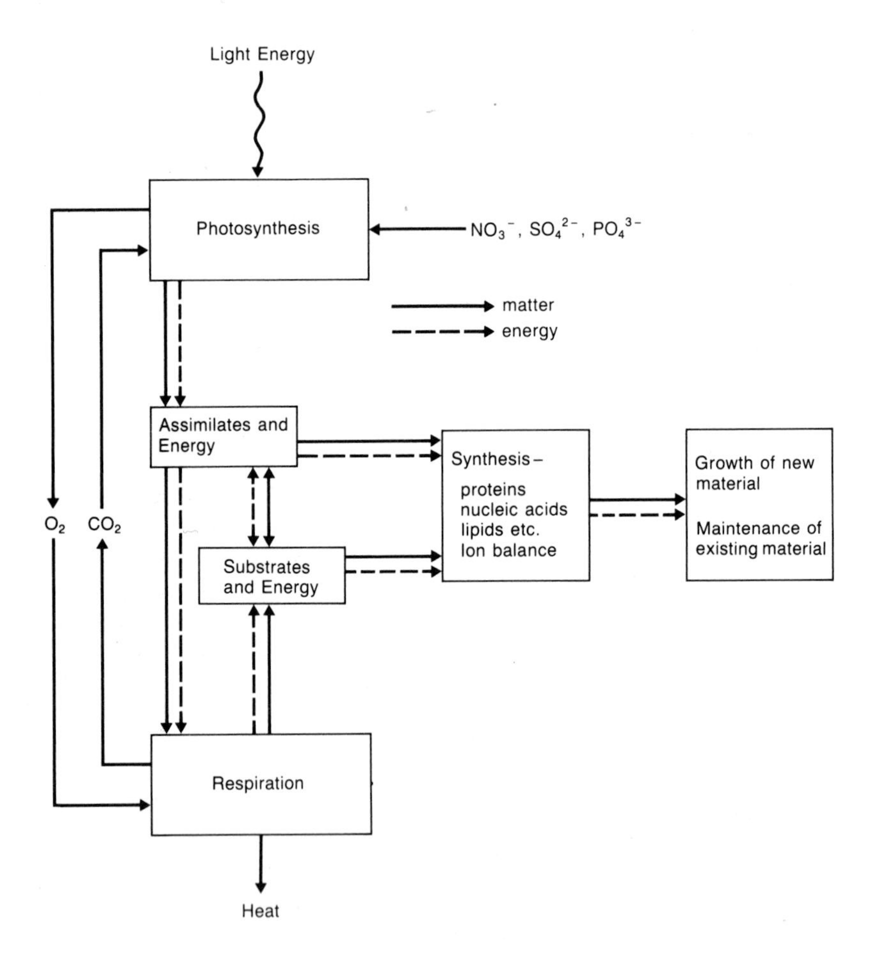

FIG. 1.2 Scheme of the interactions between the main functions of assimilation (photosynthesis) and dissimilation (respiration) and growth in the higher plant.

capable of continued growth and reproduction. Plants are composed of individual sub-systems (cells, organs and processes) which are controlled by their internal conditions and outside factors but are highly integrated. Photosynthesis is the driving factor for all processes, yet is related to the other cellular functions in complex ways which are not yet well understood.

Global photosynthesis

The earth's surface receives about 5.2×10^{21} kJ year^{-1}, about 50 per cent of which is of wavelengths used in photosynthesis, that is about 3×10^{21} kJ year^{-1}. Only about 0.05 per cent of this energy (3.8×10^{18} kJ year^{-1}) is captured in organic molecules, the rest is re-radiated into space as heat. Almost 50 per cent of the total photosynthesis is by marine organisms. The turnover of CO_2 is about 3×10^{12} tonne year^{-1} and of nitrogen about 5 per cent of this. The enormous scale of this most important photochemical reaction is comparable to geological processes, such as mineral weathering. Photosynthesis produces 5×10^{11} tonne of organic matter per year and the earth's standing organic matter is estimated at 10^{12}–10^{13} tonne dry matter. At present, the rate of destruction of forests and burning of fossil fuels is about 15 per cent of the rate of photosynthesis so that the CO_2 content of the atmosphere, which is small (~ 340 volumes per million volumes, or 34 Pa partial pressure) turns over rapidly, with a half-life of 10 years. Oxygen turnover is slower, 6500 years, as the concentration is much greater (21 volumes per 100 volumes) and water turns over in 3 million years, owing to the huge global water reserves.

History of photosynthesis research

Appreciation of photosynthesis as a process carried out by green plants in the light, consuming CO_2 and producing O_2, sugars and complex assimilates has developed over some two hundred years. The scale of understanding has moved from the whole plant and timescales directly comprehensible to the human senses, to scales of nanometres and events occurring in less than a million millionth of a second which can only be inferred from indirect measurements. In the early part of the eighteenth century, there was considerable interest in the chemical composition of the atmosphere and the behaviour of plants. Stephen Hales in 1727 suggested in his *Vegetable Statiks* that plants draw their nourishment from the air, a deduction supported by Joseph Priestly who observed that vegetation in the light could purify air rendered unfit by the respiration of a mouse. Later experiments failed to confirm this, due possibly to inadequate illumination, but Ingen-Housz in 1779 clearly showed that illuminated plants purified the air for the support of animal life. However, it was Senebier who stated that illuminated plants removed a gas produced in respiration, with the liberation of pure air. The description of fixed air (CO_2) by Black in 1754 and the discovery of Scheele that air contained an inert gas (later called nitrogen) and one supporting combustion (O_2) opened the way for a full understanding of plant gas exchange.

Lavoisier established that air was composed of different gases, with O_2 supporting burning and respiration, which were oxidative (i.e. O_2-adding) processes and that 'fixed air' composed of carbon and oxygen and produced by respiration or burning, was essential for plant nourishment. Thus the nature and source of the gas absorbed by plants was established, substantiating Hale's deductions, and the essential role of light. The involvement of water in photosynthesis was suspected by the end of the eighteenth century and DeSaussure in 1804 suggested that CO_2 was decomposed, the carbon reacting with water, that is O_2 came from CO_2, a view reversed in 1941 when Ruben, using isotopes, showed that O_2 originated from water.

Lavoisier's studies established the biological interconversions of CO_2 and O_2, the importance of energy in the processes of photosynthesis and respiration, and also the equivalence, but in reverse, of the two processes. By 1846 Liebig concluded that green plants, with sufficiently intense light, could reverse respiration, consuming the waste gases and oxygenating the atmosphere. Between 1880 and 1900 the site of O_2 release in algae was shown by Engelmann to be the chlorophyll bodies (now called chloroplasts) which contained starch. The essential biochemical role of inorganic nutrients was understood in outline and, coupled with Wöhler's demonstration of the essentially chemical nature of biological processes and Buchner's study of enzyme activity outside the living plant, the analysis of photosynthesis was made possible. Understanding of respiration developed faster and provided much insight into the chemistry of photosynthesis.

Plant pigments, particularly chlorophylls, were extracted; their analysis led Twsett to describe chromatography, a method of separating complex chemical mixtures of related substances. Willstäter and Stoll applied the methods of analytical chemistry to the structure of chlorophyll, laying the foundation of understanding of the molecular mechanisms by which light energy is captured. The concepts of light harvesting, photosystems and electron transport developed in the 1930s with the work of Emerson, and is continuing.

Carbon metabolism in photosynthesis was long a matter of controversy, dominated by the idea of CO_2 reacting with water on the chlorophyll molecule, producing compounds such as formaldehyde, a substance never satisfactorily measured in photosynthetic tissue and also toxic! Only in 1954 was Calvin's research group able to show, using radioactive carbon as a tracer, that CO_2 was fixed in a cyclic enzymatic process in chloroplasts by interconversion of sugar phosphates. Light energy was needed to produce the reductant and assimilatory power for generation of the substrate of CO_2 fixation.

Analysis of reductant formation was advanced by R. Hill's demonstration in 1937 that artificial electron donors could substitute for light. Van Niel suggested that 'reducing power' could be derived from a variety of reduced electron donors; in the photosynthetic bacteria H_2S, organic compounds, etc., but in higher plants only from water. However it was not until 1954 that the 'assimilatory power' (ATP) was shown by Arnon to be formed in photosynthesizing cells as a consequence of light reactions. This agreed with and substantiated Blackman's concept of 'light' and 'dark' processes.

Measurements of the exchange of $^{18}O_2$ suggested that respiration was inhibited in the light. However, Decker in 1957 showed that a form of respiration occurred in many plants in the light; this was called photorespiration. Later other plants were shown not to photorespire; their photosynthetic mechanism (called C4 photosynthesis) was found by Hatch and Slack to differ from the photorespiring (C3) plants. Analysis of the interaction between respiration, photosynthesis and photorespiration are currently exciting areas of research which will be discussed in detail in the following chapters.

The detailed history of photosynthesis research is fascinating and the major works of Rabinowitch (1945), the *Encyclopedia of Plant Physiology* edited by Ruhland (1956) and the review by Arnon (1977) should be consulted. A true historical perspective would show the development of abstract concepts from relatively simple initial observations. However, in this book the uncertainties, technical problems, erroneous theories and links to other disciplines must be ignored and the ideas of photosynthesis presented as if they were won without effort or difficulty.

Evolution of photosynthesis

As a biochemical process based on proteins and organic molecules which are rapidly decomposed, photosynthesis has left little direct trace in the geological record. Yet enough is now known of the comparative biochemistry, to suggest a plausible hypothesis of the development of this essential process. Here only the barest outline is attempted, articles by Schopf (1978) and Budyko (1974) should be consulted. The earth was formed (Fig. 1.3) some 4.6×10^9 years ago and for the first 0.5×10^9 years cooled and solidified. Because of earth's distance from the sun and its size, which determine the heat received and the force holding gases on the surface, both liquid water and an atmosphere were retained. The primitive atmosphere was highly reducing, with methane (CH_4), H_2, H_2S, CO_2, CO, NH_3 etc., but there was no free O_2; it was anoxic. This is considered crucial to the evolution of life because oxygen destroys organic molecules. Also there would have been no ozone, which today forms a thin layer in the upper atmosphere, absorbing ultraviolet (UV) rays. This energetic form of radiation, together with high temperatures and abundant gases evolved by volcanic activity, provided conditions for synthesis of organic molecules such that the primitive oceans resembled 'hot dilute soup' to quote Haldane. How this prebiotic state generated self-replicating biological systems is beyond the scope of this book and is still unresolved. However it occurred, organisms resembling present-day bacteria in size and cell structure are present in rocks 3.4×10^9 years old. A form of light-driven metabolism probably developed early, as derivatives of carotenoids have been detected in rocks of that age, although contamination by later material is possible. A primitive organism

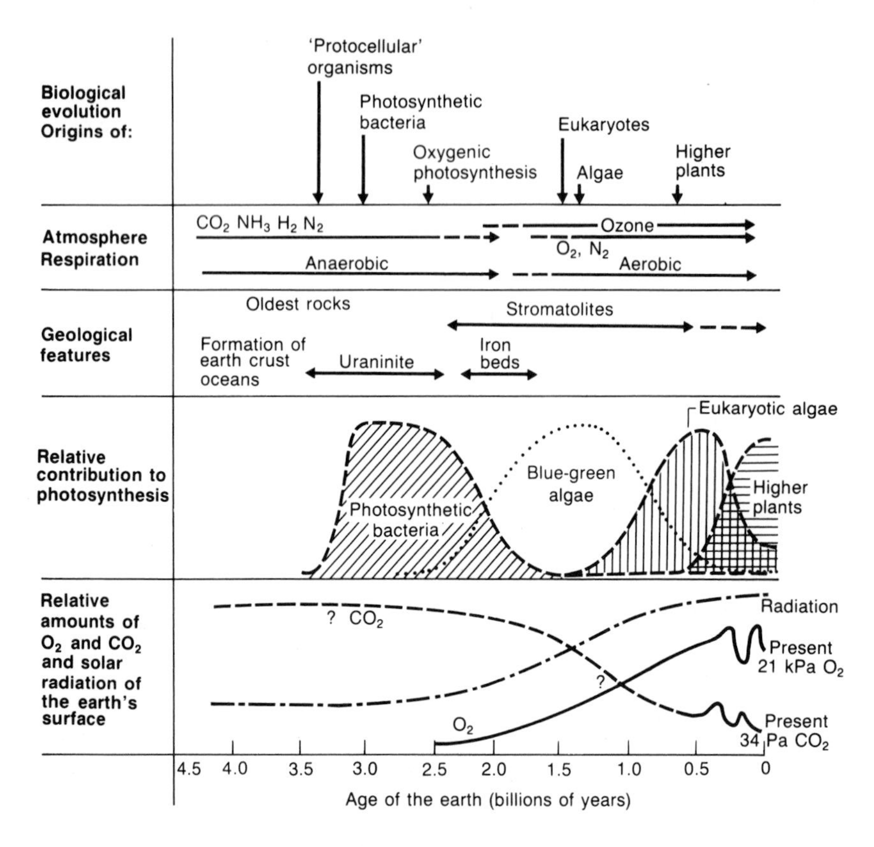

FIG. 1.3 Probable sequence of events in evolution of photosynthesis and the relation to some other geological and biological processes; highly schematic.

might have synthesized ATP by a light-driven proton pump. Such organisms would have been independent of the supply of preformed ATP and have had an evolutionary advantage by exploiting an almost limitless source of energy.

Many processes were linked to the light reactions and electron flow and ATP, including N_2, CO_2 and S assimilation. However, water was not split in early photosynthesis until 2.2×10^9 years ago so no photosynthetic O_2 was formed and the atmosphere was reducing. Linkage of two light-driven reactions or photosystems provided the necessary energy to enable water to be oxidized and O_2 (the waste product) escaped to the atmosphere. This time scale is suggested by much evidence (Schopf 1978). Extensive beds of fossil limestone, called stromatolites, were formed 3×10^9 years ago and contain blue-green algae as do present day stromatolites. However, oxygen production probably started before that time. Many geochemical processes would have consumed O_2; for example iron (Fe^{2+}) reacts to form insoluble Fe_3O_4. Oceans would have been slowly depleted of Fe^{2+} with deposition of iron ores. This may be the origin of the 'red beds' formed 2.2×10^9–1.7×10^9 years ago. To borrow Schopf's expression 'the world's oceans rusted'. A non-photosynthetic origin of O_2 by UV splitting of water was probably too slow for such a massive chemical process. Uraninite (UO_2), is a uranium ore, insoluble at O_2 concentrations above 1 per cent; no deposits younger than 2×10^9 years have been found. Thus some 2.5×10^9 years ago photosynthesis developed using H_2O as source of reductant and atmospheric O_2 increased. Deposition of reduced carbon as oil, coal, etc., also added to the O_2 content. By 1.8×10^9 years ago aerobic conditions were established as the chemical buffers were exhausted. Oxygen in the upper atmosphere formed an ozone layer, which absorbed UV radiation, allowing evolution of higher organisms and invasion of the land. Oxygen, by acting as a terminal receptor for respiratory processes, greatly (ten-fold) increased the amount of energy to be obtained by respiration of organic substances. Most of the present day biota, including man, owes its existence to the waste product of photosynthesis!

Eukaryotes, with large internally compartmented cells, evolved rapidly from their origin 1.5×10^9 years ago. Evidence from the structure and function of the nucleic acids of chloroplasts and mitochondria of higher plants suggests that these organelles arose from the invasion of non-photosynthetic eukaryotic cells by bacteria and blue-green algae. Photosynthesis evolved more complex biochemistry, separation of respiration and photosynthesis and their regulation. Photosynthesis shaped the biosphere both directly and through its effects on the earth's climate and geology. Carbon from photosynthesis was sequestered in oil, coal and gas decreasing the atmospheric CO_2 and increasing the ratio of O_2 to CO_2. This may have been unfavourable for photosynthesis as the CO_2-fixing enzyme ribulose bisphosphate carboxylase is less efficient under these conditions. On land, prevention of water loss from the plant by a thick cuticle also reduced supply of CO_2. Evolution of different types of photosynthesis, C4 and CAM (p. 192), based on the earlier metabolic forms but of greater efficiency, is probably a response to an environment of decreasing CO_2/O_2 ratio and drier atmosphere, with intense radiation. Present human activity is increasing the CO_2 concentration of the atmosphere by burning fossil fuels. This may improve plant growth in the short term but also affect the world's climate (p. 204). Humankind's future is closely linked to the photosynthetic production of food, fuel and fibre and increasing population will lead to greater demands on the efficiency of the process.

References and Further Reading

Arnon, D. I. (1977) Photosynthesis 1950–75: Changing concepts and perspectives, pp. 7–56 in Trebst, A. and Avron M. (eds), *Encyclopedia of Plant Physiology (N.S.)*, Vol. 5, *Photosynthesis I*, Springer-Verlag, Berlin.

Arnon, D. I. (1981) The discovery of photosynthetic phosphorylation, *TIBS*, **9**, 255–62.

Benedict, C. R. (1978) Nature of obligate photoautotrophy, *A. Rev. Plant Physiol.*, **29**, 67–93.

Bidwell, R. G. S. (1979) *Plant Physiology* (2nd edn), Macmillan, New York; Collier Macmillan, London.

Bonner, J. (1980) *The World's People and the World's Food Supply*, Carolina Biology Readers, Carolina Biological Supply Co., Burlington, NC.

Budyko, M. I. (1974) *Climate and Life* (English edn, ed. D. H. Miller), Academic Press, New York.

Chappell, J. B. (1977) *ATP*. Carolina Biology Readers, Carolina Biological Supply Co., Burlington, NC.

Clayton, R. K. (1980) *Photosynthesis: Physical Mechanisms and Chemical Patterns*, I.U.P.A.B. Biophysics Series, Cambridge University Press, London.

Danielli, J. F. and **Brown, R.** (eds) (1951) Carbon dioxide fixation and photosynthesis, *Symp. Soc. Exp. Biol.*, **V**, Cambridge University Press, Cambridge.

Danks, S. M., Evans, E. H. and **Whittaker, P. A.** (1983) *Photosynthetic Systems: Structure, Function and Assembly*, Wiley, New York.

Dickerson, R. E. and **Geis, I.** (1976) *Chemistry, Matter and the Universe*, W. A. Benjamin Inc., New York.

Dose, K. (1983) Chemical evolution and the origin of living systems, pp. 912–24 in Hoppe, W., Lohmann, W., Markl, H. and Ziegler, H. (eds), *Biophysics*, Springer-Verlag, Berlin.

Foyer, C. H. (1984) *Photosynthesis*, Wiley Interscience, New York.

Giese, A. (1964) *Photophysiology, 1, General Principles: Action of Light on Plants*, Academic Press, New York.

Gifford, R. M. (1982) Global photosynthesis in relation to our food and energy needs, pp. 459–95 in Govindjee (ed.), *Photosynthesis*, **2**, *Development, Carbon Metabolism and Plant Productivity*, Academic Press, New York.

Govindjee and **Govindjee, R.** (1975) Introduction to photosynthesis, pp. 2–50 in Govindjee (ed.) *Bioenergetics of Photosynthesis*, Academic Press, New York.

Gregory, R. P. F. (1979) *Biochemistry of Photosynthesis* (2nd edn), Wiley, Chichester.

Hall, D. O. (1984) Photosynthesis for energy, pp. 727–35 in Sybesma, C. (ed.), *Advances in Photosynthesis Research*, Vol. II, Martinus Nijhoff/Dr W. Junk Publishers, The Hague.

Hall, D. O. and **Rao, K. K.** (1981) *Photosynthesis* (3rd edn), Studies in Biology, No. 37, Edward Arnold, London.

Halliwell, B. (1981) *Chloroplast Metabolism: The Structure and Function of Chloroplasts in Green Leaf Cells*, Clarendon Press, Oxford.

Lehninger, A. L. (1982) *Principles of Biochemistry*, Worth, New York.

Rabinowitch, E. I. (1945) *Photosynthesis*, Interscience Publishers Inc., New York.

Reid, R. A. and **Leech, R. M.** (1980) *Biochemistry and Structure of Cell Organelles*, Blackie and Son Ltd, Glasgow.

Ruhland, W. (ed.) (1956) *Encyclopedia of Plant Physiology, 5, The Assimilation of Carbon Dioxide*, Springer-Verlag, Berlin.

Schopf, J. W. (1978) The evolution of the earliest cells, *Sci. Amer.*, **239**, 84–103.

Whittingham, C. P. (1977) *Photosynthesis* (2nd edn), Carolina Biology Readers, Carolina Biological Supply Co., Burlington, NC.

Whittingham, C. P. (1978) *The Mechanism of Photosynthesis* (2nd edn), Edward Arnold, London.

Zelitch, I. (1971) *Photosynthesis, Photorespiration and Plant Productivity*, Academic Press, New York.

CHAPTER 2

Light – the driving force of photosynthesis

Characteristics of light

Photosynthesis is driven by the energy of light, therefore the physical nature of light and its interaction with matter are described in a much simplified, qualitative way in order to understand and analyse the bases of the biological processes. The quantitative characteristics of light in relation to photochemistry are discussed in the books by Giese (1964), Clayton (1971, 1980) and Moore (1957). Light is electromagnetic radiation, emitted when an electrical dipole (a paired positive and negative charge, separated by a small distance) in an atom oscillates and causes a change in the field of force. The dipole produces an electrical and a magnetic vector, which are in phase but at right angles. Fluctuations in the field strength of these vectors are perpendicular to the direction of travel of the wave and hence light is a transverse wave. The electromagnetic wave is characterized by both wavelength, λ (in metres) which is the distance between successive positive or negative maxima on the sine wave, and by frequency, ν, the number of oscillations per unit time (s^{-1}).

Frequency is determined by the oscillations of the dipole. Wavelength and frequency are related by the velocity of propagation of the wave, v (ms^{-1}):

$$v = \lambda \nu \qquad [2.1]$$

The velocity of light (c) is 3×10^8 m s^{-1} *in vacuo*. Table 2.1 gives the approximate wavelengths and frequencies of the main groups of electromagnetic waves. Frequencies between 7.5 and 3.8×10^{15} s^{-1} (wavelengths 400 to 700 nm) are visible to the human eye and are called light. Photosynthesis is driven by radiation of 400 to 700 nm, although the response to particular wavelengths within the band is markedly different from that occurring in vision. Light is electromagnetic radiation causing a response in the human eye and strictly the term should be applied only to that spectral distribution. However the simplicity of the word and the general similarity between the wavelengths of visible light and those used in photosynthesis, justify the use of the term 'light' to describe radiation used by photosynthesis, even if different from the spectral response of human vision.

Electromagnetic radiation passes through space without matter to transmit it, in contrast to wave propagation in solids, liquids or gases. The wave form of

Table 2.1 Types of electromagnetic radiation, their wavelength, frequency and energy per photon

Type of radiation	Wavelength	Frequency (s^{-1})	Energy (eV)
Radiowaves	10^3–10^{-3} m; for 1 m	3×10^8	1.24×10^{-6}
Infra-red	800 nm	3.8×10^{14}	1.55
Visible red light	680 nm	4.4×10^{14}	1.82
Visible green	500 nm	6.0×10^{14}	2.50
Visible violet-blue	400 nm	7.5×10^{14}	3.12
Near ultraviolet	200 nm	1.5×10^{15}	6.25
Ultraviolet	10 nm	3.0×10^{16}	123.0
X-rays	0.01 nm	3×10^{19}	1.24×10^5

light is shown by interference phenomena and the slower transmission of light in dense media. Light has, however, the characteristics of both wave and particle. Emission of electrons from metal surfaces caused by light, called the photoelectric effect, and radiation of energy from atoms at distinct frequencies rather than as a continuous spectrum, shows that light is particulate.

In 1900 Planck resolved the conflict between the wave and particle concepts by considering that light behaves as discrete particles of energy, called quanta, which can only be absorbed or emitted by matter in indivisible units. Thus processes involving light are quantized, that is, 'all or nothing'. The quantum of energy is carried as the oscillating force field of the electromagnetic wave. The particle carrying a quantum of energy is called a photon, which has no rest mass. Quantum and photon are distinct concepts, the former is the energy carried by the photon. Molecular transitions caused by light involve a 'jump' or transition in energy state of electrons and can only take place if all the energy of a photon is captured; if the quantum is larger or smaller than the energy required for the transition, then the photon will not be captured. The energy of a photon (ε) depends on the frequency of the electromagnetic wave, which is related to the wavelength, and is given by:

$$\varepsilon = h\nu = hc/\lambda \qquad [2.2]$$

where h is Planck's constant (6.62×10^{-34} J s^{-1}) which has the units of energy × time or 'action'. The greater the frequency and, from eqn 2.1, the smaller the wavelength, the larger the energy content of the photon. Where the energy of light is to be related to a photochemical effect, as in spectroscopy, the wave number ($\bar{\nu} = 1/\lambda$; unit cm^{-1} by convention) is employed as it is directly proportional to the energy from Planck's law (eqn 2.2) with $\varepsilon = hc\bar{\nu}$. The Système Internationale (SI) unit of energy is the joule (symbol J). Much of the older literature used the calorie (1 calorie = 4.18 J) as the basic unit. Characteristic energy levels of molecular orbitals, ionization potentials etc., are often expressed as electron volts (eV) which is the energy acquired by an electron

Table 2.2 Characteristics of light of different wavelengths

Wavelength, λ (nm)	400	550	680
Colour to the human eye	Blue	Green	Red
Frequency, ν (s^{-1})†	7.5×10^{14}	5.5×10^{14}	4.4×10^{14}
Energy of 1 photon (J)‡	4.45×10^{-19}	3.61×10^{-19}	2.9×10^{-19}
Energy of 1 photon (eV)	3.12	2.25	1.82
Energy of 1 mole of photons (J)	29.8×10^{4}	21.7×10^{4}	17.4×10^{4}
(= 1 einstein)§ (eV)	1.87×10^{24}	1.36×10^{24}	1.09×10^{24}

†Calculated from eqn 2.1.
‡Calculated from eqn 2.2.
§Energy of 1 photon multiplied by Avogadro's number of photons.
Note: 1 nm = 10^{-9} m = 10 Angströms = 1 millimicron (mμ).

Table 2.3 Conversion factors for energy units used in the photosynthetic literature

1 electron volt (eV)	= 1.602×10^{-19} J
1 watt	= 1 J s^{-1}
1 kWh	= 3.6×10^{6} J
1 joule	= 0.239 calories
	= 6.242×10^{18} eV
1 calorie	= 4.184 joules
1 kJ mol quantum^{-1}	= 1.036×10^{-2} eV
Planck's constant	= 6.62×10^{-34} J s
	= 4.136×10^{-15} eV s

1 mole contains as many elementary particles as there are carbon atoms in 0.012 kg of ^{12}C, or Avogadro's number, 6.023×10^{23}, of particles.

Energy of 1 photon (J) = $1.986 \times 10^{-16}/\lambda$

1 mole quantum of photons (= 1 einstein) is Avogadro's number, 6.023×10^{23}, of photons.

$$\text{With } \lambda \text{ in nm 1 mol quanta} = \frac{1.986 \times 10^{-16} \times 6.023 \times 10^{23}}{\lambda}$$

$$= \frac{119.616 \times 10^{6} \text{ joules}}{\lambda}.$$

$$1 \text{ mol quanta m}^{-2}\text{ s}^{-1} = \frac{1.2 \times 10^{8}}{\lambda} \text{ J m}^{-2}\text{ s}^{-1}.$$

falling through a potential difference of 1 volt; thus the energy of a photon of red light, wavelength 680 nm, frequency $4.4 \times 10^{14}\ s^{-1}$, is 1.82 eV from eqn 2.2. Some characteristics of selected wavelengths of light are given in Table 2.2 and conversion factors for different units of energy, which have been used in the literature of radiation biology are shown in Table 2.3.

A single photon is a small unit in biological terms; at noon on a bright day the earth's surface receives a maximum of about 1.3×10^{21} photons $m^{-2}\ s^{-1}$, so a larger unit of radiation, the mole of photons (more usually referred to as mole of quanta) is used. This is often called the einstein although not allowed in SI. It is the number of photons corresponding to Avogadro's number of particles (6.023×10^{23}), and may be thought of as the number of photons required to convert a mole of a substance to another form with 100 per cent efficiency, if captured in a single discrete step. The relationship between wavelength, frequency and energy of individual photons and a mole quantum of photons is given in Table 2.2 for three selected wavelengths of light.

Confusion may arise over the many ways of measuring light and the units of expression because either the number of quanta, or their energy (or both) may be determined. There are also measures of the illuminance, that is the visual impression to the human eye. To study quantitative and kinetic aspects of the response of chemical and biological processes to light only photon number or energy in defined spectral regions are useful. Photon number incident on a surface normal to the beam, is given by photon flux (mol $m^{-2}\ s^{-1}$). Energy is given by the radiant flux (J $m^{-2}\ s^{-1}$; as 1 Js^{-1} = 1 watt (W) this is equivalent to W m^{-2}). Illuminance is given by the luminous flux, measured in lux (= lx = lumen m^{-2}). In the older literature the foot candle (1 fc = 1 lumen per square foot = 10 764 lx) was used extensively but is no longer acceptable. The biological literature uses the term photon flux density (PFD) as equivalent to photon flux, but it is not correct and should be abandoned. Radiation from about 400 to 700 nm is used in higher plant photosynthesis and is called photosynthetically active radiation or PAR and is measured by quantum sensors. Quantum sensors include germanium diodes and lead sulphide resistors; their sensitivity changes with wavelength so that at different parts of the spectrum a true photon number is counted. Energy detectors or radiometers respond to energy independent of wavelength; the instruments include thermopiles and bolometers which can measure in mono- or polychromatic light.

Spectroradiometers are instruments which measure the energy of light in narrow wavebands over the whole spectrum and are used to determine spectra of light sources. Spectrophotometers are used to determine the response of photochemical and biological processes to wavelength. Illuminance is measured by light meters, which respond to light with a similar spectral response to the human eye. Detectors must also have correct geometrical characteristics to detect light coming from different directions; most used in photosynthesis studies are cosine corrected. As the human eye is most sensitive to green light (*c.* 550 nm wavelength) measurement of illuminance and use of such units is no longer acceptable in studies of photosynthesis.

Origin of light for photosynthesis

Light is emitted by matter undergoing changes in its energy state as a result of heating, which excites electrons from the ground state to higher excited states. Radiation is emitted when the electron drops back to the lower energy state (Fig. 2.1), at discrete wavelengths corresponding to the energy difference between the ground and excited states, and giving rise to line spectra because the energy levels are distinct. Excited molecules, particularly the more complex, may emit radiation at several wavelengths, often close together giving broader emission spectra. This is due to more, closely spaced, molecular orbitals from which electronic transitions occur.

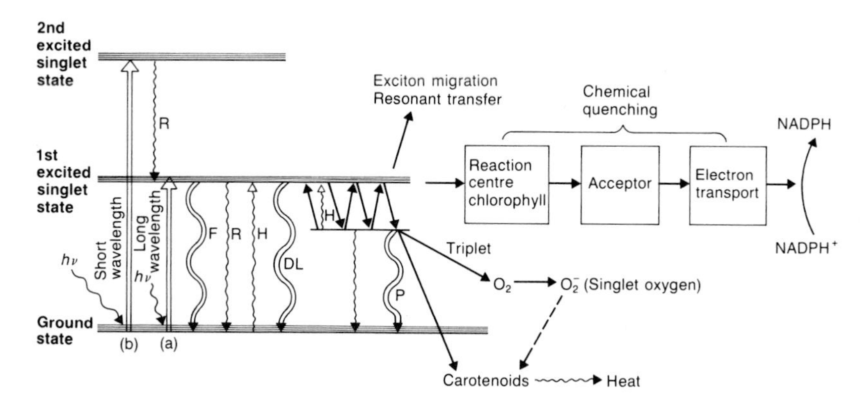

FIG. 2.1 Concept of absorption of photons ($h\nu$) by an atom, energizing an electron to an excited state (a) and its subsequent decay with release of energy. Capture of a more energetic photon (b) results in higher energy level orbitals being filled and then decay by radiationless transition (R). Heat (H) may also raise an electron to higher energy level and the energy is emitted when the electron drops back to the ground state. The main energy-dissipating processes are by radiationless transition (R), prompt fluorescence (F), delayed light emission (DL), phosphorescence (P), by chemical reactions in photosynthesis and transfer, for example of triplet energy to oxygen or carotenoids or of excitation energy to other chlorophyll and pigment molecules.

All the energy for photosynthesis comes from the sun, although light of suitable wavelength from any source, electric lamps for example, may be used. Solar nuclear fusion reactions, mainly the proton–proton cycle, consume hydrogen nuclei and produce helium nuclei, neutrinos and photons. Of the truly enormous amounts of energy generated by the sun (4×10^{26} Js^{-1}) only a minute fraction reaches the earth (6.8×10^{16} Js^{-1}) and only a very small part is used in photosynthesis. Radiation is lost from the sun's corona as if from a 'black body' (a perfect radiator and absorber of energy) with a temperature of about 5800 K, in agreement with the Stefan–Boltzmann law. The distribution of the wavelengths of the solar spectrum is concentrated between 400 and 1200 nm with the peak around 600 nm (Fig. 2.2) in the yellow-orange. At the

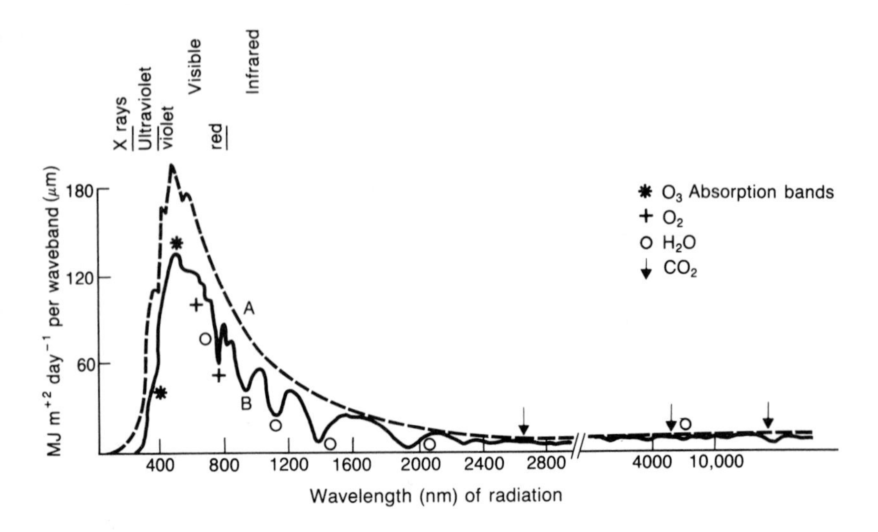

FIG. 2.2 Solar radiation above the earth's atmosphere (A) and at the earth's surface (B). The energy (in megajoules per day) for each wavelength is shown. Areas marked H_2O, O_3, O_2 show absorption of radiation by these components of the earth's atmostphere (recalculated from Gates (1962)).

earth's surface the spectrum is modified by selective absorption of wavelengths by constituents of the atmosphere such as carbon dioxide and water vapour.

Electronic state of matter and photon capture

Absorption of radiation in the visible spectrum depends on the electronic states of the atoms and molecules in the absorbing substance. Substances absorbing visible light are called pigments and are coloured according to the wavelengths absorbed; thus chlorophyll absorbs red and blue light and so appears green to the eye. As photon capture in photosynthesis is predominantly by large, complex molecules, brief consideration will be given to molecular orbitals and how electrons are arranged in them.

To capture a photon of visible light, the energy levels of the atomic or molecular orbitals must have a difference in energy corresponding to the absorbed quantum. The electronic orbitals of atoms are analogous to molecular orbitals. The main energy levels are referred to as the ground state (lowest energy), and the first and second (or higher) excited states. The energy levels are designated by a total quantum number. Within each level the number of possible orbitals depends on the magnetic motion and orientation of the electrons in relation to the nuclei, as described by quantum mechanics. Electrons spin within an orbital according to the prevailing magnetic field. Spin may be parallel to the field or antiparallel to it, and is designated by a spin quantum number, S, which has values of $+\frac{1}{2}$ or $-\frac{1}{2}$. Two electrons can only occupy the same orbital if their spins differ. When two electrons of opposite spin occupy the lowest energy orbital the configuration is a stable electronic ground state, S_0. Adding the spins of electrons together gives the total spin S and the spin multiplicity of the electrons (Fig. 2.3) is given by 2S + 1. When all spins balance out S = 0 and the spin multiplicity is 1, a singlet state. If there is spin reversal, then S equals 1 and 2S + 1 becomes 3 giving a triplet. Oxygen is an example of a ground state triplet. With an electron absent from an orbital, S equals $\frac{1}{2}$, and the spin multiplicity is 2 or a doublet. Singlet and triplet states are important in photochemistry, and the doublet in free radicals, which may cause photochemical damage in photosynthesis. The chemical and physical texts listed provide quantitative descriptions of molecular structure and quantum mechanics.

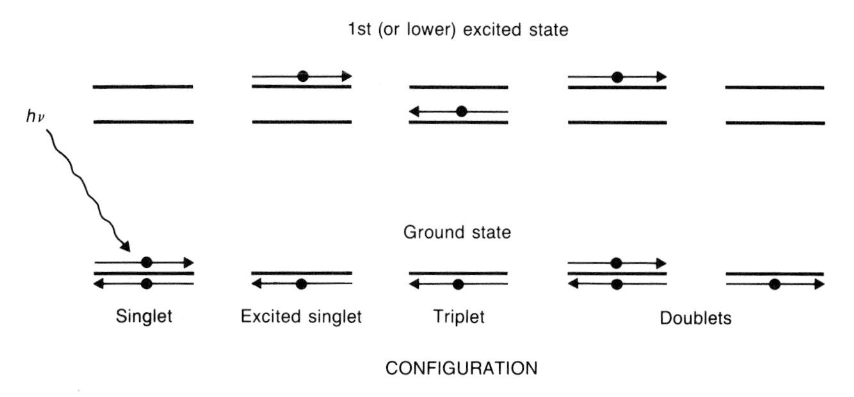

FIG. 2.3 Electron transitions and changes in electron spin induced by a photon, direction of spin is shown by →.

With a given electronic configuration, a molecule absorbs light of particular wavelength, smaller or larger quanta are not absorbed and cannot, by the Grotthus–Draper Law, cause a physical or chemical change. If the oscillating electronic vector of a photon causes an electron in a molecule to resonate (i.e. to vibrate at the same frequency) then the energy of the photon will be captured. The direction of the electrical and magnetic fields of the photon must be in correct orientation to the electronic oscillations to cause resonance. Capture of a quantum is rapid by all criteria of 'normal' biological interactions; an electron 'jumps' from one energy level to another in a time inversely proportional to the frequency of the wave, $(1/\nu)$. For red light (Table 2.1) this is about 2.3×10^{-15} s which is much faster than nuclear vibrations (10^{-13} s) and so does not influence the nuclear configuration of the molecule.

In complex molecules, with many nuclei and electrons, the possible combination of orbitals is greatly increased, and the energy levels are split into vibrational and rotational levels as electrons within the molecule are influenced by magnetic and electrical forces. Molecules absorbing visible light exhibit differences in energy of about 1 eV between ground and excited states and vibrational energy differences of approximately 0.1 eV; rotational energy levels are not observed at these wavelengths in solid state, only in gases.

Larger molecules usually have more complex energy levels because the outer electrons which provide the bonding are 'delocalized' over the whole molecule. These π electrons travel in the extended π orbitals, even in the ground state and may undergo transition to an excited orbital, π^*, in the same way that electrons in other, more 'rigid' orbitals can. The π^* orbitals are delocalized; the electrons are not orientated in the same way in π and π^* states, they are, respectively, bonding and antibonding. As π electrons are free to move in large volume the energy levels are smaller, and therefore the binding energy is greater than for a system of double and single bonds and the molecule is more stable. Also, the energy required for the ground to excited state transition is low and the capacity to absorb light of longer wavelength is increased.

Delocalization is of great importance in organic molecules, including the photosynthetic pigments which have extensive orbitals over large molecules. Extensive π systems confer high efficiency of energy capture, together with stability under normal conditions *in vivo*. Also, orbitals of different energies provide for absorption of different wavelength of light. Groups of atoms in a molecule which are responsible for absorbing light energy are called chromophores. In the 200–800 nm range chromophores always have loosely bound electrons, in π orbitals, the size of which determines wavelengths absorbed. For example, the peptide bond (found in proteins) is small and absorbs short energetic wavelengths (190 nm), the nucleic acid bases which have a larger π system absorb at 260 nm and β-carotene (absorbing at 400 to 500 nm) and chlorophyll (absorbing between 400 and 700 nm) have increasingly larger π systems. The absorption maxima become more dependent on the environment of the molecule as the size of the π system increases. Energy levels of a molecule depend on the environment, on intramolecular rearrangement, binding to other compounds etc., which alter the absorption of particular wavelengths. Chlorophyll *in vivo* is associated with protein, which 'tunes' the absorption of light over a range of wavelengths.

Chemical reactions involve reorganization of the outer shell or valency electrons which form chemical bonds. For an excited electron to be used chemically it must be held in a configuration which is stable for long enough to transfer to a chemical acceptor directly or *via* an exchange of energy (not electrons). In photosynthesis the light-harvesting pigments, which capture the energy, transfer excitation *via* other pigment molecules to special reaction centres, composed of a form of chlorophyll, which convert the excitation energy to chemical energy.

The wavelength and number of photons captured determines the maximum energy available for biological processes and therefore governs the overall energetics, although not the type of reaction possible within the limits imposed by available energy. However, the efficiency of energy capture and of all the linked conversion processes limits the actual energy which can be obtained. It is possible to overcome the inefficiency of a photochemical process if there is a suitable system allowing the energy from the capture of several photons to be 'stored' or gathered in some way and then used, by multistep processes, to generate the required energy and achieve photochemical conversions. This occurs in photosynthetic organisms as will be discussed at length in the section on photoreactions and electron transport.

Dissipation of excitation energy

Electrons remain in the excited state for a period, called the 'lifetime', dropping back to the stable ground state in an exponential decay. Processes dissipating the energy of the excited state are illustrated in Fig. 2.1. Thermal relaxation (also called radiationless transition) between the closely spaced vibrational energy levels leads to loss of excitation as heat within 10^{-12} s and is the normal pathway of energy loss from the second to first excited states. Excitation energy may be transferred to a chemical acceptor; this is the central event in photosynthesis and is considered at length later. Electrons also decay from the excited singlet state to the ground state emitting 'prompt' fluorescence in 10^{-9} s. Electrons in a lower ground state at normal temperatures reach the centre of the excited state when excited, but decay rapidly by thermal relaxation to a lower level of that excited state, before finally dropping to an energy level in the ground state of different energy from which they started. The excitation energy is therefore greater than that of the photon emitted and the fluorescence spectrum is shifted (Stoke's shift) towards longer, red wavelengths. If the molecule is excited to the second excited state then the absorption spectrum shows two bands. However, radiationless transition from the second to the first excited state allows fluorescence only from the first excited state, and only one red-shifted fluorescence band is detected (Fig. 2.4).

When electrons decay from the excited triplet before returning to ground state and releasing a photon, the process is called phosphorescence. The photon is of longer wavelength than fluorescence due to a smaller energy difference between the triplet and ground states. Phosphorescence may be much delayed as triplet states have a very long lifetime, from milliseconds to tens of seconds, due to the change in spin which accompanies the triplet transition and occurs with low probability. Electrons in the triplet, when energized to the excited singlet by thermal energy, decay to ground state releasing a photon in the process of delayed light emission (also called luminescence, delayed fluorescence and 'after glow'); the wavelength of the photon is shifted like fluorescence as both come from the same energy level. Delayed light emission is strongly dependent on temperature and on a small energy gap between triplet and excited singlet. Another type of fluorescence originates from triplet–triplet annihilation by two molecules giving only one excited singlet state molecule which decays. This fluorescence is proportional to the square of the exciting light. Fluorescence of chlorophyll in the thylakoid membrane is an important characteristic. Prompt fluorescence comes from

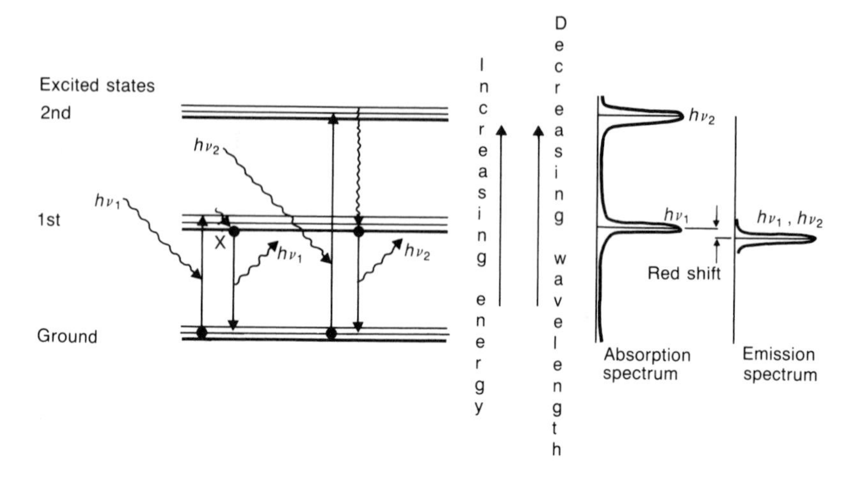

FIG. 2.4 Capture of photons of low ($h\nu_1$) or high ($h\nu_2$) energy, shown by the absorption spectrum, raises an electron to the first and second excited states. Electrons drop from the second to first excited state by radiationless decay, and photons of similar wavelength are emitted from the first excited state, giving a single peak in the emission spectrum. Radiationless decay from the higher to lower energy levels of the first excited state (at X) redshifts the emitted photons.

chlorophyll which cannot pass excitation onto another molecule, either because it is not joined to it or because the system is already overexcited by light. This fluorescence may arise because the reduced acceptor of energy in the reaction centres (page 37) passes energy back to chlorophyll which then fluoresces. However, energy is lost from the chlorophyll matrix over a long period by delayed fluorescence and the rate depends on the state of the complete system.

The rate of decay of the excited state depends on radiationless transitions, fluorescence and photochemistry. The processes have rate constants for de-excitation, respectively K_d, K_f and K_p so that the overall rate constant is $K = (K_d + K_f + K_p)$. With n_0 excited states initially, the decrease to n excited states in time, t, is given by $n = n_0 \, e^{-Kt}$ where e is the base of natural logarithms. The number of excited states decreases exponentially to $(1 - 1/e)$ of the initial number of excited states in a time $= 1/K$ which is the lifetime, the time required for 63 per cent of the excited electrons to decay. The exponential decay process also can be characterized by the 'half-life', the time needed for half the original population of excited electrons to decay. Singlet excited states are relatively short lived, that of chlorophyll is about 5×10^{-9} s.

Transfer of excited states

Decay of excited electrons to the ground state, either by radiationless decay or fluorescence, wastes the energy for photosynthesis and *in vivo* the main mechanism which captures the energy of the excited pigment is transfer to a special form of chlorophyll (RC). This can donate electrons to a chemical acceptor. Energy does not pass directly from excited pigment (usually called the donor, D*) to the final chemically reactive acceptor (CA) but *via* other pigment molecules (A) within the groups of light capturing but chemically unreactive pigments which form the light harvesting 'antenna' (see Fig. 3.1):

$$\begin{aligned} &D + h\nu + A \rightarrow D^{\star} + A \rightarrow D + A^{\star} \\ &A^{\star} + A \rightarrow A + A^{\star} \rightarrow \text{repeated} \rightarrow \\ &A^{\star} + RC \rightarrow A + RC^{\star} \rightarrow RC^{\star} + CA \rightarrow RC^{+} + CA^{-} \end{aligned} \qquad [2.3]$$

The mechanisms of energy transfer depend on the types of molecules, their size, energy levels of the electronic orbitals etc., and on concentration and orientation. In solids or concentrated solutions, orbitals 'overlap' and form extensive 'superorbitals'. When excited, an electron enters the delocalized conduction bands and leaves a 'hole' (a positive charge) in the ground state; electrons migrate through the pigment matrix. This semiconductor type of mechanism leads to photoconductivity, which is observed in dried chloroplasts but is not thought to play an important role *in vivo*. However it has been suggested to occur in closely bound chlorophyll groups. Another 'strong' interaction occurs between closely packed molecules at 1–2 nm spacing; it is rapid (10^{-12} s) and depends on (1/distance) and there is no radiationless decay. At very close spacing (< 1 nm) energy is rapidly lost, preventing photochemistry. None of these mechanisms is thought to be important in photosynthesis. More important is repeated energy transfer between donor and acceptor leading to energy migration between groups of the same and different molecules at distant spacing. Dipole interaction (Fig. 2.5) causes inductive resonance in the acceptor, so that the excitation is passed to it. There is no mass or electron transfer, the excitation migrates as a spin-coupled electron-hole pair and is localized on a definite molecule. This 'weak interaction' involves rates of transfer of 10^{-12}–10^{-14} s and is called a Förster mechanism, after the discoverer. The interaction decreases as R_0^{-6}, where R_0 is the Förster distance, defined as the distance between molecules at which energy transfer rate equals the losses due to fluorescence and radiationless decay. R_0 is the mean distance between the dipole oscillating centres in each molecule and, for photosynthetic systems *in vivo*, is between 4 nm and 10 nm for chlorophyll with an average of about 6 nm for efficient transfer, and minimizes the number of vibrations undergone by the donor; it is temperature dependent. Transfer is primarily *via* singlet excited states. Orientation between dipoles is important; transfer is zero with perpendicular orientation and maximal with parallel orientation. Energy levels are most critical; transfer only occurs if the fluorescence bands overlap because the electron drops by radiationless decay to the lowest vibrational level of the excited state from which it decays as a fluorescence photon if its energy is not transferred. The acceptor absorption band is not at the lowest vibrational level but in the centre of the energy band and will take excitation if of the correct energy, that is, similar to fluorescence.

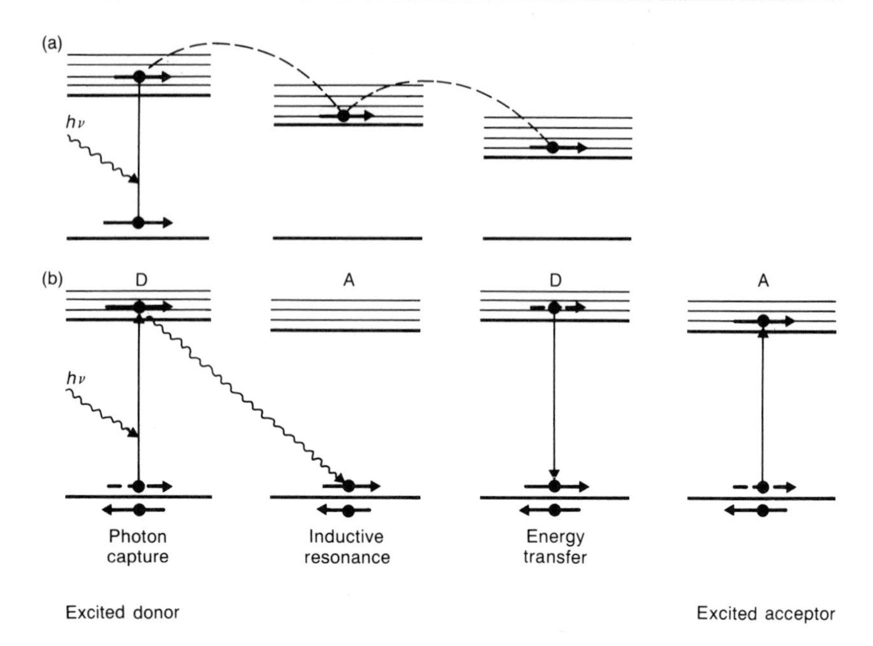

FIG. 2.5 Energy transfer between molecules by rapid excitation transfer (a) from closely spaced molecules of similar energy level or (b) by inductive resonance between donor (D) and acceptor (A) molecules.

As energy is lost at each transfer, a donor–acceptor chain of decreasing ground state to first excited state energy difference favours funnelling of energy in a particular direction and increases the rate of exciton transfer. There may be transfer between similar molecules (homogeneous) by random walk with excitation passing in the pigment matrix at random but, as this leads to energy loss, transfer between different types of molecules (heterogeneous) is more likely. It is faster, requiring fewer steps (perhaps 10–100), and very efficient. However *in vivo* there may be different transfer mechanisms in different parts of the pigment antenna.

Light absorption and absorption spectra

When light passes through a solution of a compound, some wavelengths are transmitted, that is, passed through without alteration, some scattered and the remainder absorbed in proportion to the concentration of absorbing substance. Absorbed light is of the greatest importance in photosynthesis. The number of photons captured at different wavelengths is the difference between incident and transmitted light (allowing for scattering etc.). The number of photons absorbed at different wavelengths constitutes an absorption spectrum, which gives much information on molecular configuration, electronic transition and energy levels. Substances have characteristic spectra from which they may be identified and quantified. Absorption spectra are measured with a spectrophotometer; radiation of the required wavelength band or monochromatic light is passed through a known thickness of solvent (in which the substance to be studied will be dissolved) contained in a cuvette of material transparent to the wavelength to be measured (e.g. glass for visible light, quartz for ultraviolet light). This is a reference or standard solution to correct for reflectance and transmission characteristics of the cuvette. A solution of the substance is then substituted for the solvent and the decrease in transmitted light is measured. The books by Clayton (1980), Willard *et al.* (1965) and several contributions in the volume edited by Hoppe (1983) consider techniques.

With I_0 photons per unit area and time passing into the cuvette of thickness l (cm) and a fraction I not absorbed and passing through it, the fraction absorbed is $I_a = I_0 - I$. Thus dI photons are absorbed in a thin layer dl of area A by n molecules, uniformly distributed per unit volume. The average cross section, σ, of each molecule is the area of the molecule for photon capture. With a probability, p, that a photon will be captured by a molecule, the absorption cross section, k, is equal to $p\sigma$ (units, cm^{-2}). Thus the absorption of photons is given by:

$$-\mathrm{d}I = k\,n\,\mathrm{d}l\,I \qquad [2.4]$$

which integrates to give the Beer–Lambert law:

$$I = I_0\,\mathrm{e}^{-knl} \qquad [2.5]$$

where *knl* is the absorbance. This may be expressed as

$$I = I_0\,10^{-\varepsilon cl} \qquad [2.6]$$

where c (units, molar) is the molar concentration of absorber and ε is the molar extinction coefficient or molar absorbtivity (unit, $M^{-1}\ cm^{-1}$), characteristic of the absorber in a particular solvent at a given wavelength. Transmittance, I/I_0, is related to optical density (OD), also called absorbance or extinction as:

$$\mathrm{OD} = \log I_0/I = \varepsilon\,c\,l \qquad [2.7]$$

Absorbance increases linearly with concentration in dilute solution and is often quoted as $A^{1\%}_{1\,\mathrm{cm}}$ or absorbance of a 1 per cent concentration in a 1 cm layer.

It is possible to measure the concentration of several components (which do not react) in a solution if their extinction coefficients are known at different wavelengths, as at any one wavelength

$$\mathrm{OD} = \varepsilon^{\mathrm{A}}\,c^{\mathrm{A}}\,l + \varepsilon^{\mathrm{B}}\,c^{\mathrm{B}}\,l \qquad [2.8]$$

and by measuring at different wavelengths (at least as many as there are components) the equations can be combined to give the concentrations. This is the basis for Arnon's much used method of measuring chlorophylls *a* and *b* in the same extracts of leaves (see Neubacher and Lohmann 1983; Arnon 1949).

References and Further Reading

Arnon, D. I. (1949) Copper enzymes in isolated chloroplasts. Polyphenoloxidase in *Beta vulgaris*, *Plant Physiol.*, **24**, 1–15.

Broda, E. (1978) *The Evolution of the Bioenergetic Processes*, Pergamon Press, Oxford.

Clayton R. K. (1971) *Light and Living Matter: A Guide to the Study of Photobiology*, Vol. 1. *The Physical Part*, McGraw-Hill, New York. Reprinted in 1977 by R. E. Krieger, Huntington, New York.

Clayton, R. K. (1980) *Photosynthesis: Physical Mechanisms and Chemical Patterns*, I.U.P.A.B. Biophysics Series, Cambridge University Press, London.

Gates, D. M. (1962) *Energy Exchange in the Biosphere*, Harper and Row, New York.

Giese, A. (1964) *Photophysiology, 1, General Principles: Action of Light on Plants*, Academic Press, New York.

Govindjee and **Whitmarsh, J.** (1982) Introduction to photosynthesis: Energy conversion by plants and bacteria, pp. 1–18 in Govindjee (ed.), *Photosynthesis*, Vol. 1, *Energy Conversion by Plants and Bacteria*, Academic Press, New York.

Hallett, F. R., Speight, P. A. and **Stimson, R. H.** (1982) *Physics for the Biological Sciences. A Topical Approach to Biophysical Concepts*, Methuen/Chapman and Hall, Toronto.

Hoppe, W., Lohmann, W., Markl, H. and **Ziegler, H.** (1983) *Biophysics*, Springer-Verlag, Berlin.

Malkin, R. (1977) Delayed fluorescence, pp. 473–91 in Trebst, A. and Avron M. (eds), *Encyclopedia of Plant Physiology* (N.S.), Vol. 5, *Photosynthesis I*, Springer-Verlag, Berlin.

Moore, W. J. (1957) *Physical Chemistry* (3rd edn), Longman, London.

Neubacher, H. and **Lohmann, W.** (1983) Applications of spectrophotometry in the ultraviolet and visible spectral regions, pp. 100–9 in Hoppe, W., Lohmann, W., Markl, H., and Ziegler, H. (eds), *Biophysics*, Springer-Verlag, Berlin.

Nobel, P. S. (1974) *Biophysical Plant Physiology*, W. H. Freeman, San Francisco.

Parson, W. W. and **Ke, B.** (1982) Primary photochemical reactions, pp. 331–85 in Govindjee (ed.), *Photosynthesis*, Vol. I, *Energy Conversion by Plants and Bacteria*, Academic Press, New York.

Smith, K. C. (ed.) (1977) *The Science of Photobiology*, Plenum, New York.

Sybesma, C. (1977) *An Introduction to Biophysics*, Academic Press, New York.

Wells, C. H. J. (1972) *Introduction to Molecular Photochemistry*, Chapman and Hall, London.

Willard, H. H., Merritt, L. L. and **Dean, J. A.** (1965) *Instrumental Methods of Analysis* (4th edn), Van Nostrand Reinhold Co., New York.

CHAPTER 3

Light harvesting and energy capture in photosynthesis

Three functions of the light-harvesting apparatus contribute to the utilization of light quanta to produce a chemical intermediate of higher energy state:

- light absorption: the energy of a photon is captured by an antenna pigment molecule and an electron is excited,
- energy transfer: excitation energy moves through the antenna to a reaction centre, a special form of pigment, in which it excites an electron,
- electron transfer: an energized electron from the reaction centre passes to a chemical acceptor and the oxidized reaction centre is reduced by an electron from water.

These processes are controlled by the physical and chemical characteristics of the pigments, particularly chlorophyll (chl).

Light-harvesting pigments

Several types of pigments harvest the energy of light (Table 3.1). Only special forms of chlorophyll *a* (chl *a*) and bacteriochlorophyll (bchl) form reaction centres in higher plants and photosynthetic bacteria respectively. All other pigments are therefore accessory pigments, forming an antenna for capturing photons which donates excitation to the reaction centres. A group of many antenna molecules of different type, donates energy to a single reaction centre complex. Pigments absorb in different parts of the spectrum and, in combination, enable an organism to absorb light of different wavelengths. This is of great importance ecologically for organisms may then exploit a greater energy supply or grow in different radiation environments. For example phycobilins of red algae absorb blue light, which predominates in deep water where the plants grow. Figure 3.1 illustrates, for different organisms, the pigments, wavelengths absorbed and composition of the light-capturing system.

Photosynthetic bacteria have several forms of pigments donating energy to the reaction centre bacteriochlorophyll. In higher plants chlorophyll *b* (chl *b*) is an auxiliary pigment passing excitation to chl *a*. In all O_2-evolving organisms excitation moves to reaction centres composed of forms of chl *a* which pass the energy on as an excited electron to chemical reactions.

Table 3.1 Pigments of photosynthetic organisms: only some of the more important pigments and groups of plants in which they occur are given. Primary pigments are those involved in the photochemical process, accessory pigments function only in light harvesting

Organism	Primary pigment	Accessory pigment(s)	Wavelength absorbed (nm) (approximate)
Prokaryotes			
Purple bacteria	bchl *a*	bchl *a*	Blue-violet–red 470–750
		bchl *b*	Blue-violet–red 400–1020
Green sulphur bacteria	bchl *a*	bchl *c* = *Chlorobium* chl	Blue–red 470–750
Blue-green algae	chl *a*	Phycocyanin Phycoerythrin Allophycocyanin	Orange 630 Green 570 Red 650
Eukaryotes			
Red algae	chl *a*	Phycocyanin Phycoerythrin Allophycocyanin	Orange 630 Green 570 Red 650
Brown algae	chl *a*	chl *c*	Violet-blue–red
Higher plants	chl *a*	chl *b*	Violet-blue–orange-red 454–670
Most higher plants, algae and bacteria		Carotenoids α, β in different groups Xanthophylls	Blue-green 450

Chlorophylls

Chlorophylls are probably the most abundant biological pigments and the world appears green to the human eye due to their absorption of blue and red light. Leaves may contain up to 1 g of chlorophyll m^{-2} of surface area but this varies with species, nutrition (particularly nitrogen fertilization), age, etc. Chlorophyll is extracted with fat solubilizing solvents (e.g. ether, acetone) for it is a lipophilic molecule only found in membranes containing lipid and hydrophobic protein. Chl *a* and chl *b* may be separated by chromatography.

Chemically chlorophylls are chlorin macrocycles with four-fold symmetry, derived from porphyrin. Chl *a* (Fig. 3.2) is a conjugated macrocyclic molecule (mass 894) with a planar 'head' of four pyrrole rings; it is about 1.5 × 1.5 nm but the overall size is much greater due to a phytol group, a terpene alcohol chain some 2 nm long, which positions the molecule in membranes. The

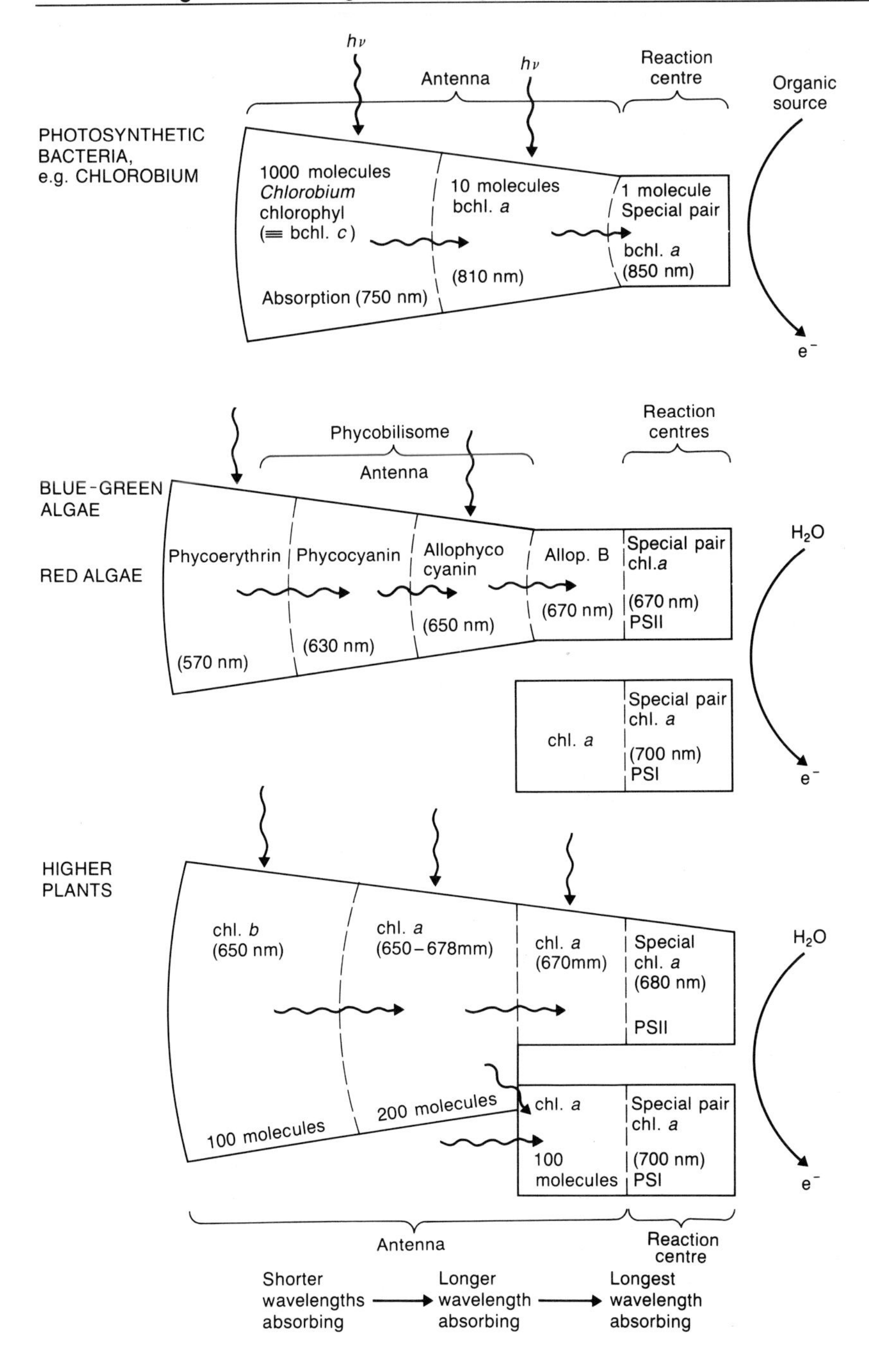

FIG. 3.1 Energy absorption and excitation transfer between pigments in the light-harvesting antenna and to the photochemical reaction centres of different photosynthetic organisms.

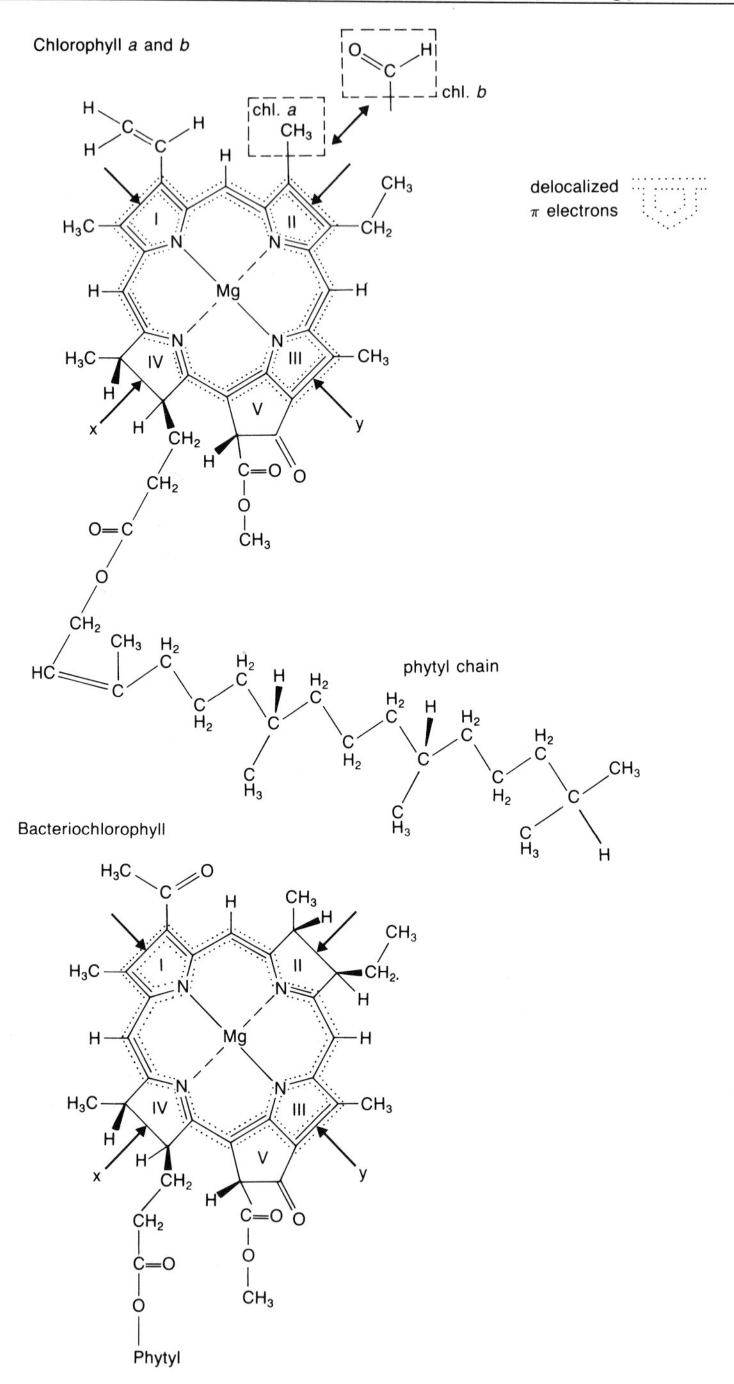

FIG. 3.2 Structure of chlorophyll *a* and bacteriochlorophyll *a*. The side groups on ring II are responsible for the difference between chlorophyll *a* and *b*. Axes x and y are the principal electronic transitions.

chemical groups and H^+ on the outer edge of the pyrrole unit confine the electrons to a single plane, increasing absorption of red wavelengths, an advantage for plants of land or surface water. A non-ionic magnesium atom, bound by two covalent and two co-ordinate bonds in the centre of the molecule, co-ordinates the rings. Magnesium is crucial to capture of light energy, an iron atom (as in related pigments) cannot do this. Magnesium is a 'close shell' divalent cation which changes the electron distribution and produces powerful excited states; chl *a* is a good donor of electrons (at a large negative potential), and excited chl *a* at the reaction centre is a very strong oxidizing substance able to accept electrons (indirectly) from water. The large size and extensive ring structures of the chlorophylls, with ten double bonds, allows electrons to delocalize in the π orbitals over the 'head' of the molecule, increasing the area for capture of a photon, and giving many redox levels which are important for efficient energy capture and transfer, and produce a complex absorption spectrum. In chl *b* a formyl (—CHO) group replaces the methyl (—CH_3) group on ring II (Fig. 3.2), which increases the blue and decreases the red absorption maxima and alters solubility; chl *b* is less soluble in petroleum ether than chl *a* but more soluble in methyl alcohol for example. Bchl *a* (Fig. 3.2) is similar to chl *a* but has an acetyl instead of a vinyl group on ring I, and ring II is saturated with hydrogen instead of unsaturated. This loss of the double bond alters the π system. Such differences, which are probably later evolutionary developments, increase absorption at longer wavelengths, but bchl cannot generate a strong oxidant to remove electrons from water.

Absorption spectra of chlorophyll

Measured in organic solvents after extraction from the plant (Fig. 3.3), chl *a* absorbs most strongly at 430 nm (Soret band) and 660 nm and chl *b* at 450 and 640 nm. The absorption maxima 'shift' with the solvent; in forty different solvents the red absorption of chl *a* is between 660 and 675 nm. Polar solvents, such as acetone, cause strong dipole–dipole interactions, weaken London dispersion forces and change hydrogen bonding, altering the electronic configuration of the molecule and hence absorption. Aggregation of chlorophyll also causes a shift; crystalline chl *a* has its long wave absorption minimum at 740 nm.

Differences in absorption spectra and molar extinction coefficients enable chlorophylls to be distinguished and measured spectrometrically in unpurified solutions. Chl *a* has a molar extinction coefficient of $1.2 \times 10^5\ M^{-1}\ cm^{-1}$ (p. 29) at 430 nm; a 10^{-5} M solution is intensely coloured and absorbs some 80 per cent of the incident light. Chlorophyll is a very efficient pigment with a cross-section absorbing area per molecule of $3.8 \times 10^{-16}\ cm^2$. On a bright noon day at the earth's surface (a photon flux of 2×10^{-3} mol quanta $m^{-2}\ s^{-1}$) a chlorophyll molecule will capture about 45 photons s^{-1}.

The maxima in light absorption correspond to different energy levels in the molecule (Fig. 3.3). The highest energy level is the second (or higher) singlet,

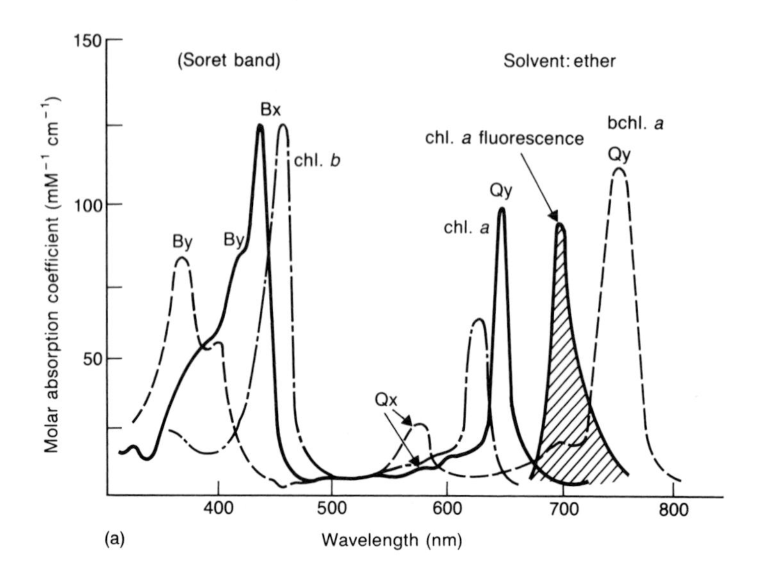

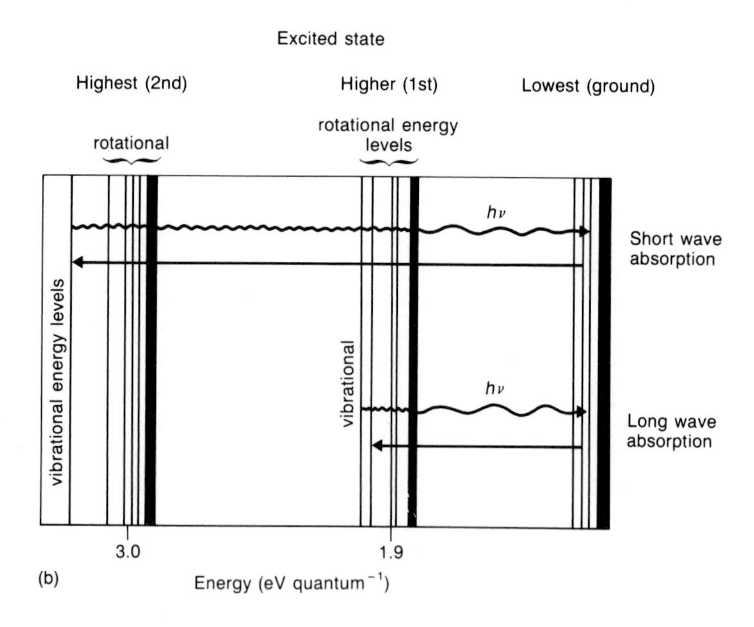

FIG. 3.3(a) Absorption spectra of chlorophylls *a* and *b* and bacteriochlorophyll *a* in ether showing principal electronic transitions in blue (Soret) and red wavelengths. An energy level diagram (b) for chl *a* is related to the main absorption peaks.

excited at 430 nm. The excited electrons drop in 10^{-12} s from the second to the first excited singlet state, to which electrons are excited by red light. Thus the energy of blue light is dissipated and a blue photon is no more effective in photosynthesis than a red.

The lifetime of the lowest excited singlet state is about 5×10^{-9} s and decays to the singlet ground state with a probability of $\frac{1}{3}$ or to the lower triplet state with the probability of $\frac{2}{3}$. In a solution of monomeric chlorophyll there is little radiationless dissipation to the ground state but in aggregated chlorophyll it dominates. Absorption spectra of chlorophylls show the electronic transitions along axes of the molecule. Polarized light and paramagnetic or electron spin resonance (ESR) are used to analyse the transitions, which are very important in understanding the mechanism of energy transfer (Clayton 1980). The x-axis of chlorophyll is through the nitrogen (N) atoms of rings II and IV and the y-axis through the N–N atoms of I and III (Fig. 3.2). The two main absorption bands in the blue and two in the red are called B and Q respectively, due to $\pi \rightarrow \pi^\star$ transitions. The polarization of the transition along the axes are called x and y, and may be from the lowest vibrational energy state (called 0) or the next higher energy state (called 1). Thus absorption at 430 nm is a Bx (0,0) transition and at 660 nm it is a Qx (0,0). The Qy transition is most altered by solvent (e.g. water bound to Mg) and association in membranes which increases the red absorption.

The chlorophylls at the reaction centre

Normal absorption spectra cannot measure small changes in absorption occurring in a few molecules if the bulk of the pigments have overlapping absorption bands. Most chl *a* and *b* molecules in higher plants form the antenna, and only few chl *a* form reaction centres which pass electrons to acceptor molecules. Antenna chlorophyll 'swamps' the absorption even in preparations enriched in reaction centre chlorophyll. To overcome this, difference spectra compare absorption between light and dark with repetitive flashes or with chemical oxidation and reduction (see Witt 1975) enabling the chemically reactive form of chlorophyll to be studied. ESR and electron nuclear double resonance (ENDOR) techniques are also used to detect changes in paramagnetism caused by the transfer of electrons to an electron acceptor, independent of changes in the non-paramagnetic bulk chlorophylls and other pigments.

In isolated spinach chloroplasts from which most of the chlorophyll had been removed, Kok (1976) observed from difference spectra, a decrease in absorption (called photobleaching) near 700 nm following illumination (Fig. 3.4). The signal was altered by the redox state of the chlorophyll; when oxidized there was a loss of absorption at 680 and 690 nm and increase at 686. The form of chl *a* responsible is called P700 (P for pigment and the wavelength of the absorption change) and it is the reaction centre for photosystem I (PSI)

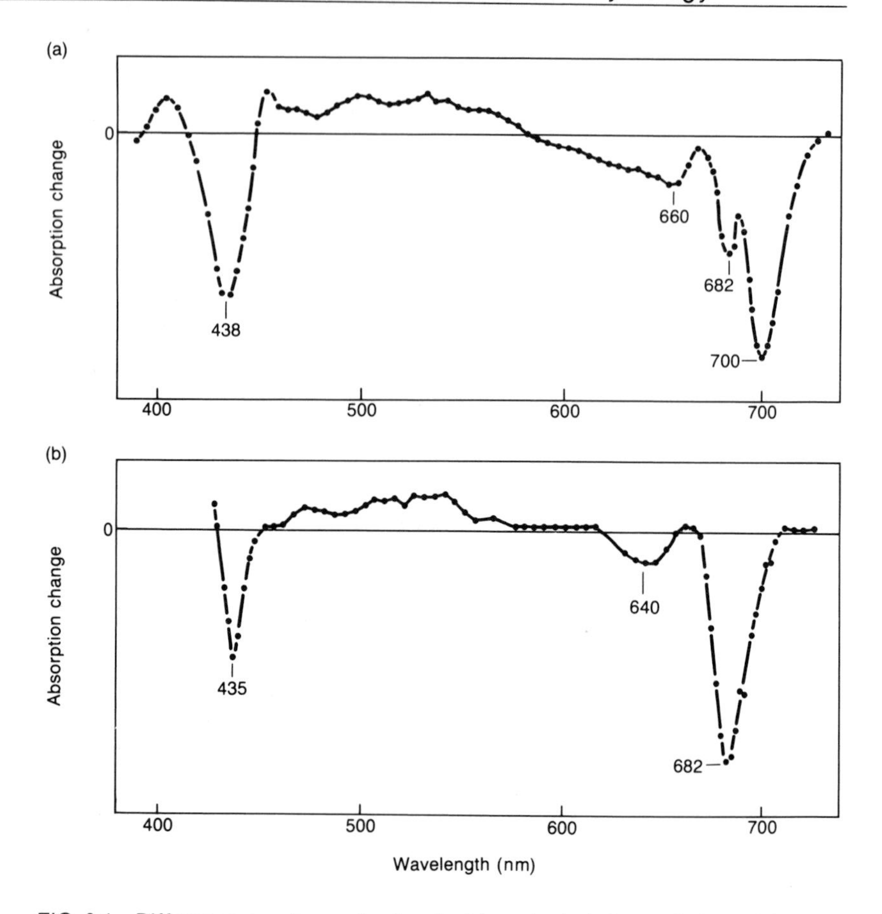

FIG. 3.4 Difference spectrum of spinach chloroplasts between light and dark; (a) the change at 700 nm is due to absorption by PSI reaction centre chl *a* (after Kok (1961)). (b) The reaction centre chl *a* absorption of PSII (P680). (After Doring *et al.* (1969)).

(p. 47). Only one in 300 to 400 chlorophyll molecules is the special form, P700. Difference spectra also show fast change (60 ps) in absorption at 680 nm, separate from but equivalent to the P700; it is associated with the reaction centre chl *a* of photosystem II (PSII) (p. 47) and is called P680.

Duysens observed, in bacterial chromatophores, a small (2%) decrease in absorption at about 870 nm wavelength caused by illumination or chemical oxidation; the absorption then increased slowly in the dark. By removing the major part of the bchl with detergent the signal from P870, the bacterial reaction centre, was enhanced (see Clayton 1980).

Changes in absorption with illumination or chemical modification show charge separation during transfer of electrons from reaction centre chl *a* (or bchl) to an acceptor, A

$$\text{chl } a + \text{A} \xrightarrow{\text{light}} \text{chl } a^{+} + \text{A}^{-} \qquad [3.1]$$

P700, P680 and P870 convert light energy to chemical form. This step is the true 'photochemical act' in photosynthesis and thus has a unique role, not simply a special one. The organization and function of higher plant reaction centres and comparison with bacterial reaction centres, which are similar to PSI, is discussed by Clayton (1980). The structure of the higher plant reaction centre is also considered by Katz, Shipman and Norris (1979) and Okamura, Feher and Nelson (1982).

The molecular structure and arrangement of P700 and the other reaction centres are unknown in detail. Photo-oxidation is accompanied by an ESR signal identified as an unpaired electron delocalized over a large π-system. The ESR signal and photobleaching of P700 is 1:1, that is, an excitation produces one free radical and ejects one electron from chl *a*. The structure of P700 is, on evidence from ESR and ENDOR, probably a 'special pair' of chl *a* molecules joined covalently as a dimer in a specific way. In photosynthetic purple bacteria a special bacteriochlorophyll dimer complex (linked to protein and electron acceptors) acts as electron donor. The electron is distributed over the π system of both molecules of the dimer. Several models of the PSI reaction centre have been proposed, an example (Fig. 3.5) has Mg of one chlorophyll linked to the ketocarboxyl group on ring V of the other, possibly by water, for it has a lone pair of electrons on the O to join to Mg and H bonds to form the other link. The two chl *a* molecules are then parallel with the NI–NIII axes aligned with a small distance (0.36 nm) between the Qy transitions bringing the orbitals into an overlapping position and forming a very extended π orbital system.

Chemical groups other than water could be the ligand for the dimer, for example the —OH groups of serine or threonine, the $—NH_2$ on lysine and —SH groups. Thus the protein matrix could be responsible for orientation of the chlorophylls and P700 would then be a protein–chlorophyll special pair. However, the electronic state transition involves only chlorophyll, not the protein. In membranes the Qy transition of special pair chl *a* might be orientated, probably by the protein, almost parallel to the plane but Qx may be at a greater angle. Antenna chl *a* is less closely orientated and chl *b* is possibly at angles greater than 35°.

The characteristics of such a special pair chl *a* dimer would be that the Qy transition is shifted to the required 700 nm, the energy levels are correct for sharing the unpaired electron and that it is stable, allowing an excited electron to transfer to an acceptor (with which it forms a radical pair) whilst holding the electron for sufficient time for reaction to occur. Also, the $P700^{+}$ cation free radical formed on oxidation is not as reactive as the monomer, thus reducing the chances of back reactions with the acceptor. However, the dimer model is not fully accepted for the PSI reaction centre, as more refined ESR and ENDOR spectroscopy favours a monomer of chl *a*, which has a more realistic redox

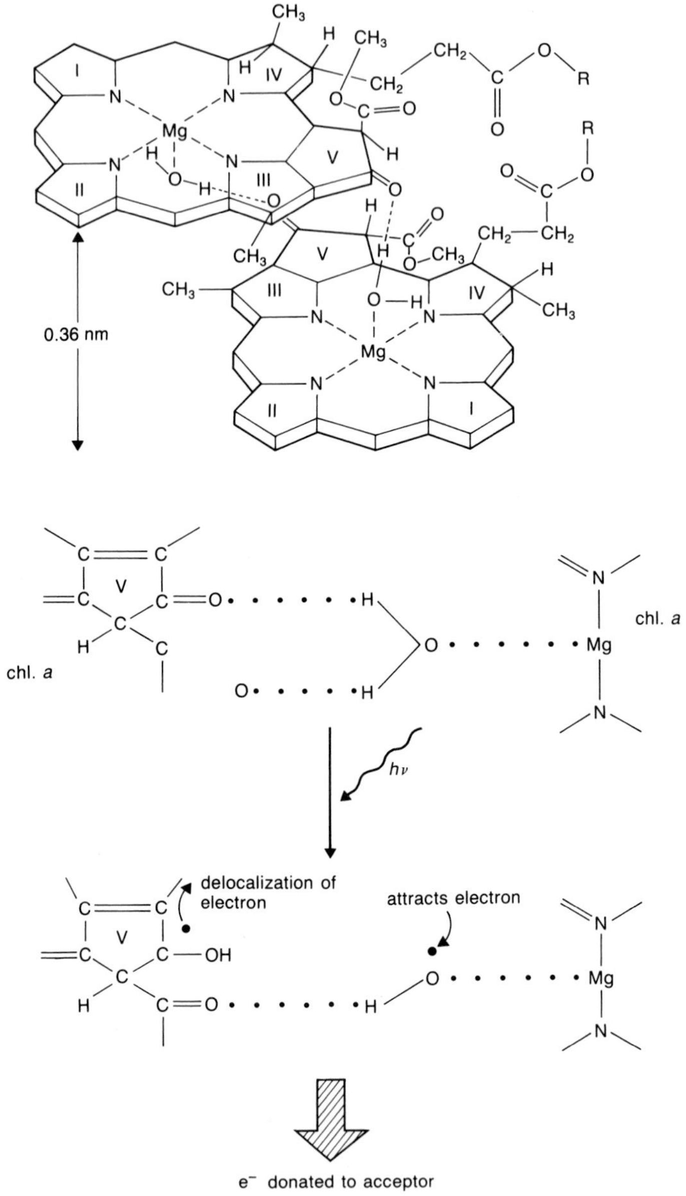

FIG. 3.5 Possible configuration of photosystem I chl special pair, P700, at the reaction centre. The ligand is shown as water although other chemical groups may function. A scheme of electron movement, resulting in charge separation, shows how the ligand may function. Adapted from Katz, Shipman and Norris (1979) with permission.

potential. Thus there is no generally accepted model of PSI reaction centre structure or function. The molecular arrangement of the PSII reaction centre, P680, is also not yet understood. Events at the reaction centre are of great importance and interest. The mechanism of the oxidation processes is one of exciton dissociation, with the dissociated negative charge transferred (*via* electron tunnelling) over distances of 2–3 nm, into an acceptor pool. This pool must be reduced at a rate two orders of magnitude greater than the rate of photon absorption and the back reactions must be of the same order as the photon capture. These conditions give an efficiency of the PSI reaction centre of about 20 per cent of the solar radiation energy, although losses, outside the reaction centre, in the electron transport and related processes give an efficiency of 5 per cent; such efficiency has been measured for rapidly growing plants under optimal conditions. Detailed understanding of the mechanism may help to improve the efficiency of energy transduction in photochemical processes of many types.

Absorption spectra of chlorophylls in thylakoids

Aggregation of chlorophylls in membranes is analysed mathematically (deconvoluted) by fitting curves of a normal (Gaussian) distribution to absorption spectra of thylakoids (Fig. 3.6). Chl *a* shows up to ten absorption maxima, the main ones absorbing at about 660, 670, 678, 685 and 689 nm. A 650 nm

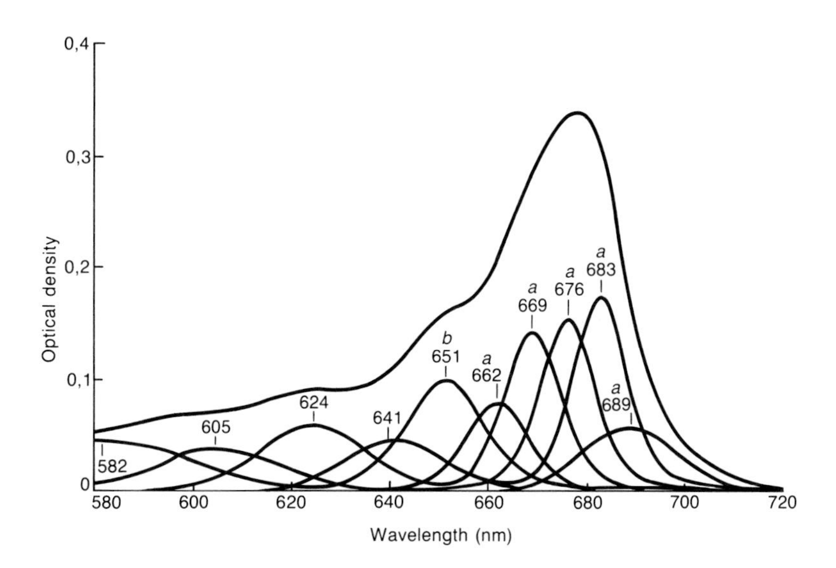

FIG. 3.6 Aggregates of chlorophyll in membranes absorb light at different wavelengths; the number and importance of the aggregates are analysed mathematically by deconvolution of the spectra. Spinach chloroplasts at 25 °C; numbers on curves are absorption maxima of individual complexes. (Modified from van Ginkel and Kleinen Hammans 1980).

absorption is due to chl *b*. Light-harvesting chlorophyll–protein (LHCP) complex (p. 75) has six bands, including that at 678 nm. Chl *b* in LHCP may be in groups exchanging by the strong excitation mechanism and linked to chl *a* by the Förster mechanism. The chl *b* antenna shows strong interaction but not chl *a*.

Chlorophyll fluorescence

Electrons in the higher levels of the first excited singlet, S, state, (energy E_a) decay by radiationless transition (R) to the lower levels and, if not used in photochemistry or transferred to other molecules, decay to the singlet ground state, S_0, by emission of 'prompt' fluorescence of lower energy (E_f) than the exciting light, as $E_f = E_a - R$. Thus a solution of chlorophyll irradiated with blue light emits red fluorescence. Phosphorescence is light emission occurring many seconds or minutes after illumination due to the triplet (T) to singlet (S) ground state transition. Delayed fluorescence involves the $T \rightarrow S_i \rightarrow S_0$ transition (with small, $T \rightarrow S_i$, energy gap) and has the phosphorescence decay time but fluorescence spectrum, or triplet annihilation or $T_1 + T_1 \rightarrow S_i \rightarrow S_0$ transition with different relationship to the energy absorbed. In chl *a* of the thylakoids prompt fluorescence is emitted at a peak of 685 nm. It shows the accumulation of excitation energy in the antenna and is inversely related to the use of electrons; it indicates the state of electron transport and biochemical processes relative to energy capture (Ch. 11). Delayed fluorescence, also called delayed light emission, was first observed by Strehler and Arnold and occurs in all photosynthetic organisms; it is from excited singlet states of chlorophyll and is of similar wavelength to fluorescence, but shorter wavelength than phosphorescence. Fluorescence and delayed light emission have similar action and emission spectra and saturation characteristics to photosynthesis. Delayed light is emitted from chlorophyll re-excited over a long period from a store of energy produced in the light, so that the time course is long compared to prompt fluorescence.

Accessory light-harvesting pigments

Pigments (Table 3.1) other than chl *a* form a major part of the light gathering antenna. Chl *b*, abundant in higher plants, has been discussed. In other organisms, such as the algae, accessory pigments are important for they capture light of wavelengths not absorbed by chlorophyll and pass the energy to the reaction centre.

Phycobiliproteins of the red algae and blue-green algae are composed of a chromophore, a bilin pigment (phycocyanin, phycoerythrin or allophycocyanin) attached to a protein, characteristic of the organism. The chromophores are straight chain tetrapyrroles related to porphyrins and therefore chlorophyll, but are water soluble. Phycocyanin and phycoerythrin absorb mainly at 630 and 550 nm respectively and allophycocyanin at 650 nm; they have high molecular extinction coefficients. Phycobiliproteins transfer energy to reaction centres with almost 100 per cent efficiency.

Carotenoids, that is the carotenes and xanthophylls which are synthesized by the same synthetic pathway except for the final steps, have the dual functions of accessory pigments in energy capture and of dissipating energy and excited states of O_2 in plant membranes. The main carotene of leaves is β-carotene and lutein is the principal xanthophyll, although violaxanthin and neoxanthin are very important. Carotenes are fat soluble orange pigments (maximum absorption at 530 nm); they are long (3 nm) chain hydrocarbons of isoprene units (five-carbon) with alternating double bonds, nine or more in different positions in the different photosynthetic carotenoids. Xanthophylls contain O_2 and are therefore not hydrocarbons, they differ in the position of the oxygen in the terminal ring structure.

The most important function of carotenes in higher plants is not as accessory pigment, for they are only 30 to 40 per cent efficient in transferring energy to reaction centres, but in dissipation of the excess energy of chlorophyll and of 'detoxifying' reactive forms of O_2. Carotenoids occur in all photosynthetic organisms that evolve O_2, and are closely associated with the reaction centres. Plants lacking carotenoids, because of chemical inhibition of synthesis or mutation, are damaged during photosynthesis. The photosynthetic bacterium *Rhodopseudomonas sphaeroides* is also destroyed in the light in the presence of oxygen when a mutation prevents carotenoid production. In the light, ground state oxygen reacts with excited state chlorophyll giving singlet oxygen which is very reactive, oxidizing (bleaching) chlorophyll, purines in nucleic acids and polyunsaturated fatty acids, which form lipid peroxides. Superoxide ($O_2^{\circ -}$) with a free radical (an unpaired electron) is produced in photosynthesis; it is both a reducing and oxidizing agent. Hydrogen peroxide, which is formed when O_2 reacts with electrons from photosynthesis, reacts with $O_2^{\circ -}$ to form the hydroxyl (OH) radical, which is very reactive indeed and damages most biological material (see p. 102). It is therefore essential to prevent formation of these compounds or destroy them quickly, before the photosynthetic system is damaged. In light, oxygen becomes toxic and photosynthesis self destructive! Only one set of chlorophylls is formed in a cell so their destruction must be avoided. Peroxide and $O_2^{\circ -}$ are destroyed by catalase and superoxide dismutase (SOD) respectively but carotenoids dissipate $O_2^{\circ -}$ and also quench excited chlorophyll, preventing reaction with O_2. Carotenoids form excited triplets, by energy migration from the chlorophyll triplet state, which are dissipated by radiationless decay (Fig. 3.7), transferring energy to the medium. For the triplet energy migration the distance between carotenoid and chlorophyll must be very small, one order of magnitude smaller than the Förster distance so that close association between pigments is essential for π-electron systems to overlap. Excited O_2 is destroyed in an epoxidation reaction involving the ring structure of the carotenoids and an enzyme. Singlet O_2 is also removed, *via* the carotenoid triplet state to give ground state oxygen. Carotenoids also absorb 380 to 520 nm radiation, which is rather energetic and may damage biological

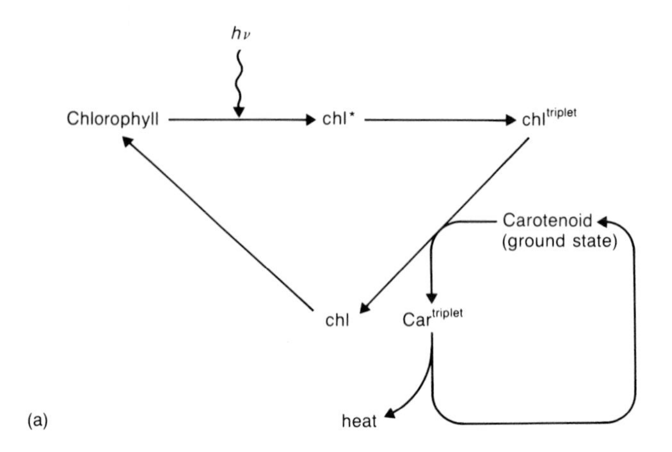

(a)

HO
O_2*
O
Carotenoid
Epoxycarotenoid
Carotenoid
de-epoxidase

(b)

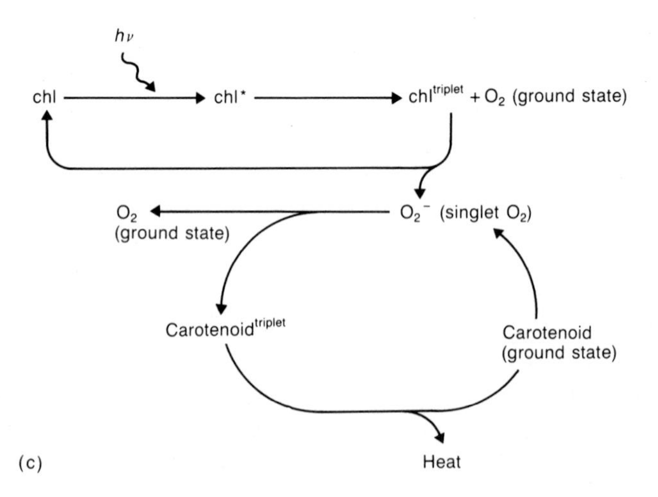

(c)

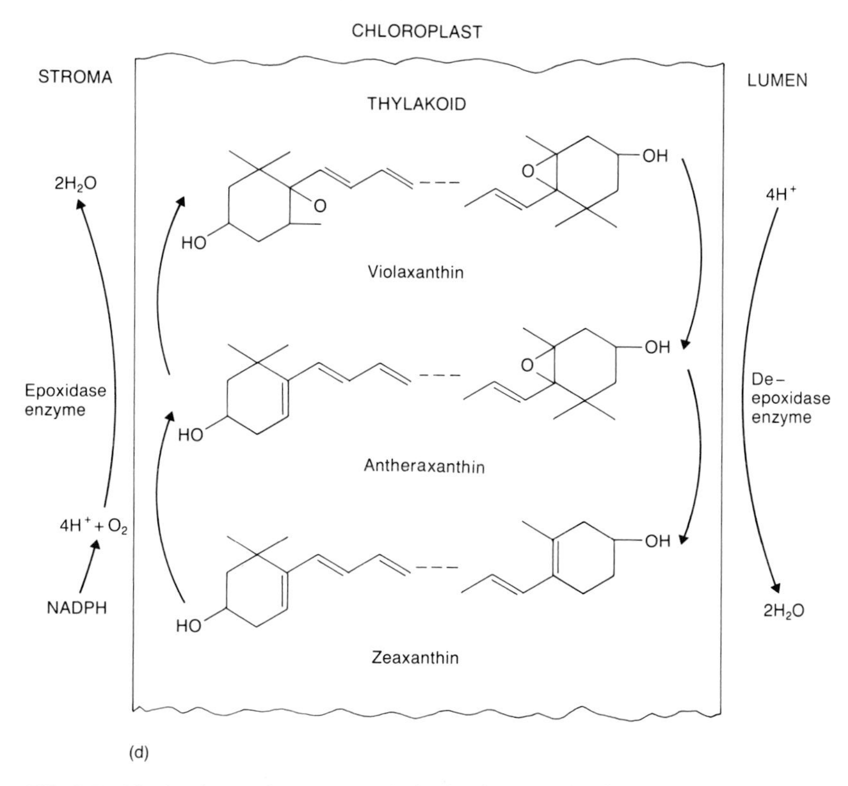

FIG. 3.7 Mechanisms of energy regulation in photosynthesis by carotenoids. (a) the direct reaction with excited chlorophyll; (b) the reaction with excited states of oxygen by epoxidation; (c) reaction of carotenoids with excited states of oxygen; (d) reaction with NADPH and O_2 in the violaxanthin cycle.

systems. Carotenoids themselves are destroyed only with excessive energy load. The structure with nine or more double bonds is essential for efficient energy dissipation, seven double bonds or less being ineffective.

Violaxanthin and zeaxanthin may regulate the amount of NADPH and the redox state of chloroplasts *via* an epoxidation reaction. The violaxanthin cycle (Krinsky 1978) involves two enzymes on the thylakoid membrane which convert zeaxanthin by NADPH or NADH and O_2 (catalysed by epoxidase enzyme) to violaxanthin. This shuttles across the membrane to be de-epoxidized by the required enzyme in reaction with protons in the lumen; the enzyme is activated by the light reactions. Interestingly chloroplast envelopes are rich in violaxanthin in the dark and in zeaxanthin in the light, as expected from the cycle (see Douce and Joyard 1979).

Carotenoids are thus an important 'safety valve' dissipating the excess energy of the excited pigments and the products of the reaction with O_2. This limits the danger of photodestruction of tissues in intense light, when O_2 concentration is high and the normal acceptor of electrons is deficient.

Action spectra and two light reactions

The absorption spectrum is the amount of light captured by photosynthetic pigments as a function of wavelength and the action spectrum is the rate of photosynthesis (e.g. O_2 evolution or CO_2 absorption) or other response resulting from the capture. An action spectrum for O_2 release by algal cells is given in Fig. 3.8. Action spectra show the efficiency of energy use. From the Stark–Einstein Law of equivalence the amount of a reaction is proportional to the number of photons absorbed, so that efficiency may be determined from an action spectrum. However if the action spectrum is the product of several reactions, as in photosynthetic CO_2 assimilation or O_2 release, the action spectrum of the basic reactions may be obscured. Ideally a process directly linked to energy consumption is measured, independent of other processes. For example, measurement of photosynthetic action spectra by O_2 or CO_2 exchange of whole leaves, algae etc., is complicated by respiration and photorespiration which must be assessed in deriving the true action spectrum. Also the response at different wavelengths can only be compared when light intensity is limiting (i.e. the process is linearly dependent upon the amount of light captured) not at saturating light intensity where other factors limit.

If all pigments captured energy and delivered it to reaction centre chlorophyll with equal efficiency, and each wavelength was equally effective in promoting

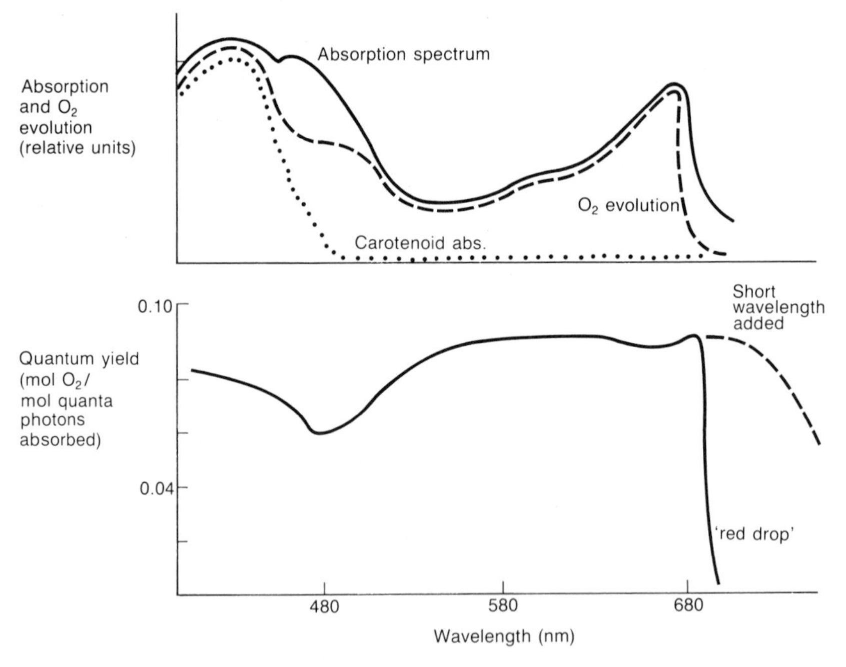

FIG. 3.8 Absorption and action spectra for O_2 evolution by the alga *Chlorella*. Red light is ineffective in O_2 evolution ('red drop'). Supplementation of red light ($>$ 690 nm) with shorter wavelength light ($<$ 690 nm) increases the O_2 evolution (-----). In the blue region, where carotenoids absorb, efficiency of O_2 production is poor.

photosynthesis then the chemical process would be the same at all wavelengths, and the action and absorption spectra would have the same shape. However, several pigments may capture energy with different efficiency at different wavelengths so that absorption and action spectra do not match. Thus in Fig. 3.8 there is close agreement between the action and absorption spectra for chloroplasts between 570 and 680 nm, but in the blue region where carotenoids absorb, less O_2 is produced per unit of light absorbed, as energy transfer from carotenoids to chlorophyll is inefficient.

Emerson and Lewis observed in 1943, that algae evolved O_2 at different wavelengths of light (Fig. 3.8) with an almost constant quantum yield (mol O_2 per mol photons absorbed) up to about 650–690 nm, at which point the rate decreased rapidly despite absorption of photons by chl *a*. This phenomenon is known as the 'red drop'. In 1957 Emerson and co-workers discovered that adding short wavelengths together with red light greatly increased (enhanced) O_2 evolution compared with the two wavelengths given separately (Figs 3.8 and 3.9) – this is the enhancement or Emerson effect. It shows that there is co-operation between two pigment systems, now called photosystems (abbreviated PS), one absorbing short, the other long wavelengths. Separation in time of the activity of two photosystems was demonstrated by giving short flashes (milliseconds) of each wavelength separated by darkness. With long wavelength and a 6 s dark interval, O_2 was evolved. With the opposite wavelength sequence up to 60 s darkness was required before O_2 was evolved. This is evidence that two photosystems are arranged in series, the short-wave absorbing photosystem (PSII) preceding the long (PSI). However, storage and transfer ('spill-over') of energy occurred from the shorter- to the longer-wavelength absorbing photosystems. The order of wavelengths was important. With long given to red algae before the short, O_2 was evolved in a sudden gush, which then decreased before increasing to a steady state; with short wavelength before the long, O_2 evolution decreased and then increased to a steady state.

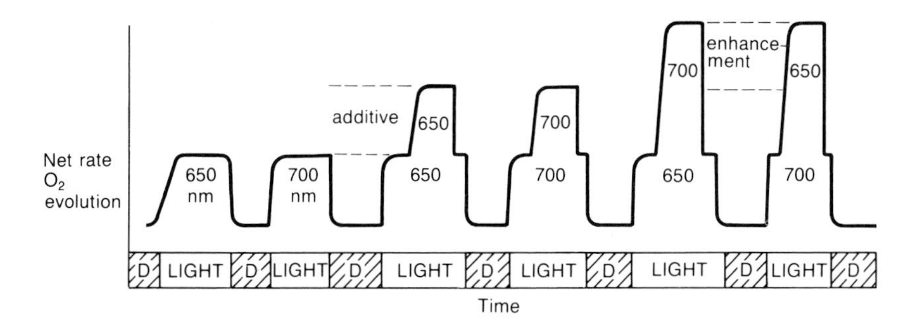

FIG. 3.9 Photosynthetic enhancement illustrated schematically. Algae or chloroplasts are illuminated with short (650 nm) or longer wave (700 nm) light giving equal rates of O_2 evolution. Addition of the same wavelength, i.e. 650 + 650 nm, doubles O_2 evolution, but 650 + 700 nm enhances O_2 evolution. (Modified from Myers 1971).

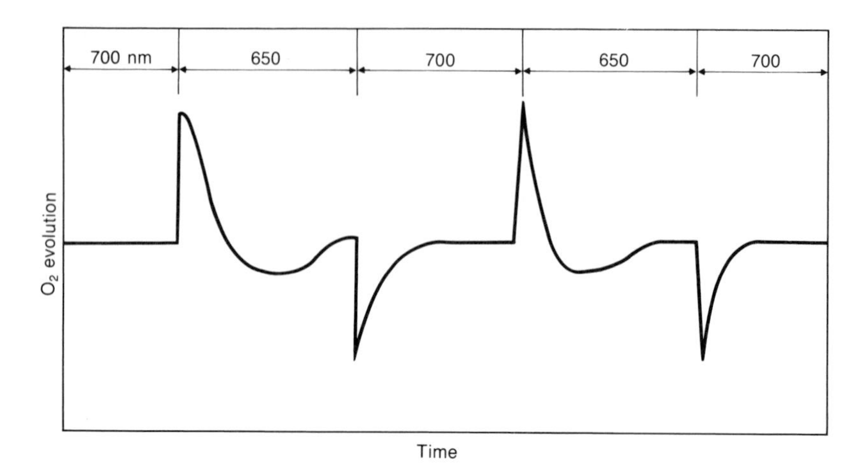

FIG. 3.10 Chromatic transients are observed as an increase in O_2 evolution when chloroplasts are illuminated with short wave light followed by long wave light, but when the order of wavelength is reversed O_2 evolution decreases. These transients induced by light quality indicate the operation of two photosystems, shown by the enhancement effect, in series. (Modified from Myers 1971).

These chromatic (= colour) transients (Fig. 3.10) were discovered by Blinks who suggested that chlorophyll was inactivated and then recovered, or that the transients were caused by respiration. However, Myers and French showed that the transients were due to the activity of two photosystems. Where the 'red drop' occurs only PSI operates and the PSII does not. The long-wavelength absorbing PSI 'primes' the system and the shorter-wavelength absorbing PSII delivers energy to it, with the photosystems acting in series. Evidence from fluorescence induction curves also supported this model as did measurements of the reduction/oxidation state of different components of the electron transport chain under different light regimes. Identification of the two photosystems was a major achievement and forms the basis of much of our understanding of the biophysics of photosynthesis. Before discussing the characteristics of PSI and II, evidence for groups of chlorophylls functioning as semi-discrete, energy gathering units, called photosynthetic units, will be considered.

The concept of photosynthetic units developed from Emerson and Arnold's studies of oxygen production. Algae (*Chlorella*) were illuminated with brief (10^{-5} s) flashes of intense light to saturate photosynthesis. This permitted only a single photon capture per chlorophyll molecule ('single turnover flash') and prevented the energy being used chemically during the flash. A dark period of about 40 ms at room temperature was required for maximum O_2 production per flash. Thus photosynthesis was divisible into light and dark reactions which together gave a maximum efficiency of light utilization. In the cold a longer (0.4 s) dark period was needed for maximum O_2 evolution due to slower chemical processes. However, the maximum O_2 yield per flash was the same despite the different rates. When O_2 evolution was measured following flashes of different energy, separated by a dark period greater than 40 ms (at 25 °C) O_2 per flash increased to a maximum of one O_2 molecule for about 2400 molecules of chlorophyll activated. The ratio of O_2 evolved to photons absorbed (the quantum yield) in dim light is about one O_2 molecule per eight or ten photons absorbed, but at saturating light 2400 chlorophylls are excited to produce one O_2. Therefore to evolve one O_2, 2400 chlorophyll molecules co-operated to collect eight photons, *c.* 300 chlorophylls per photon, and supply the energy to the chemical processes.

The group of chlorophylls, reaction centre and other pigments have been called a 'photosynthetic unit'. The eight quanta are captured by two photosystems, so the energy of four photons per photosystem must be transferred to each reaction centre for 1 O_2 molecule, 4 e^- and 4 H^+ to be evolved and liberated from 2 H_2O molecules. A series coupling of two photosystems is suggested; each e^- is energized twice, first by a short wave photon (PSII) and again by a longer wave photon (PSI). Some antenna chlorophylls gather photons and transfer the excitation to the reaction centre, increasing the effective area per chlorophyll and allowing quanta to be absorbed in very dim light. This explains the gush of O_2 in experiments where the chance of capturing a photon was small, the eight quanta must first be 'accumulated' before electrons can be removed from water. However, energy gathered in one photosynthetic unit can be passed to other units or photosystems ('spill-over') under normal conditions, so that units are not functionally distinct. The photosynthetic unit corresponds loosely to the chlorophyll–protein complexes in thylakoid membranes (p. 72) associated with an individual photosystem. A thylakoid disc 500 nm in diameter may contain 10^5 chlorophylls associated with some 200 electron transport chains, each with one PSI and one PSII unit, 250 chlorophylls per photosystem. Chlorophyll *b* passes excitation energy on to chl *a* and then to P680 (PSII) or P700 (PSI) reaction centres. However present concepts of photosynthetic organization are more dynamic; the size of the antenna and number of reaction centres and the proportion of PSI and II per chloroplast thylakoid varies with species and conditions. Also, the photosynthetic unit is not the functional unit for all thylakoid processes; photophosphorylation, for example, is related to the whole thylakoid.

Quantum yield (also called quantum efficiency) measures the ability of photons to produce chemical change and is the number of O_2 molecules evolved (or CO_2 fixed) per quantum of light absorbed. Its reciprocal is the quantum requirement, that is, the number of quanta needed per O_2 of CO_2 produced or consumed. The quantum yield (see Radmer and Kok 1977) of photosynthesis is controversial; Warburg and associates claimed one O_2 evolved per four quanta absorbed but from eight to ten quanta is now accepted. The value of quantum yield is very important as it provides a basis for understanding the energetics of photosynthesis and to interpret the mechanism. To reduce one CO_2 requires 460 kJ mol^{-1} of quantum energy. Three quanta of red light (680 nm, 174 kJ mol^{-1}) would suffice if efficiency of capture and conversion approached 90 per cent, which is thermodynamically impossible (Clayton 1980). Even four quanta of red light would require 66 per cent efficiency at minimum. Blue light, although more energetic, is no more effective than red light. Four quanta per CO_2 fixed or O_2 evolved conflicts with experiment and the concept of two light reactions. It is now accepted that eight photons are needed per O_2 as a theoretical minimum and more may be required, depending on conditions.

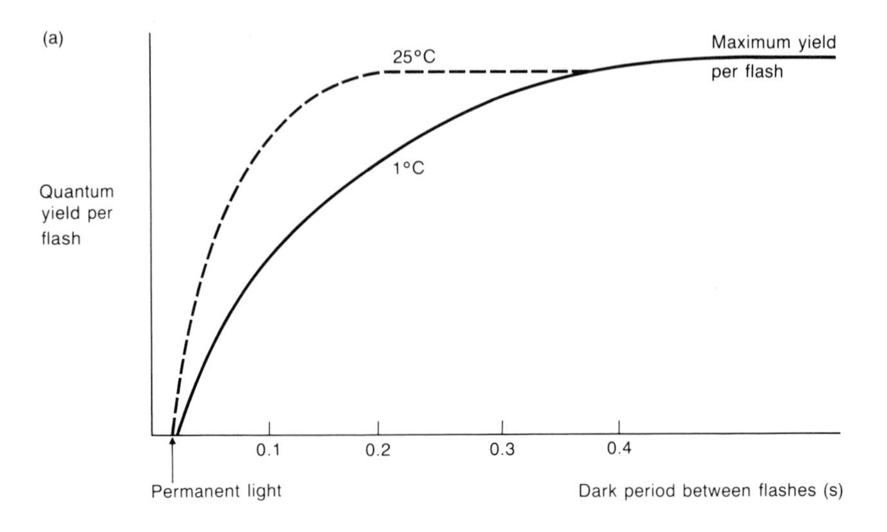

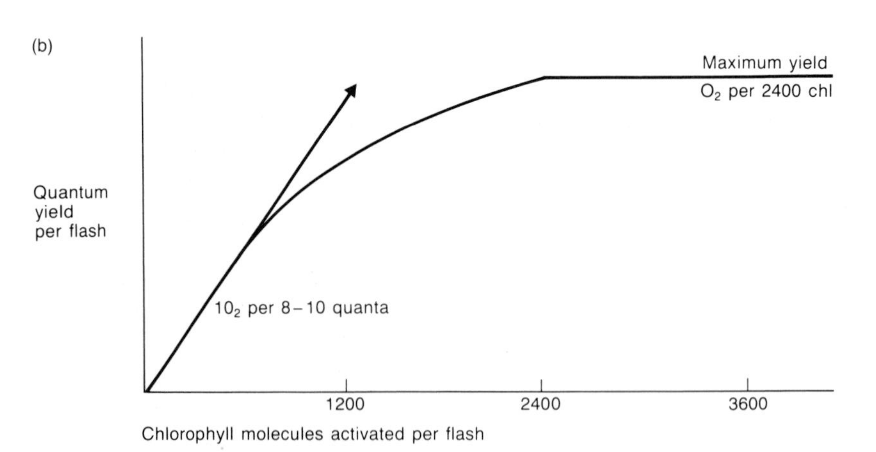

FIG. 3.11(a) Quantum yield (O_2 evolved/number of photons absorbed) of photosynthesis for *Chlorella* with 10^{-5} s flashes of saturating light separated by dark period of different duration. Measurements made in cold (1 °C) and warm (25 °C) conditions to change the rate of chemical reactions. (b) Quantum yield per flash for *Chlorella* with the energy of the flashes increased and a dark period between greater than 0.04 s, at 25 °C.

The energy available from eight red photons is $8 \times 2.9 \times 10^{-19}$ J or 1.4 MJ mol^{-1}. Energy fixed in carbohydrates is 470 kJ mol^{-1} so the efficiency is $(0.47 \times 10^6)/(1.4 \times 10^6) \times 100$ or 34 per cent; ATP formulation increases efficiency to 38 per cent, a good efficiency for many energy conversions. However, under intense natural illumination with only about 50 per cent of radiation as PAR (400–700 nm), overall efficiency is less than 5 per cent; algae in dim light achieve 10 per cent efficiency.

Energy transfer

The pigment antenna is a mechanism for accumulating quanta in dim light to drive a process requiring several quanta. The effective cross sectional area of the reaction centre is increased more than 100-fold, and by combining pigments, different wavelengths of light are used. In bright light excited chlorophyll can 'hold' energy for the next unoxidized reaction centre. Energy movement between closely aggregated chlorophylls is possibly by the 'rapid' (exciton) mechanism in homogeneous transfer (p. 28). These groups interact with less organized pigments by the slow Förster mechanism. Fluorescence occurs in 10^{-9} s and the Förster mechanism in 10^{-10} or 10^{-11} s and radiationless relaxation in 10^{-12} s. Therefore excitation may pass by random walk between many molecules before it reaches a reaction centre, which can accept an electron. Theoretically many steps (perhaps up to 1000) may occur between antenna and reaction centres, but heterogeneous transfer directs excitation movement so there are many fewer; 200–300 steps. This is shown by, for example, loss of fluorescence polarization induced by polarized light. In dim light practically all photons are tunnelled to reaction centres and transfer is about 100 per cent efficient. To achieve this the chlorophyll molecules are arranged at 4–10 nm distances. Energy transfer in the orientated pigments close to the reaction centre may be by electron tunnelling, but the mechanisms are not understood. Reaction centres of PSI trap the excitons from the antenna in 60–70 ps but PSII is slower, 200–500 ps; these are derived from fluorescence lifetime measurements. Models of energy transfer between groups of chlorophylls are discussed in relation to the light-harvesting chlorophyll–protein complexes in thylakoids (p. 75).

Formation of a reduced acceptor is the most important way of dissipating chlorophyll excited states and requires particular structure of the reaction centre–acceptor complex. Passage of the electron from higher energy orbital of the reaction centre chlorophyll to lower energy orbital of the acceptor requires correct molecular orientation and overlap of molecular orbitals in order that net transfer is faster than back reactions caused by thermal excitation. Transfer between stable states of two different molecules is not achieved directly (the transition is 'forbidden', i.e. of very low probability) but occurs *via* an intermediate state of higher energy. Activation energy is needed for an electron to cross this energy barrier but back transfer to donor *via* the intermediate is energetically unfavourable. In photosynthesis excitation drives an electron from the reaction centre over an energy barrier to an acceptor, the pigment acting as a catalyst or sensitizer.

References and Further Reading

Arnez, J. and **Van Gorkom, H. J.** (1978) Delayed fluorescence in photosynthesis, *A. Rev. Plant Physiol.*, **29**, 47–66.

Bolton, J. R. and **Warden J. T.** (1976) Paramagnetic intermediates in photosynthesis, *A. Rev. Plant Physiol.*, **27**, 375–83.

Breton, J. and **Vermeglio, A.** (1982) Orientation of photosynthetic pigments *in vivo*, pp. 153–94 in Govindjee (ed.), *Photosynthesis*, Vol. 1, *Energy Conversion by Plants and Bacteria*, Academic Press, New York.

Butler, W. L. (1978) Energy distribution in the photochemical apparatus of photosynthesis, *A. Rev. Plant Physiol.*, **29**, 345–78.

Clayton, R. K. (1980) *Photosynthesis: Physical Mechanisms and Chemical Patterns*, I.U.P.A.B. Biophysics Series, Cambridge University Press.

Cogdell, R. J. (1983) Photosynthetic reaction centers, *A. Rev. Plant Physiol.*, **34**, 21–45.

Döring, G., Renger, G., Vater, J. and **Witt, H.T.** (1969) Properties of the photoactive chlorophyll-a_{II} in photosynthesis, *Zeitschrift für Naturforschung*, **B24b**, 1139–43.

Dörr, F. (1983) Photophysics and photochemistry. General principles, pp. 265–88 in Hoppe, W., Lohmann, W., Markl, H. and Ziegler, H. (eds), *Biophysics*, Springer-Verlag, Berlin.

Douce, R. and **Joyard, J.** (1979) Structure and function of the plastid envelope, *Adv. Bot. Res.*, **7**, 1–116.

Glazer, A. N. (1983) Comparative biochemistry of photosynthetic light-harvesting systems, *A. Rev. Biochem.*, **52**, 125–57.

Golbeck, J. H., Lien, S. and **San Pietro, A.** (1977) Electron transport in chloroplasts, pp. 94–116 in Trebst, A. and Avron M. (eds), *Encyclopedia of Plant Physiology* (N.S.), Vol. 5, *Photosynthesis I*, Springer-Verlag, Berlin.

Katz, J. J., Shipman, L. L. and **Norris, J. R.** (1979) Structure and function of photoreaction-centre chlorophyll, pp. 1–40 in *Chlorophyll Organisation and Energy Transfer in Photosynthesis*, Ciba Foundation Symposium 61 (N.S.). Excerpta Medica, Amsterdam.

Kok, B. (1961) Partial purification and determination of oxidation-reduction potential of the photosynthetic chlorophyll complex absorbing at 700 mμ, *Biochem. Biophys. Acta*, **48**, 527.

Kok, B. (1976) Photosynthesis: the path of energy, pp. 846–86 in Bonner, J. and Varner, J. E. (eds), *Plant Biochemistry* (3rd edn), Academic Press, New York.

Krinsky, N. I. (1978) Non-photosynthetic functions of carotenoids, *Phil. Trans. R. Soc. Lond. B.*, **284**, 581–90.

Malkin, R. (1977) Primary electron acceptors, pp. 179–86 in Trebst, A. and Avron M. (eds), *Encyclopedia of Plant Physiology* (N.S.), Vol. 5, *Photosynthesis I*, Springer-Verlag, Berlin.

Mathis, P. (1981) Primary photochemical reactions in photosystem II, pp. 827–37 in Akoyunoglou, G. (ed.), *Photosynthesis III. Structure and Molecular Organisation of the Photosynthetic Apparatus*, Balaban International Science Services, Philadelphia.

Mathis, P. and **Paillotin, G.** (1981) Primary processes of photosynthesis, pp. 98–161 in Hatch, M. D. and Boardman, N. K. (eds), *The Biochemistry of Plants*, Vol. 8, *Photosynthesis*, Academic Press, New York.

Mauzerall, D. (1977) Porphyrins, chlorophyll and photosynthesis, pp. 117–24 in Trebst, A. and Avron, M. (eds), *Encyclopedia of Plant Physiology* (N.S.), Vol. 5, *Photosynthesis I*, Springer-Verlag, Berlin.

Milgrom, L. (1984) Vampires, Plants and Crazy Kings, *New Scientist*, **26**, 9–13.

Moore, W. J. (1957) *Physical Chemistry* (3rd edn), Longman, London.

Myers, J. (1971) Enhancement studies in photosynthesis, *A. Rev. Plant Physiol.*, **22**, 289–312.

Okamura, M. Y., Feher, G. and **Nelson N.** (1982) Reaction Centers, pp. 195–272 in Govindjee (ed.), *Photosynthesis*, Vol. 1, *Energy Conversion by Plants and Bacteria*, Academic Press, New York.

Radmer, R. J. and **Kok, B.** (1977) Light conversion efficiency in photosynthesis, pp. 125–35 in Trebst, A. and Avron, M. (eds), *Encyclopedia of Plant Physiology* (N.S.), Vol. 5, *Photosynthesis I*, Springer-Verlag, Berlin.

Renger, G. (1983) Photosynthesis, pp. 515–42 in Hoppe, W., Lohmann, W., Markl, H. and Ziegler, H. (eds), *Biophysics*, Springer-Verlag, Berlin.

Sauer, K. (1981) Charge separation in the light reactions of photosynthesis, pp. 685–99 in Akoyunoglou, G. (ed.), *Photosynthesis III. Structure and Molecular Organisation of the Photosynthetic Apparatus*, Balaban International Science Services, Philadelphia.

van Ginkel, G. and **Kleinen Hammans, J. W.** (1980) Action spectra of photophosphorylation-II. ATP formation catalysed by phenazinemethosulphate suggesting the involvement of long wavelength pigment forms in the light-harvesting process for PSII and PSI. *Photochemistry and Photobiology 31*, 385–395, Pergamon Press, Oxford.

Witt, H. T. (1975) Energy conservation in the functional membrane, pp. 495–554 in Govindjee (ed.), *Bioenergetics of Photosynthesis*, Academic Press, New York.

CHAPTER 4

Architecture of the photosynthetic apparatus

Structural components of photosynthesis form a hierarchy of organizational 'levels' of different size and complexity which co-operate:

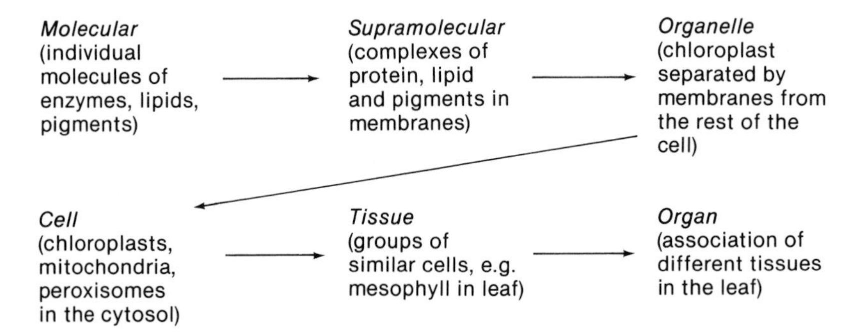

Leaves are the site of most higher plant photosynthesis and provide the necessary conditions to maintain it, for example epidermis and cuticle minimize water loss. Leaf structure, shape and cell distribution are genetically determined but change, within limits, with growth conditions, allowing adjustment to environment. A semi-quantitative estimate of the number and size of component parts of an 'average' higher plant leaf (Table 4.1) is based on data for spinach, tobacco and wheat given by Esau (1958), Nobel (1974), Heath (1969) and Mühlethaler (1977).

The mesophyll tissue is differentiated in many species (Fig. 4.1) into palisade and spongy mesophyll cells, both with extensive air passages contiguous with the sub-stomatal cavities. Rate of exchange of gases between cells and the atmosphere is related to their surface area and to the geometry of the intercellular spaces of the mesophyll (see p. 208); capture of light also depends on mesophyll structure, for example the number and density of cell layers. Photosynthetic processes appear to be the same in palisade and spongy mesophyll cells of C3 plants but in C4 plants (Ch. 9) the mesophyll is clearly differentiated into morphologically distinct tissues with different photosynthetic functions.

Individual photosynthetic cells in algae and most higher plants perform all the primary, and many secondary, reactions of photosynthesis, requiring only water and nutrients. Mesophyll cells may be isolated from leaves by enzymes

Table 4.1 A semi-quantitative analysis of the photosynthetic system in an 'average' C3 plant leaf

Leaf characteristics	
Area of leaf	1 m^2
Thickness of leaf	3×10^{-4} m
Volume of (1 m^2) leaf	3×10^{-4} m^3
Fresh mass of (1 m^2) leaf	0.17 kg
Dry mass of (1 m^2) leaf	0.04 kg
Volume of cells (assuming density of dry matter = 1.2×10^3 kg m^{-3})	2.2×10^{-4} m^3
Air space in leaf = total-cell volume	0.8×10^{-4} m^3
As % of total volume	27%
Leaf cell number per m^2 leaf	
Leaf has: 1 layer of palisade mesophyll cells in transverse section and in parallel section contains	4×10^9 cells
5 layers of spongy mesophyll cells in transverse section and in parallel section each layer contains	0.6×10^9 cells
Total spongy mesophyll cell number	3×10^9 cells
Total mesophyll cell number	7×10^9 cells
Cell volumes	
Palisade cell	
'average' diameter	2×10^{-5} m
'average' length	8×10^{-5} m
total volume (for cylinder with hemispherical ends)	2.9×10^{-14} m^3
total volume of all palisade cells	1.2×10^{-4} m^3
Spongy mesophyll cell	
'average' diameter of cell	3×10^{-5} m^3
volume of cell	1.4×10^{-14} m^3
total volume of all spongy mesophyll cells	4.2×10^{-5} m^3
Total volume of mesophyll cells	1.6×10^{-4} m^3
Chloroplasts in cells	
Assume that each palisade cell contains 100 chloroplasts: total chloroplasts m^{-2}	4.0×10^{11}
Assume each spongy mesophyll cell contains 50 chloroplasts: total chloroplasts m^{-2}	1.5×10^{11}
Total number of chloroplasts m^{-2} leaf	5.5×10^{11}

Table 4.1 A semi-quantitative analysis of the photosynthetic system in an 'average' C3 plant leaf (continued)

Chloroplast	
A chloroplast approximates a hemisphere of diameter	5×10^{-6} m
Volume of 1 chloroplast	3.3×10^{-17} m^3
Total chloroplast volume in 1 m^2 of leaf	1.8×10^{-5} m^3
Chloroplasts occupy $\approx$ 8% of cell volume	
Vacuoles occupy 80% of the cell volume, therefore the cytoplasm volume is and the chloroplasts occupy 38% of cytosol volume	4.8×10^{-5} m^3 m^{-2} leaf
Chloroplast envelope area chloroplast^{-1}	5.9×10^{-11} m^2
Chloroplast envelope area m^{-2} leaf	32 m^2
Chloroplast envelope area mg chlorophyll^{-1}	0.064 m^2
Total volume of chloroplast stroma	1.2×10^{-5} m^3
Thylakoid system	
A typical chloroplast contains 60 grana stacks and each has 15 thylakoids/granum, i.e. 30 membranes. Diameter of 1 'end' of granum is	4.5×10^{-7} m
Area of vesicle (not allowing for ends)	1.6×10^{-13} m^2
Total area of membranes in one stack	5.0×10^{-12} m^2
Total area of grana membranes in one chloroplast	3×10^{-10} m^2
If four stromal thylakoids the width of the granum pass across the diameter of chloroplast the stromal thylakoid area is	2.4×10^{-10} m^2
Total thylakoid area, grana + stroma per chloroplast	5.4×10^{-10} m^2
Total thylakoid area m^{-2} leaf	300 m^2 (other estimates 835 m^2)
Thickness of thylakoid membranes (approx.)	7.5×10^{-9} m
Volume of thylakoid membranes	4.1×10^{-18} m^3
Volume of thylakoid lumen	
area of end surface of a granum	1.6×10^{-13} m^2
space between membranes is approx.	8×10^{-9} m
volume of individual vesicle	1.3×10^{-21} m^3
volume of lumen in one stack	1.9×10^{-20} m^3
volume of grana lumen in a chloroplast	1.2×10^{-18} m^3
volume of stromal thylakoids chloroplast^{-1}	9.0×10^{-19} m^3
total volume of thylakoid lumen chloroplast^{-1}	2.1×10^{-18} m^3

Table 4.1 A semi-quantitative analysis of the photosynthetic system in an 'average' C3 plant leaf (continued)

total volume of thylakoid lumen m^{-2} leaf	$1.2 \times 10^{-6}\ m^3$
total volume of membrane + lumen in a chloroplast	$6.2 \times 10^{-18}\ m^3$
or ~ 20% of chloroplast volume	
Chlorophyll content of leaves	
Average content of chlorophyll m^{-2}	0.5 g
molecular mass of chlorophyll *a*	894
chlorophyll content	5.6×10^{-4} mol m^{-2}
or	1.9 mol m^{-3} leaf
Chlorophyll/chloroplast	9×10^{-13} g
or	1.0×10^{-15} mol
Vol. thylakoid lumen (g chlorophyll)$^{-1}$	$2.3 \times 10^{-6}\ m^3\ g^{-1}$
Vol. chloroplast (g chlorophyll)$^{-1}$ or vol. mol^{-1}	$3.6 \times 10^{-5}\ m^3\ g^{-1}$ $0.033\ m^3\ mol^{-1}$
Vol. thylakoid membrane (g chlorophyll)$^{-1}$	$4.6 \times 10^{-6}\ m^3\ g^{-1}$
Area thylakoid membrane (g chlorophyll)$^{-1}$	$600\ m^2\ g^{-1}$ (other estimates $1670\ m^2\ g^{-1}$)
Conc. chlorophyll in thylakoid	$2.2 \times 10^5\ g\ m^{-3}$
or	2.4×10^2 mol m^{-3} (0.24 M)
No. chlorophyll molecules chloroplast^{-1}	6.7×10^8
Area membrane (chlorophyll molecule)$^{-1}$	$1.2 \times 10^{-18}\ m^2$
Area per 'head' of chlorophyll molecule	$2.2 \times 10^{-18}\ m^2$
Miscellaneous values	
Chlorophyll molecules per photosystem I & II	300
Conc. of NADP reductase in stroma	8×10^{-5} M
Coupling factor (CF_1)	0.45 g (g chlorophyll)$^{-1}$
Molecular mass of CF_1	325 kD
CF_1 granum^{-1}	200
ATP content illuminated chloroplasts (nmole ATP mg chlorophyll^{-1})	40
ADP content (nmole ATP mg chlorophyll^{-1})	12
ATP conc.	1.5 mM

Table 4.1 A semi-quantitative analysis of the photosynthetic system in an 'average' C3 plant leaf (continued)

ADP conc.	0.5 mM
ATP/ADP ratio	3
NADPH conc.	0.1 mM
RuBP carboxylase/oxygenase conc. of enzyme sites	4 mM
Electron transport chains per thylakoid 'disc'	200
Dry mass per chloroplast	20×10^{-12} g
Vol. water to total vol.	75%
Protein content (% dry mass of chloroplast)	60%
Lipid content (% dry mass of chloroplast)	20%
Chlorophyll (% dry mass of chloroplast)	4%
Carotenoids (% dry mass of chloroplast)	0.9%
Nucleic acids (% dry mass of chloroplast)	2.5%
Soluble products of photosynthesis	7.5%
Mg conc. in chloroplast stroma (light)	30 mM
Mg conc. in chloroplast stroma (dark)	15 mM
Inorganic phosphate in chloroplasts (moles P_i mg chlorophyll^{-1})	3
Inorganic phosphate concn	100 mM
RuBP conc.	0.1–2 mM

which break the bonds holding together the walls of adjacent cells. Protoplasts may also be freed from the cell wall by enzymes which lyse bonds in the wall. Cells and protoplasts will photosynthesize actively when given CO_2, light and nutrients, if the osmotic potential of the medium is correct. Cells and protoplasts from C3 plants function independently. However, in C4 plants co-operation has evolved between photosynthetic cells and their autonomy has been lost.

The ultrastructure of photosynthetic cells shown by electron microscopy is discussed by Gunning and Steer (1975a and b). In the thin layer of cytoplasm around the large central vacuole are chloroplasts, mitochondria and peroxisomes, which are all involved in aspects of photosynthetic metabolism, together with inclusions such as pigment granules and starch grains. The number of chloroplasts per cell is very variable, approximately 20–100 in higher plants. Chloroplast distribution in cells differs with species and changes with conditions, such as illumination. C3 plants have typically flattened spherical or lens-shaped chloroplasts some 3–10 μm in greatest dimensions

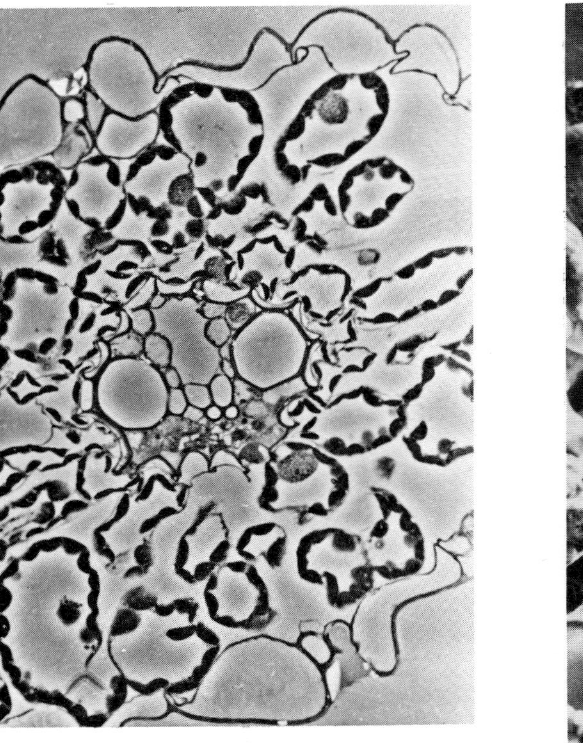

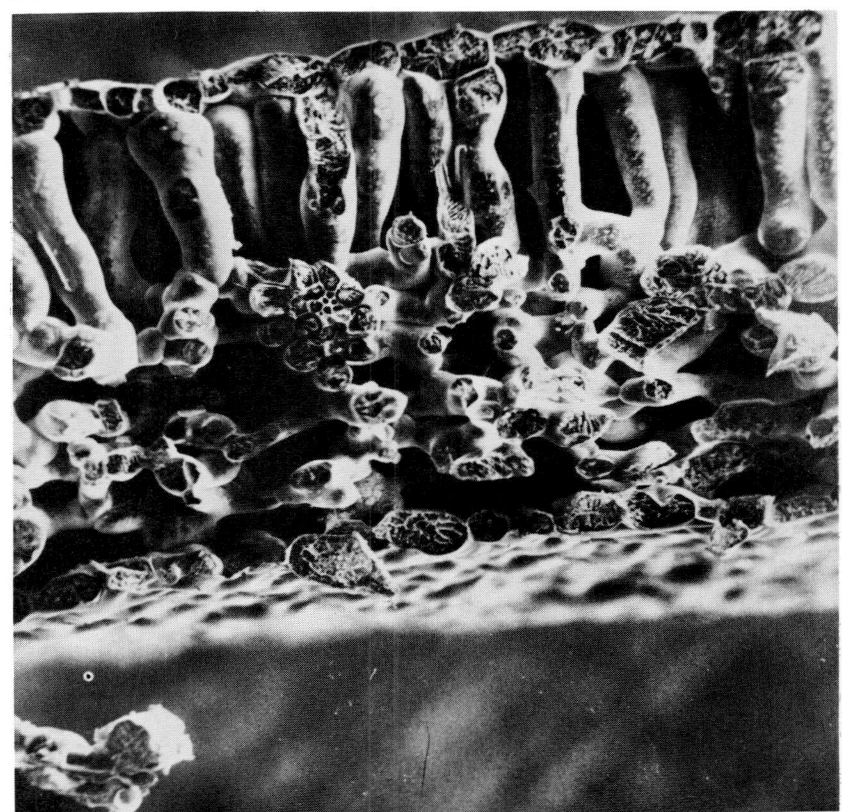

FIG. 4.1(a) Transverse section of a wheat (*Triticum aestivum L.*) leaf (× 400) showing the loosely packed mesophyll cells arranged around a central vascular bundle. The cells contain chloroplasts lining the cell wall. (b) Scanning electron micrograph of a transverse section of a bean (*Vicia faba*) leaf (× 216). Electron micrographs courtesy of R. Turner, Rothamsted Experimental Station, Harpenden, Herts, UK.

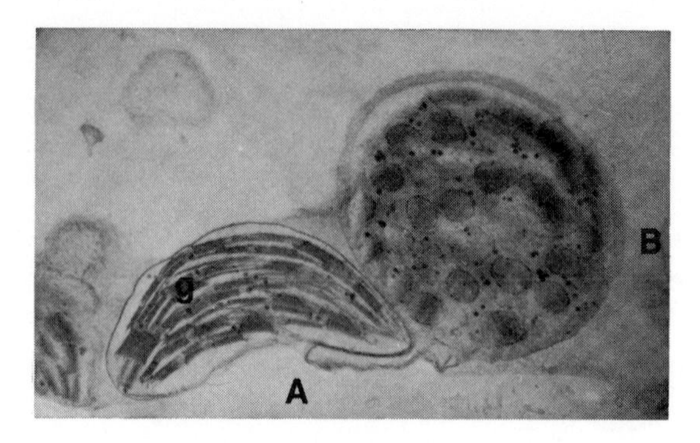

FIG. 4.2 Structure of higher plant chloroplasts, illustrated by an electron micrograph (× 10 000) of bean (*Vicia faba*). (A) shows in transverse section the double chloroplast envelope enclosing the stroma and the thylakoid membranes forming granal stacks (g) and unstacked stromal thylakoids; (B) has been sectioned at right angles to (A). The granal discs are clearly visible. Courtesy of Dr A.D. Greenwood, Imperial College, London.

(Fig. 4.2). Mitochondria are prominent, with characteristic double membranes, the inner often folded. Peroxisomes have a single limiting membrane and granular contents, without distinctive features in electron micrographs, and are often closely associated with chloroplasts. The number of mitochondria and peroxisomes varies and may depend on the function of the cell and on the environment during growth. Peroxisomes metabolize products from the chloroplast in co-operation with mitochondria, for example in photorespiration (Ch. 8).

Chloroplasts perform all the primary (e.g. light capture and electron transport leading to NADPH and ATP synthesis) and most of the secondary processes (e.g. synthesis of 3-carbon phosphorylated compounds from CO_2) in photosynthesis and their structure will be considered in detail. They are bounded by a continuous envelope and contain all the membrane bound light-harvesting chlorophyll and other pigments, proteins and redox compounds involved in transport of electrons, and the associated proteins, plus the soluble enzymes and substrates required for CO_2 NO_3^- and SO_4^{2-} assimilation by photosynthesis together with its products.

Chloroplasts are isolated by breaking the cell wall mechanically, in a buffer solution (pH about 7) with non-permeating osmotic substances (e.g. sorbitol or sucrose, 0.3 M) to maintain the osmotic potential and ions, for example phosphate (30 mM) and magnesium (5 mM). Cell debris is removed by filtration through layers of muslin. Chloroplasts are separated from other organelles by centrifugation (90 s at 3000 *g*) and then suspended in buffered solution (see Leegood and Walker (1983) for methods). However, mechanical damage and unphysiological conditions may impair the envelope membranes. Intact chloroplasts assimilate CO_2 faster than damaged ones because they retain enzymes and co-factors. The envelope is detected by the bright appearance of isolated chloroplasts in phase contrast microscopy. However the envelope may appear intact yet have broken and resealed. Ferricyanide, an electron acceptor in electron transport, cannot penetrate intact membranes, so measuring its reduction $(Fe(CN)_6)^{3-} + e^- \rightarrow (Fe(CN)_6)^{4-}$ at 420 nm indicates the state of the envelope. Hall (1972) has classified chloroplasts as:

Chloroplast	Description	Classification
Complete	Outer envelope intact, stroma retained, rapid CO_2 fixation. No $NADP^+$ or ferricyanide penetration	Type A
Unbroken	Torn, but resealed envelope, some stroma lost, slow CO_2 fixation. $NADP^+$ and ferricyanide penetrate	Type B
Broken	Envelope and stroma lost in medium, thylakoids naked, no CO_2 fixation	Type C
Free lamellar	Lamellae only, no CO_2 fixation (only if required factors added)	Type D
Chloroplast fragments	Stromal protein removed	Type E, F

Manipulation of isolated chloroplasts has been very important in analysing their functions. Separated thylakoids and envelope may be disrupted by detergents or ultrasonically; the components are separated by column chromatography, density gradient fractionation, gel electrophoresis etc., and chemically analysed to determine the components of light-harvesting systems and electron transport chains. The position of proteins or other components within the membranes is found by using antibodies produced by injecting chemically pure components of the membrane into animals. If an antibody only affects, for example, electron transport in disrupted thylakoids, then the antigen is inside the membranes. Similarly, chemical probes of known function are used to penetrate the membranes, changing processes and indicating the position of components. X-ray studies may be made on particles or membrane pieces to study the orientation of components. These techniques in combination provide a picture of chloroplast membrane and thylakoid vesicle structure and function.

Chloroplast internal structure

Electron microscopy shows the chloroplast to consist of an envelope enclosing a complex of membranes, the thylakoid system often joined or stacked into grana (Figs 4.2 and 4.3); the lipid membranes contrast with the background when stained with lipophilic electron dense osmium. The space between envelope and thylakoid membranes is the chloroplast stroma. The envelope is composed of two membranes each about 10 nm thick separated by the intra-envelope space (*c.* 10 nm). The membranes are lipid bilayers, of galactosyl glycerides and phosphatidyl choline, containing carotenoids but no chlorophyll. The membranes are not identical in function. The outer cytoplasmic membrane allows many substrates to pass freely, whereas the inner (stromal) membrane is highly selective, allowing passage of only some solutes by special enzyme systems, called translocators. The cytoplasmic membrane has 9 nm diameter particles in both halves of the lipid bilayer at a density of 1.5×10^2 particles μm^{-2}. The stroma membrane has 7 nm and 9 nm diameter particles at a density of 1×10^3 and 1.8×10^3 μm^{-2} in the halves of the lipid bilayer next to the stroma and intermembrane space respectively. The particles in both membranes are probably protein complexes, some associated with the translocators.

The stroma contains indistinct granules and particles, mainly of proteins; the enzyme ribulose bisphosphate (RuBP) carboxylase is the major soluble protein and may crystallize in unfavourable conditions such as water stress or air pollution. Other inclusions are products of the photosynthetic processes; for example starch granules up to 2 μm long accumulate in the stroma and

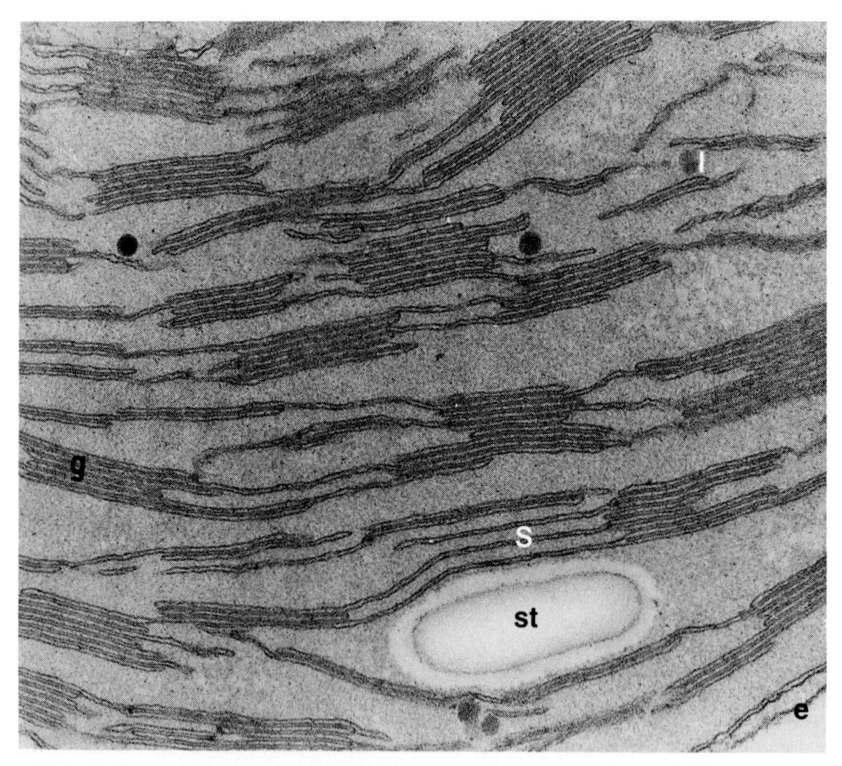

FIG. 4.3 Transverse section of spinach (*Spinacea oleracea*) chloroplast by electron microscopy (× 103 000); thylakoid membranes are associated into grana (g) joined by the stromal thylakoids (S). Osmophilic lipid globules (l) are apparent and a starch granule (st) also. The envelope (e) appears as a single membrane at the lower right hand corner. Courtesy of Dr A.D. Greenwood, Imperial College, London.

disturb the thylakoid membranes, and globules of lipids and plastoquinone accumulate (see Coombs and Greenwood (1976) for a review); RNAs and DNA occur in chloroplasts which synthesize many of their constituent proteins.

The most noticeable feature of chloroplasts in electron micrographs (Fig. 4.2) is the thylakoid (from the Greek *θυλακοειδζσ* for 'sack like') membrane vesicle system. In transverse section the thylakoids appear as parallel pairs of continuous membranes separated by a space, the thylakoid lumen (Fig. 4.3) which is 5–10 nm wide and contains few identifiable features. Thylakoid membranes frequently associate into granal stacks, interconnected by pairs of membranes, called stromal thylakoids (or alternatively intergranal connections or frets), which are in contact with the stroma on both sides. The interface between the appressed membranes is the partition region. In C3 plants over 60 per cent of the thylakoid surface is typically in the grana. The end membranes of stacked thylakoids and the ends of the grana, but not the partition regions, have direct contact with the stroma (Fig. 4.4).

Several models of the three-dimensional structure of the thylakoid system have been suggested from analysis of serial sections of chloroplasts; a stylized view is shown in Fig. 4.4a together with an electron micrograph of the membrane interconnections (Fig. 4.4b). Thylakoid membrane vesicles in grana are stacked and flattened, but not closed, sacs (Fig. 4.2) interconnected with the other membranes. The vesicles join the stromal lamellae at different points around the periphery of a granum (Fig. 4.4b). The structure derives from folding and joining of separate sheets of lamellae which are interconnected and probably originate from a single point (Fig. 4.5), the prolamella body, in the developing chloroplast (Staehelin and Arntzen 1979). The thylakoid system appears to be a single interconnecting giant closed vesicle with continuous lumen, a feature of great importance in electron transport and ATP generation. Composition of the lumen is not known; but proteins, of the water-splitting complex and the light-harvesting complex for example, may occupy part of the volume and it is unlikely to be an homogeneous aqueous solution of small molecules. Grana differ in extent and size between species, and with conditions during growth, for example, with bright illumination there is less granal stacking. A semi-quantitative summary of the size of an average thylakoid system is given in Table 4.1. Grana in isolated thylakoids stack and unstack, according to the ionic concentration and light quality.

The internal composition and structure of thylakoid membranes

A membrane is 5–7 nm thick and consists of a lipid bilayer (50 per cent of the mass) together with proteins, pigments and other major components which are vital for photosynthesis (Kirk and Tilney-Bassett 1978). Thylakoid lipids are a complex mixture; some 80 per cent is glycolipid containing galactose, such as monogalactosyl-diglyceride (50 per cent of total lipid on a molar basis) and digalactosyl-diglyceride (25 per cent) which have neutral hydrophobic heads. The remainder is mainly phospholipid (10 per cent) and sulpholipid (5 per

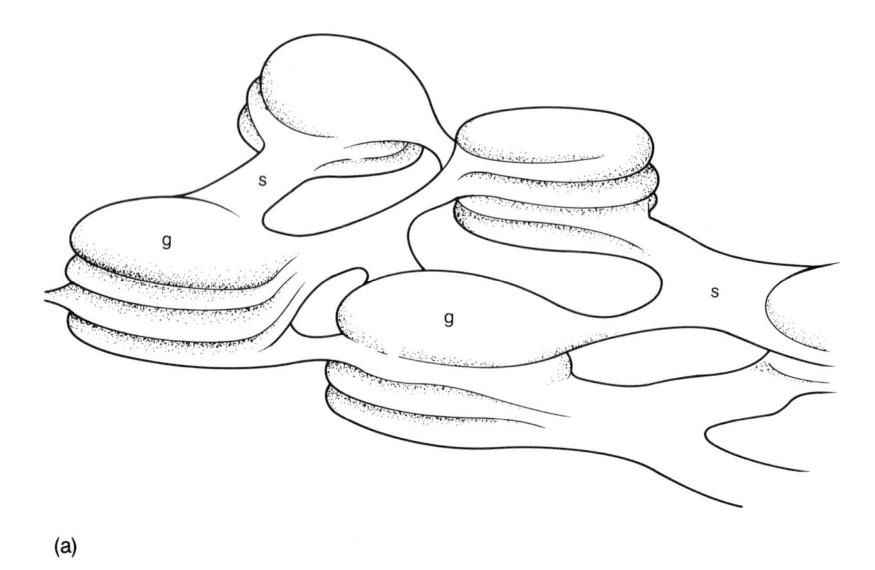

(a)

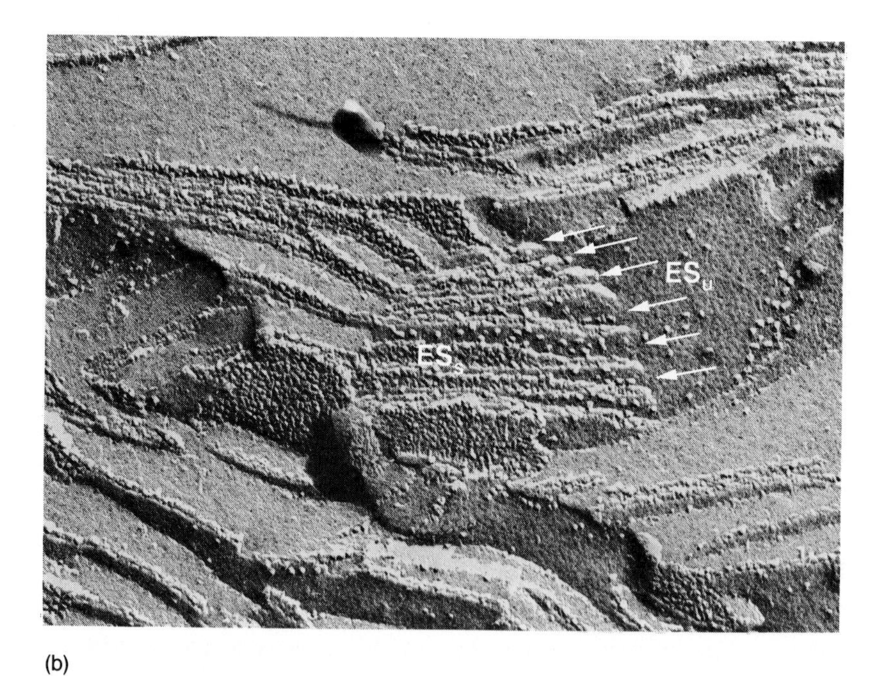

(b)

FIG. 4.4(a) Diagram of the connections between granal (g) and stromal (s) thylakoids. (b) Spinach chloroplasts showing interconnections (→) between the granal thylakoids (ES_S) and a single stromal thylakoid (ES_U). Electron micrograph ($\times$ 185 000) courtesy of Professor L.A. Staehelin, University of Colorado, Boulder, Colorado, USA.

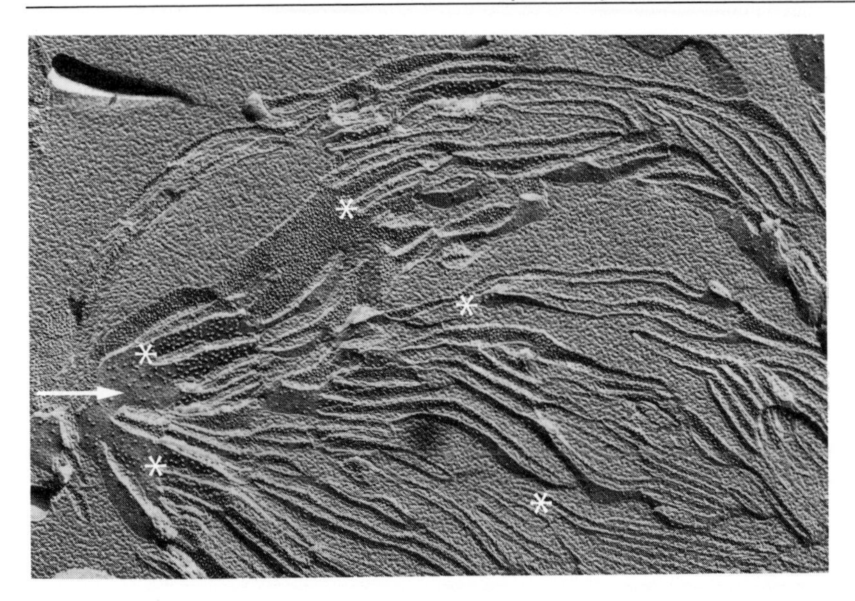

FIG. 4.5 Thylakoid membranes of an isolated spinach chloroplast after freeze-fracture, showing progressive branching of membranes (*) from a common point (→). Electron micrograph (×50 000) by courtesy of Professor L.A. Staehelin, University of Colorado, Boulder, Colorado, USA.

cent), charged at pH 7. Synthesis and structure of lipids is considered in Chapter 8. The fatty acids of lipids are highly unsaturated. Linolenic acid (C18:3) is the predominant fatty acid and trans-3-hexadecanoic acid (C16:1) acylated to phosphatidyl glycerol is specific to thylakoid membranes; its function may be structural. The fatty acid tails form a non-aqueous, hydrophobic, central core to the membrane, whilst the hydrophilic heads are at the surface. The outermost half of the membrane, next to the stroma is some 3–5 nm thick; the inner, next to the lumen is 2–3 nm thick. As the two most abundant lipids are highly unsaturated, the membrane is very fluid at physiological temperatures with little cholesterol or other sterol to cause rigidity. Fluidity allows movement of pigment–protein complexes laterally through the membrane. The lateral diffusion coefficient of lipids is 10^{-10} m^2 s^{-1} and of proteins 5×10^{-11} m^2 s^{-1}. As the distances over which pigment–protein complexes move are small, displacements of the order of 10–100 nm will occur rapidly, particularly if the proteins are charged, enabling the thylakoid to change its structure and function as necessary for optimizing ion and water fluxes across the membrane and energy distribution between light transducing complexes within it.

Particles in thylakoid membrane (Fig. 4.6)

Surface structure of membranes is observed by electron microscopy of isolated membranes. Inner membrane structure may be seen after freeze-fracturing

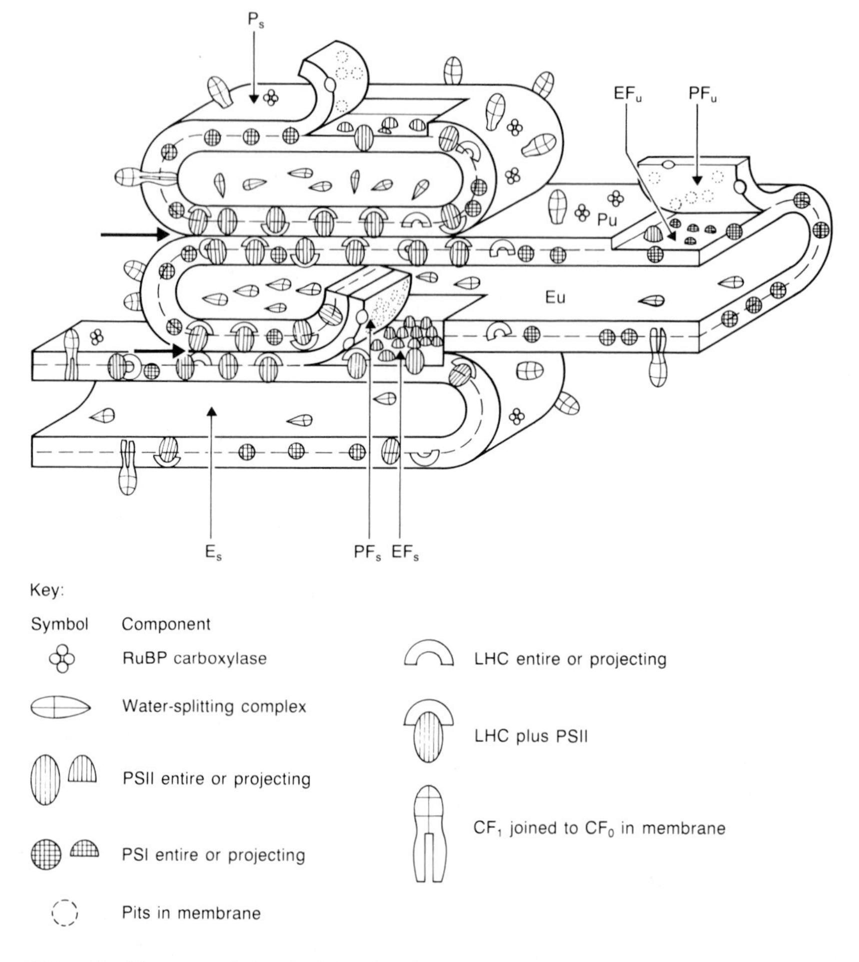

FIG. 4.6 Diagram of the thylakoids of a higher (C_3) plant chloroplast. Part of a granal stack and stromal thylakoid is shown. Coupling factor (CF_1) and ribulose bisphosphate carboxylase (RuBPc) are attached but not in the partition region (→). In the lumen of the granal thylakoids are particles of the photosystem II, water-splitting complex. The partition region contains particles on the endoplasmic fracture face (EF_S) identified with PSII and the light-harvesting protein complex. Smaller particles on other membranes may be PSI and the base of coupling factor. See text for explanation of membrane surfaces and particles.

thylakoids. Membranes are frozen and cut, during which the membrane bilayer separates along the line of weakness caused by the hydrophobic tails of the membrane lipids, exposing particles within the membrane.

The surfaces of membranes in electron micrographs of freeze-fractured thylakoids are denoted by their contact with the stroma, i.e. protoplasmic surface (PS) or the lumen, i.e. endoplasmic surface (ES) and their fractured surfaces are PF and EF respectively. Membranes from stacked (granal) or unstacked (agranal) regions are shown by subscript s and u respectively (Branton *et al.* 1975). On the outer surface of stromal (PS_u) and of granal thylakoids (PS_s) in contact with the stroma, are particles of RuBP carboxylase loosely attached and easily removed; most prominent are club-shaped 15 nm diameter particles of coupling factor, called CF_1 which synthesizes ATP. Neither RuBP carboxylase nor CF_1 ('extrinsic', i.e. external proteins) occur in the partition region. Smaller (9 nm diameter) particles on PS are exposed parts of intrinsic proteins (i.e. occurring within the membrane) some possibly the base part of CF_1, called CF_0. Also, particles from within the membrane may project rather indistinctly out of the surface, and form a lattice with rows 8–9 nm apart and interparticle distances of 10 nm. On the very smooth inner lumen surface, ES_s, of granal thylakoids are rectangular (10 × 15 nm) particles of four (sometimes two or six) subunits each *c.* 5 nm in diameter (Fig. 4.7). In artificially unstacked thylakoids these large particles may be arranged in a lattice when they always have four units of 18 × 20 nm spacing on the ES_s face. The stromal thylakoid inner surface (ES_u) is much more textured than ES_s and has only a few smaller (4–6 nm diameter particles) just projecting above the surface. More than 80 per cent of the large particles of EF faces occur in granal regions and 20 per cent or less in stromal areas.

Particle distribution on fractured membranes has been analysed mainly on spinach (see Mühlethaler (1977) and Staehelin *et al.* (1977)). Figure 4.7 shows

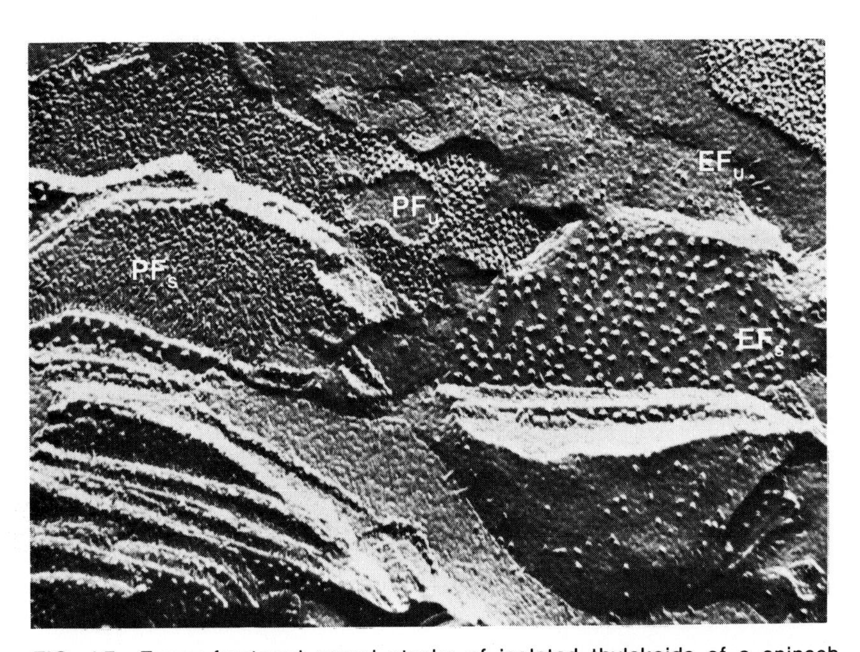

FIG. 4.7 Freeze-fractured granal stacks of isolated thylakoids of a spinach chloroplast, showing the EF and PF faces of stacked and unstacked areas. Note the characteristic large (15 nm) particles on EF_s. The stacks are linked by a sheet of unstacked membrane (↓). Electron micrograph (× 185 000) by courtesy of Professor A.L. Staehelin, University of Colorado, Boulder, Colorado, USA.

electron micrographs of the freeze-fracture faces. On EF_s are many (*c.* 1500 particles μm^{-2}) large particles in two populations of 15 and 11 nm diameter, 60–70 per cent and 30–40 per cent respectively of the total population. Depending on the conditions (salt concentrations etc.) these particles, which appear lobed, may be arranged in very regular arrays or lattices forming a uniform population about 16 nm diameter. There are also about 700 smaller particles μm^{-2}, 10 nm diameter, of two subunits. On the stromal face, EF_u, there are fewer (450–570 μm^{-2}) smaller (10 nm diameter) particles which are arranged not in a lattice, and are more scattered than on the EF_s face.

The PF faces in both stromal and granal thylakoids contain particles: PF_s has 3500–4500 particles μm^{-2}, of 8 nm average diameter (range 5–12 nm). PF_u has 3600 particles μm^{-2}, in two groups of 8 and 11 nm average diameter and less deeply embedded than those in the PF_s faces. Deep pits in the PF_s are left by the large particles on the EF_s face tearing out of the PF_s lipid bilayer during fracturing (Figs 4.6 and 4.7). Four (sometimes two or three) smaller PF_s particles are regularly arranged around pits and would be closely associated with large EF particles. The groups, separated by about 2.5 nm, form the lattices in granal areas when they penetrate the ES and PS faces.

Particles in membranes have been identified with PSI and II, and the light-harvesting chlorophyll-protein and cytochrome *b-f* complexes (see p. 93). Probably the large EF particle has a core of a PSII complex (8 nm diameter) associated with two, four or six units of light-harvesting complex in granal thylakoids, but only one in stromal membranes. Smaller particles in the PF_u and PF_s surfaces may be chlorophyll-protein complexes of PSI (see p. 72), light-harvesting complexes and CF_0. Variations in size may be due to association of light-harvesting complexes with different numbers of other complexes or with components of the electron transport chain.

Large particles span the lipid bilayer from inside to outside surface in stacked membranes. Unstacked membrane particles are not arranged in regular lattice arrangements and do not span the membrane. Perhaps forces on or in the membrane cause the large particles to stand upright, make the lipid layer thinner, and produce arrays. However, the smaller PF particles do not project into the thylakoid lumen.

Thylakoid membranes are 'sided' in construction, with the water-splitting complex in the lumen, a PSII chlorophyll–protein complex, a cytochrome *b/f* complex and light-harvesting complex spanning the membrane interspersed with the PSI chlorophyll–protein complex on the outer side, and finally enzymes of carbon metabolism and ATP synthesis on the outer surface. This sidedness allows thylakoids to transport electrons to the stroma from water in the lumen and accumulate protons in the lumen.

Membrane structure of C4 plants

Considerable differences occur between C4 species in chloroplast structure; here maize is taken as an example. Its photosynthetic tissue is divided into bundle sheath and mesophyll (Fig. 4.8a and b) with different forms of chloroplast (Fig. 4.9). Mesophyll cell chloroplasts are comparable to those of C3 plants

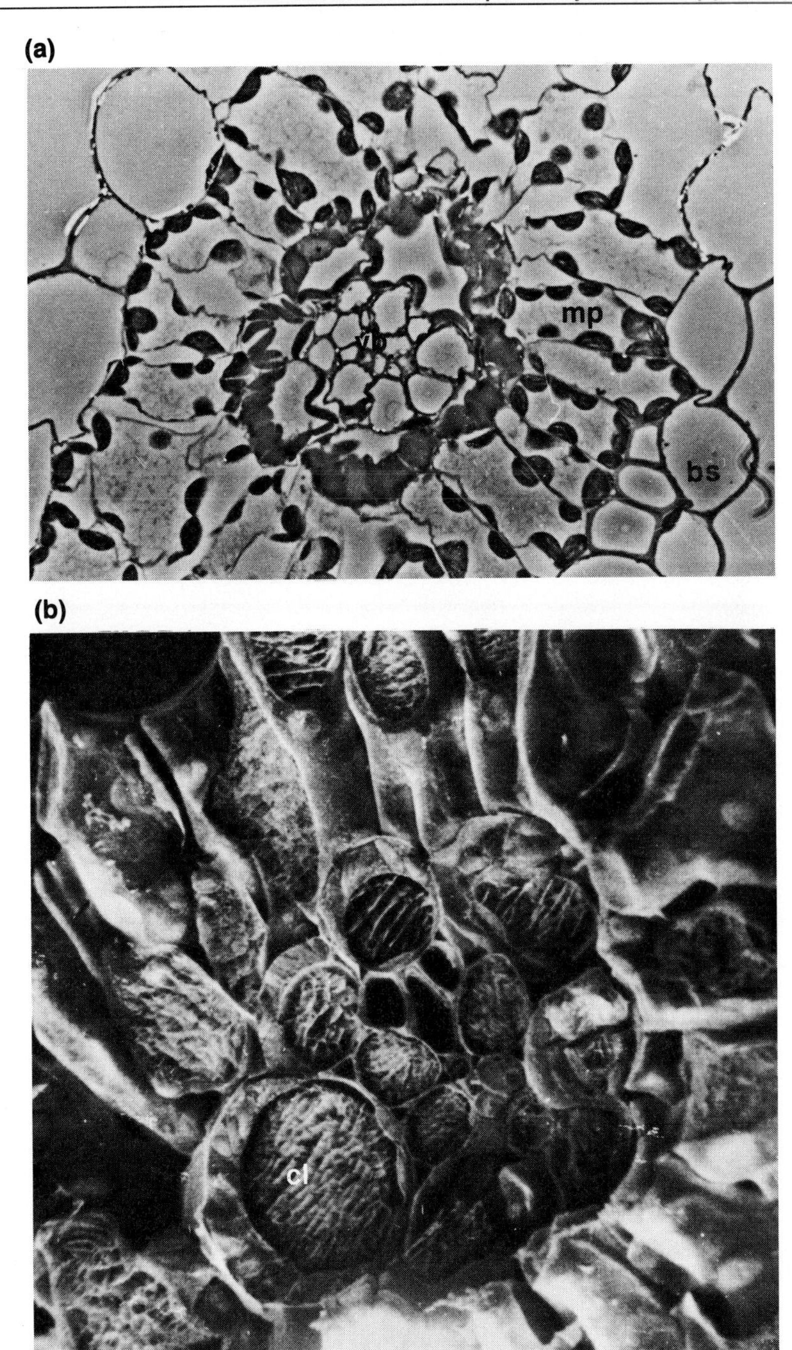

FIG. 4.8(a) Transverse section of a maize leaf (× 440) with the vascular bundle (vb) surrounded by a bundle sheath (bs) of large cells with prominent chloroplasts. Cells of the mesophyll parenchyma (mp) have smaller, less densely packed chloroplasts. (b) Bundle sheath of maize (scanning electron micrograph × 640) with closely packed chloroplast lamellae (cl). Courtesy of Mr R. Turner, Rothamsted Experimental Station, Harpenden, Herts, UK.

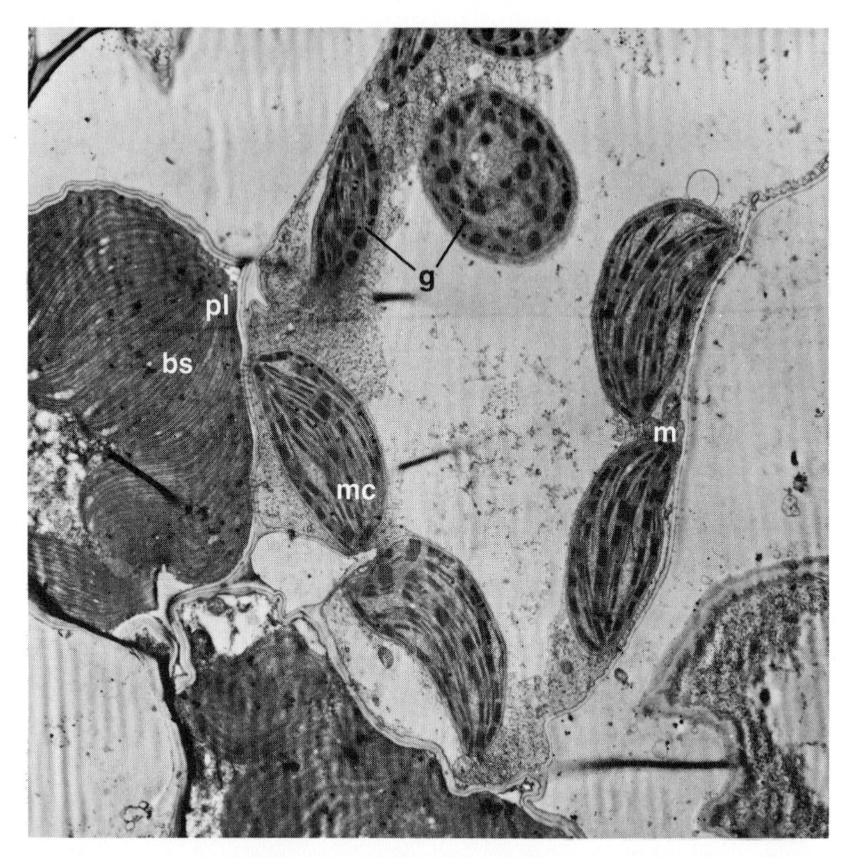

FIG. 4.9 Section of maize bundle sheath cell with chloroplast (bs) and mesophyll cell (m) with chloroplast (mc) showing chloroplast dimorphism. Plasmodesmata (pl) traverse the wall between the two cell types (g = grana). Electron micrograph (× 3560) by courtesy of Mr R. Turner, Rothamsted Experimental Station, Harpenden, Herts, UK.

already described, with many discoid chloroplasts about 8 μm diameter, double membrane envelope, thylakoids arranged in parallel and many obvious grana. Bundle sheath cells contain few, large (15–20 μm diameter) chloroplasts with many parallel, densely packed thylakoids with no obvious grana, although some contact occurs. A dense layer of tubules, the chloroplast reticulum, lines the envelope which is closely appressed to a dense layer of cytoplasm next to the cell wall; groups of plasmodesmata connect the bundle sheath and mesophyll cells (Fig. 4.10).

Thylakoid membranes of the two types of chloroplast differ in macromolecular construction (Fig. 4.11). Granal and stromal areas of the C4 mesophyll chloroplast are similar to the equivalent areas of C3 plants. The EF_s face is extensive with large particles (15 nm diameter) on a smooth background. On the PF_s face the particles are smaller and more numerous. EF_u faces lack 15 nm particles and have 10 nm diameter particles, similar to the PF_u particles, but at much lower density.

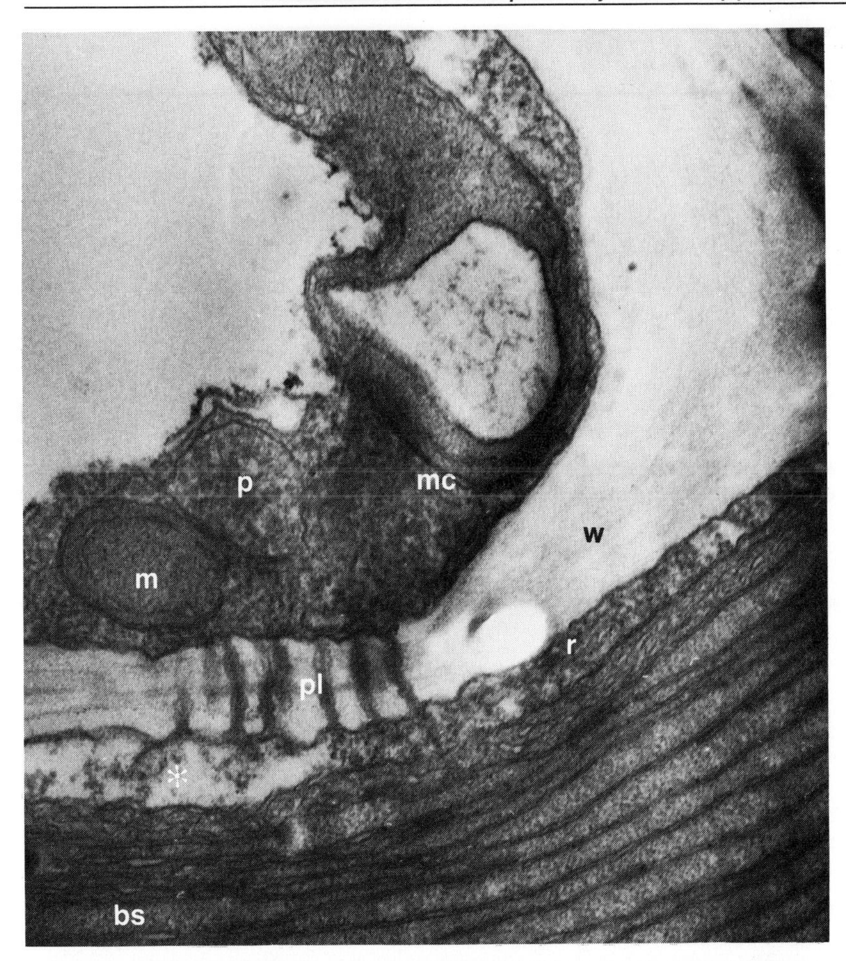

FIG. 4.10 Detail (×35 000) of connections between the bundle sheath cell (bs) and mesophyll cell (mc) of maize. Bundle sheath chloroplasts have a dense reticulum (r) within the envelope and dense cytoplasm connecting the envelope to the outer membrane. At * the chloroplast appears to have pulled away from the wall (w). Plasmodesmata (pl) link bundle sheath and mesophyll cell, which has mitochondria (m) and peroxisomes (p) close to the plasmodesmata. Courtesy of Mr R. Turner, Rothamsted Experimental Station, Harpenden, Herts, UK.

Bundle sheath thylakoids lack obvious grana. However, the restricted contact areas have 15 nm EF_s particles and PF particles of 8.5 nm, only slightly larger than the mesophyll particles, and appear to be rudimentary grana, of comparable subunit structure to mesophyll and C3 chloroplasts. Particle density is similar in stacked mesophyll and in contact areas in the bundle sheath, but as there is much less stacked area in bundle sheath compared to mesophyll chloroplasts, the number of larger particles per thylakoid is only one tenth. However, the PF particles are similar in number. Stromal lamellae of bundle sheath and mesophyll chloroplasts have 10 nm particles on EF_u and PF_u surfaces. The small PSII activity (p. 190) and few large EF particles in bundle

MESOPHYLL THYLAKOIDS

EF_s PF_s P_s E_s

CPI/PSI

LHC

CP_aPSII

P_u

PF_u

E_u

EF_u

Partition region

Lumen

Water-splitting complex

CF_1

Granal thylakoids

Stromal thylakoids

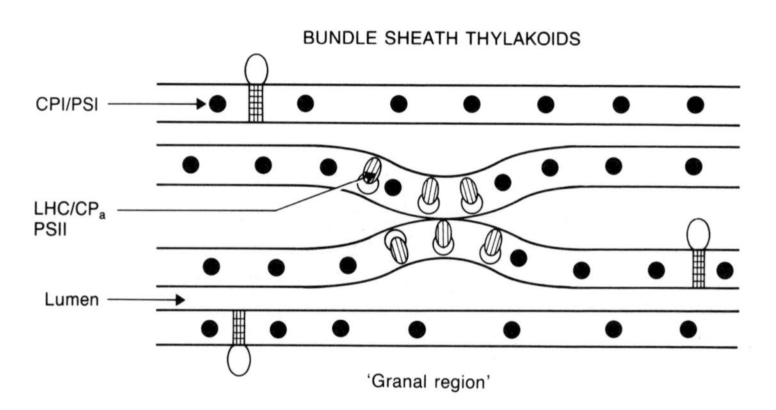

FIG. 4.11 Diagram of maize mesophyll and bundle sheath thylakoids. Only small areas of contact (grana) occur between bundle sheath lamellae compared with extensive grana in the mesophyll. Particles of PSII and associated light-harvesting complex are in grana of both types.

sheath thylakoids suggest that the EF particle is a PSII unit together with the light-harvesting chlorophyll–protein complex which is responsible for chloroplast stacking. Thus the basic structure of thylakoids is similar in different plants and tissues but with quantitative differences in composition related to function.

Chlorophyll and protein complexes in thylakoids

Chl *a* and chl *b* occur only in thylakoid membranes and may form 5 per cent of their total mass (Table 4.1). Chlorophyll is complexed with proteins, but they are not covalently bonded; the hydrophobic phytyl group of chlorophyll may be between the membrane proteins and lipids, and the hydrophilic part of the

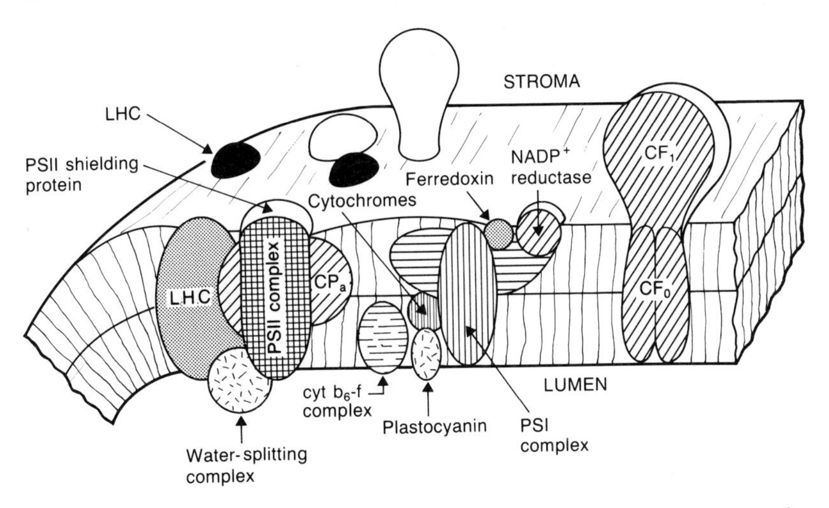

FIG. 4.12 Diagram of the light-harvesting pigment complexes and associated electron transport substances in the thylakoid membrane. Particles are mobile in the membrane depending on light and other conditions.

prophyrin ring in the protein. This would orientate pigment molecules for efficient energy capture and transfer.

The macromolecular structure of thylakoids (Fig. 4.12) has been determined from studies (Thornber 1975) in which chlorophyll–protein complexes are removed from membranes, with the anionic detergent sodium dodecyl sulphate (SDS) for example, and separated by electrophoresis on polyacrylamide gels containing SDS (Fig. 4.13). As conditions of extraction and electrophoresis

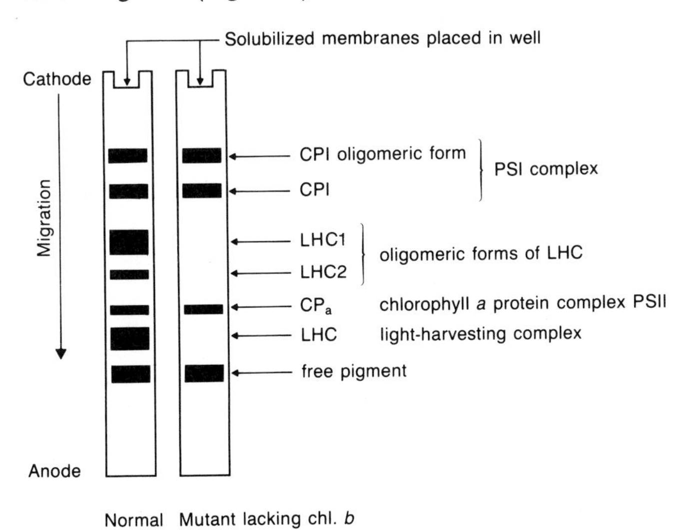

FIG. 4.13 Electrophoretic separation of thylakoid pigment-protein complexes after solubilizing membranes in SDS detergent; semi-schematic, based on studies with normal and mutant barley.

Table 4.2 Chlorophyll–protein (CP) complexes and carotenoids of higher plant thylakoids

Name	Abbreviation	No. sub-units	Chlorophylls (chl *a/b* ratio)	Chlorophyll content (% total)	Mol chl/ mol protein	Main carotenoids (α/β carotenoid ratio)	Mass of complex (kD) (mass sub-units, kD)
CP complex I	CPI, CPI_a (also called P700-chl *a* complex)	2	chl *a*, P700 ($a/b \approx 20$)	30	12	β-carotene ($\alpha/\beta \approx 0.7$)	100 (50, 60, 70)
CP complex II	CP_a (also called CPIV)	1	chl *a*, P680 ($a/b \approx 5$)	10	6	β-carotene ($\alpha/\beta \approx 0.5$)	40–45 (43)
Light-harvesting CP complex	LHCP 1 LHCP 2 LHCP 3	4	chl *a*, chl *b* ($a/b \approx 1.1 \rightarrow 1.3$)	10 10 30	 6	lutein, neoxanthin ($\alpha/\beta \approx 50 \rightarrow 100$)	23–30 (24–26) forms oligomers

(1 dalton = 1.6605×10^{-27} kg, the mass of 1 hydrogen atom. Protein mass is frequently expressed in Daltons; a mass of 20,000 D (20 kD) is equivalent to a protein of mole mass 20 kg.)

affect separation, a variable number of bands results, and size and composition of membrane components is difficult to interpret – for example polypeptide monomers may associate into larger units (oligomers) which give distinct bands in gel separation. Although there is uncertainty about the chlorophyll–protein complexes in the membrane and their correspondence to membrane particles, three main complexes contain 90 per cent of the chlorophyll (Table 4.2). One corresponds to PSI and its antenna chl *a*; it is called P700 chlorophyll *a* complex or chlorophyll–protein complex I, CPI for short. A second is light-harvesting chlorophyll *a/b*–protein complex, now called light-harvesting complex, LHC, which has only antenna function and no photochemical activity. The third is less well resolved but contains PSII and its antenna chlorophyll *a*. It is called chlorophyll–protein complex IV (CPIV) or CP_a.

CPI has a mass of 110 kilodaltons (kD), composed of two polypeptides of 50 or 60 and 70 kD, and contains about 30 per cent of the total chlorophyll, (no chl *b*), 10 mole chl *a* per mole of protein and 1 P700 molecule to every 100 chl *a* molecules. P700 is probably surrounded by a tightly bound inner and loosely bound outer chl *a* antenna, with about 5 β-carotenes and 2 xanthophylls. A quinone (napthoquinone, phylloquinone or tocopherylquinone but no plastoquinone) and perhaps iron–sulphur centres are bound in the complex, and transport electrons. Some CPI may be composed of P700 and 5–10 monomers of protein–chlorophyll, a functional unit. CPI occurs throughout the plant kingdom in all organisms; photosynthetic bacteria have an equivalent and very similar photosystem complex. CPI predominates in membranes in contact with the stroma.

The CP_a complex is formed independently of the LHC but closely linked to it; it occurs in all oxygen-evolving plants and contains 10 per cent of the total chlorophyll, mainly or only chl *a*. P680 forms the core with 6 mole of chl *a*, β-carotene and electron carriers bound per mole of protein, a 40–45 kD polypeptide which does not form oligomers. Probably an antenna of chl *a* links CP_a to the LHC, thus preventing easy or consistent isolation. It has been suggested that there may be two types of CP_a units called $PSII_\alpha$ and $PSII_\beta$. $PSII_\alpha$ contains more chl *a* in the antenna than $PSII_\beta$ and several $PSII_\alpha$ complexes may join together but the $PSII_\beta$ complexes are not joined together or with PSII. The more $PSII_\alpha$ complex the greater the *a/b* ratio and the less grana stacking occurs. Possibly $PSII_\beta$ complexes are in the stromal membrane and $PSII_\alpha$ in the stacked regions.

Light-harvesting complex, mainly found in granal thylakoids of plants with chl *b*, contains 50 per cent of chl *a* and all chl *b* (ratio *a/b*, 1:3) and 1 carotenoid per 6 chlorophylls. Over 50 per cent of the total thylakoid protein (and most of the structural protein) is in LHC with a molecular ratio of 6–12 chlorophyll per protein. LHC is not essential for photosynthesis as mutants lacking it can photosynthesize efficiently whereas loss of CPI or CP_a inhibits photosynthesis. The amount of LHC varies with growth conditions, for example in dim light the chl *a/b* and P700/chl *b* ratios decrease as LHC increases relative to CPI or CP_a, whereas water stress and iron deficiency decrease LHC formation.

Within the LHC complex, groups of chl *b* molecules, probably arranged within exciton transfer distance, deliver energy with high efficiency to groups of chl *a* molecules more loosely arranged and transferring energy by the Förster mechanism. LHC passes energy mainly to CP_a and hence to PSII but as light changes so does the association of LHC with CPI and CP_a, altering the energy transfer between LHC and CPI and therefore the distribution ('spill-over') of energy between PSI and PSII. The protein is composed of two abundant polypeptides, 25 and 23 kD, found with chl *b* and chl *a* (it is not known if there is specific binding between polypeptides and pigments). LHC is not found in mutants or plants grown in intermittent light which lack chl *b*. If each polypeptide contains 6 chlorophyll molecules there will be 30 or more monomers aggregated to form LHC *in vivo*. Most of the LHC is hydrophobic and buried in the membrane. However, a segment (2 kD) of 20 amino acids, on the major polypeptides is exposed to the stroma, and a segment of each is exposed to the lumen. The exposed portion in the stroma may be removed by adding protease enzyme (trypsin). The amino acid sequence at the C-terminal of the peptide segment is (lysine$^+$ or arginine$^+$) lysine$^+$, arginine$^+$, serine, alanine, threonine$^+$, threonine$^+$, lysine$^+$, lysine$^+$, with positive charges as indicated. Despite these positive charges the thylakoid surface has a net negative charge (1 charge per 6 nm^2) so that the membranes are mutually repelling and dissociate unless cations (e.g. Mg^{++}) are present to shield the negative charges. This allows the positive charged portion of LHC to join onto some negative charges and cause thylakoid stacking. The threonine residues on the 2 kD peptides can be phosphorylated by a specific kinase, abolishing the positive charge and so altering the balance of charges on the membrane surface, changing stacking and distribution of particles in the granal and stromal thylakoids and also energy distribution.

Arrangement of complexes in thylakoids

Current models of particle organization illustrated in Fig. 4.12 are reviewed by Anderson (1981) and Hiller and Goodchild (1981). A large LHC antenna of chl *a* and *b* is linked closely to the antenna chl *a* of CP_a, which adjoins the PSII core. LHC delivers energy preferentially to PSII and the CPI antenna transfers energy to P700. PSI may be almost restricted to membranes exposed to the stroma and absent from the interior of stacked membranes. Probably 80 per cent of PSII is in the granal regions, sheltered from the stroma (Staehelin *et al.* 1977; Barber, 1982). Illumination causes PSI to associate with PSII and LHC, from which it receives more energy, at the edge of the granum. Groups of LHC, CPI and CP_a, perhaps 200 per granum, may function independently, only distributing energy within a single group (a 'puddle' model) or co-operatively (a 'lake' model) with excitation energy passing between groups, but efficient energy distribution under different conditions may require that complexes interact in various ways.

Photosynthetic units and quantasomes

The functional 'photosynthetic unit', a group of 200–300 chlorophyll molecules, co-operating in all the light reactions (considered in Ch. 3), has been identified with large particles, called 'quantasomes', seen in electron micrographs (see Park and Sane 1971). When isolated they contained 230 molecules chlorophyll per unit, carotenoids, quinones (e.g. plastoquinone) and proteins. However, separation of yet smaller pigment–protein complexes and their occurrence in different proportions in parts of the thylakoid has undermined the concept of a 'quantasome' of fixed size and composition. Now the 'photosynthetic unit' is identified with groupings of LHC, CPI and CP_a in the membrane, which allow greater flexibility in photosynthetic response to environmental, and possibly physiological changes.

Granal stacking and excitation energy distribution

Granal stacking and arrangement of particles within the membranes are influenced by the thylakoid's environment. Low concentrations of cations causes grana to dissociate into individual thylakoids. Dissociation is independent of type or concentration of anion and is not affected by osmotic concentration which, however, changes thylakoid volume and membrane thickness. Thylakoids restack if cations are added (Fig. 4.14); cations are most effective in the order trivalent, divalent, monovalent, provided they do not bind to the surface. Stacking is associated with accumulation of large particles on EF_s and decrease on EF_u, and these large particles of LHC are thought necessary for grana formation. Significantly maize mesophyll cells with LHC are granal but bundle sheath chloroplasts without LHC do not stack (page 70).

Changes in the arrangement of photosystems in thylakoids are related to changes in the intensity of fluorescence emission at 685 nm from chl *a* of PSII

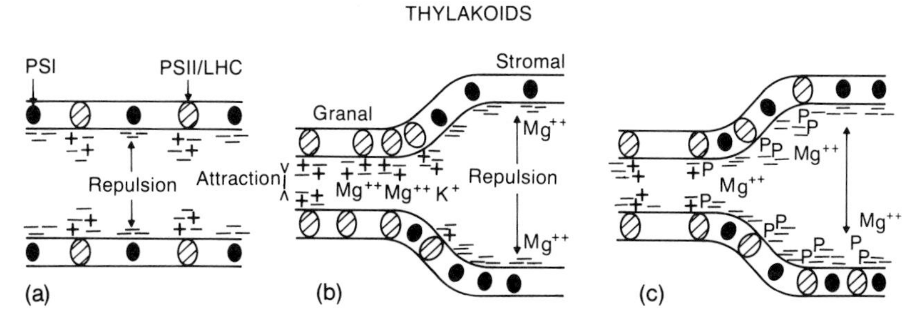

FIG. 4.14 Hypothetical state of thylakoids; (a) unstacked, random PSI and PSII particles with repulsion due to excess negative charges; (b) cations (Mg^{2+}, K^+) screen the charges, PSII/LHCP complexes associate, with grana formation, and PSI moves to unstacked areas; (c) with illumination PSII and PSI mix at the edges of grana, which shrink.

at room temperature. Bonaventura and Myers (1969) showed that darkened *Chlorella* when illuminated with light that does not saturate photosynthesis but excites PSII predominantly (650 nm red light), gave large fluorescence and low photosynthetic O_2 evolution (similar results have been obtained by others with chloroplasts and isolated thylakoids); this is called state 1 and is caused by the 'trap' of PSII being 'closed', that is unable to accept an electron because the chain is reduced, and excitation decays by fluorescence. With time, however, energy redistributes or 'spills over' from PSII (which has a slow reaction centre process and therefore is often 'closed') to PSI, which is always 'open trap', so electrons flow in the chain, O_2 is evolved, and excitation in PSII decreases, giving low fluorescence or state 2. This is not caused by changes in the light absorbing area of the antenna, but by different degrees of association between PSI and PSII and LHC. In isolated thylakoids low cation concentration gives state 2 (maximum 'spill-over') but increasing ionic strength decreases 'spill-over' and increases fluorescence (state 1). Low cation concentration causes membranes to unstack and random distribution of protein complexes in them. Stacking causes redistribution of large membrane particles (LHC and CP_a) into the partition areas of thylakoids, and is related to increased fluorescence. As electron transport is from PSII to PSI in series, lack of excitation of PSI will reduce electron flow, slow photosynthesis and increase fluorescence. However, increasing the energy to PSI increases the overall rate of electron flow and decreases fluorescence. This redistribution of energy to reaction centres of PSI increases efficiency of photosynthesis by about 20 per cent.

Increased efficiency related to changes in thylakoid composition is also observed in intact leaves. When PSII is stimulated after light which is absorbed by PSI (730 nm, far red) or darkness, fluorescence from PSII is initially great but decreases to a constant value in about 10 minutes (the Kautsky effect, page 236) due to spill-over. Adding PSI light and thus increasing the ratio of PSI to PSII light does not increase fluorescence, showing that spill-over activates PSI efficiently, allowing electron flow. If the PSI light is removed there is an immediate increase in fluorescence (state 1), which again drops to state 2 in 10 minutes. Provision of PSI light eliminates the capacity for spill-over although it may be re-established. Such changes in fluorescence have been related to redistribution of LHC and stacking, which alters the distance between LHC and the PSI complex. In the intact chloroplast system, cation effects observed with isolated thylakoids are unlikely to play a major role in the state 1 to state 2 transitions, and light quality and spectral distribution are more important. However, both these and the redox state of the plastoquinone pool in the thylakoid interact to regulate the position of the membrane particles which determine energy transfer (Barber 1982). Spill-over is regulated by conditions in the membranes. Thylakoid surfaces are negatively charged above pH 5, due to dissociation of carboxyl (—COOH) groups of glutamate and aspartate in proteins. CPI has a greater negative charge than LHC, which carries positive charges on the threonine residues of the 2 kD polypeptide portion of LHC exposed to the stroma. In dilute ionic solutions (Fig. 4.14) grana dissociate as negative charges are not shielded and there is electrostatic repulsion, and CPI and LHC/CP_a are randomly distributed to achieve charge equalization, so that spill-over from PSII to PSI is efficient and state 2 is observed. Cations (e.g. 3 – 10m M Mg^{2+}) screen the negative charges and the positively charged LHC can link to other membranes so that grana form. CPI migrates into unstacked areas and large LHC particles are observed in stacked, and small PSI particles in unstacked, membranes. Spill-over decreases as PSI and II are separated, giving high (state 1) fluorescence and decreasing electron flow.

Under conditions *in vitro* cations screen the charges and allow stacking, but state 1 to 2 transitions are still observed, due to enzymatic phosphorylation of LHC (Bennett 1984). Light activates, (*via* electron flow and reduced plastoquinone, PQH_2) a protein kinase enzyme, located on the stromal surface (Fig. 4.15). The kinase phosphorylates the threonine residues on the exposed LHCP polypeptide, using ATP (plus Mg^{2+}) and requires about 10 minutes to phosphorylate a portion of the LHC. This timescale is similar to the observed state 1 to state 2 fluorescence changes. It has also been shown that this is

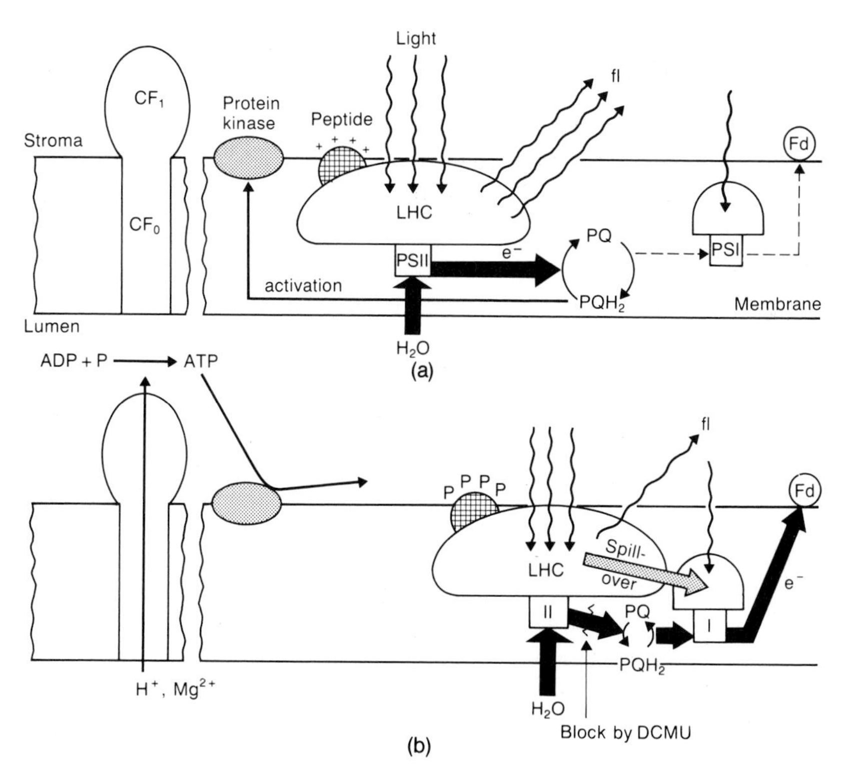

FIG. 4.15 Control of energy distribution in thylakoids by phosphorylation of a peptide on LHC. (a) initially light is absorbed mainly by PSII, electron (e^-) transport to ferredoxin (Fd) is small but reduction of plastoquinone (PQ) activates a protein kinase which (b) increases association of PSII and I, spill-over of energy and electron flow to Fd and decreases fluorescence (fl).

related to the reduction of the PQ pool and provides a mechanism for regulating spill-over. Rapid PSII activity compared to PSI reduces PQ and so increases phosphorylation, masking positive charges on about 20 per cent of LHC which migrate, due to charge, into unstacked regions. Hence grana become smaller in the light (in algae from 60 to 40 per cent of total membrane) and LCH/CP_a and CPI associate, giving greater electron transport, faster O_2 evolution and state 2. The importance of such a mechanism is that it allows flexibility in energy use. In darkness or when the reduction of PQ is low and the kinase is not active, a phosphoprotein phosphatase (not light activated but functional in light and dark) reverses the activation in 10 minutes. Currently the nature of kinase and phosphatase enzymes, amounts in different species, function and regulation are active and exciting research areas (Barber 1982; Horton 1983). This is not the only phosphorylation of possible importance in the thylakoids. There is evidence that an 8 kD polypeptide of CF_0 undergoes phosphorylation and regulates the H^+ flux through CF_1 (see Ch. 7).

Why should energy distribution be inefficient on illumination and what is the advantage of variable association between particles? They may optimize energy transduction. Possibly it is necessary to pump H^+ into the thylakoid (from water splitting) quickly to generate ATP and avoid excessive formation of reductant. On sudden illumination the high energy supply to PSII may permit rapid H^+ accumulation with minimum e^- transport to avoid over-reduction of $NADP^+$ when ATP is limiting. If NADPH production is slow, the e^- transfer is slowed, PQ reduced, the kinase is activated and more energy passes to PSI, stimulating electron flow. If production of ATP is limiting then the kinase does not phosphorylate LHC, energy builds up, and PQ is reduced, so increasing the H^+ pump and ATP synthesis. Spill-over may allow 'fine tuning' of the ATP/NADPH ratio and flexibility under conditions of changing light quality, such as in dense vegetation where red light may overstimulate PSI. Changes in thylakoid granal surface area have also been suggested to regulate ATP production in dim light but are probably only a minor function of stacking grana. Stacking may permit ATP synthesis and regulate reductant formation particularly in dim light, where extensive grana formation is characteristic, but in bright light ATP synthesis may not be limiting and so PSI is stimulated and stacking is decreased. ATP is known to dissociate grana and decrease fluorescence. The kinase and phosphatase act as a light regulated switch mechanism for effective thylakoid function. This complex feedback system involves the thylakoid components and their redox state, cations, and regulatory proteins, for optimization of energy use in different radiation environments. Perhaps differences in thylakoid structure between C3 and C4 plants and between sun and shade plants (Ch. 9) determine the overall efficiency of the thylakoid in relation to demand for ATP and NADPH in a particular light environment, and complex control of a portion of the light-harvesting apparatus allows rapid response to relatively small changes in the environment.

References and Further Reading

Anderson, J. M. (1975) The molecular organisation of chloroplast thylakoids, *Biochim. Biophys. Acta*, **416**, 191–235.

Anderson, J. M. (1981) Consequences of spatial separation of photosystem 1 and 2 in thylakoid membranes of higher plant chloroplasts, *FEBS Lett.*, **124**, 1–10.

Arntzen, C. J. and **Briantais, J.-M.** (1975) Chloroplast structure and function, pp. 52–133 in Govindjee (ed.), *Bioenergetics of Photosynthesis*, Academic Press, New York.

Barber, J. (1976) Ionic regulation in intact chloroplasts and its effect on primary photosynthetic processes, pp. 89–134 in Barber, J. (ed.), *The Intact Chloroplast*, Elsevier/North-Holland Biomedical Press, Amsterdam.

Barber, J. (1982) Influence of surface changes on thylakoid structure and function. *A. Rev. Plant Physiol.*, **33**, 261–95.

Bennett, J. (1984) Chloroplast protein phosphorylation and the regulation of photosynthesis, *Physiol. Plant.*, **60**, 583–90.

Boardman, N. K., Anderson, J. M. and **Goodchild D. J.** (1978) Chlorophyll–protein complexes and structure of mature and developing chloroplasts, pp. 36–109 in *Current Topics in Bioenergetics*, Academic Press, New York.

Bonaventura, C. and **Myers, J.** (1969) Fluorescence and oxygen evolution from *Chlorella pyrenoidosa*, *Biochim. Biophys. Acta*, **189**, 336–83.

Branton, D. *et al.* (1975) Freeze-etching nomenclature, *Science*, **190**, 54–6.

Coombs, J. and **Greenwood, A. D.** (1976) Compartmentation of the photosynthetic apparatus, pp. 1–51 in Barber, J. (ed.), *The Intact Chloroplast*, Elsevier/North-Holland Biomedical Press, Amsterdam.

Douce, R. and **Joyard, J.** (1979) Structure and function of the plastid envelope, *Adv. Bot. Res.*, **7**, 1–116.

Douce, R. and **Joyard, J.** (1981) The chloroplast envelope: structure, composition and biological properties, pp. 187–98 in Akoyunoglou, G. (ed.), *Photosynthesis III. Structure and Molecular Organisation of the Photosynthetic Apparatus*, Balaban International Science Services, Philadelphia.

Esau, K. (1958) *Plant Anatomy*, Wiley, New York. Chapman and Hall, London.

Gunning, B. E. S. and **Steer, M. W.** (1975a) *Plant Cell Biology – An Ultrastructural Approach*, Edward Arnold, London.

Gunning, B. E. S. and **Steer, M. W.** (1975b) *Ultrastructure and the Biology of Plant Cells*, Edward Arnold, London.

Hall, D. O. (1972) Nomenclature for isolated chloroplasts, *Nature New Biol. (Lond.)*, **235**, 125–6.

Heath, O. V. S. (1969) *The Physiological Aspects of Photosynthesis*, Stanford University Press, Stanford, CA.

Hiller, R. G. and **Goodchild, D. J.** (1981) Thylakoid membrane and pigment organization, pp. 1–49 in Hatch, M. D. and Boardman, N. K. (eds), *The Biochemistry of Plants*, Vol. 8, *Photosynthesis*, Academic Press, New York.

Horton, P. (1983) Control of chloroplast electron transport by phosphorylation of thylakoid proteins, *FEBS Lett.*, **152**, 47–52.

Kaplan, S. and **Arntzen, C. J.** (1982) Photosynthetic membrane structure and function, pp. 65–151 in Govindjee (ed.), *Photosynthesis*, Vol. 1, *Energy Conversion by Plants and Bacteria*, Academic Press, New York.

Kirk, J. T. C. and **Tilney-Bassett, R. A. E.** (1978) *The plastids, their chemistry, structure, growth and inheritance*, Freeman, London.

Laetsch, W. M. (1971) Chloroplast structural relationships in leaves of C4 plants, pp. 323–49 in Hatch, M. D., Osmond, C. B. and Slatyer, R. O. (eds), *Photosynthesis and Photorespiration*, Wiley Interscience, New York.

Leegood, R. C. and **Walker, D. A.** (1983) Chloroplasts (including protoplasts of high carbon dioxide fixation ability), pp. 185–210 in Hall, J. L. and Moore, A. L. (eds), *Isolation of Membranes and Organelles from Plant Cells*, Academic Press, London.

Miller, K. R., Miller, G. J. and **McIntyre, K. R.** (1977) Organisation of the photosynthetic membrane in maize mesophyll and bundle sheath chloroplasts, *Biochim. Biophys. Acta*, **459**, 145–56.

Mühlethaler, K. (1977) Introduction to structure and function of the photosynthesis apparatus, pp. 503–21 in Trebst, A. and Avron, M. (eds), *Encyclopedia of Plant Physiology* (N.S.), Vol. 5, *Photosynthesis I*, Springer-Verlag, Berlin.

Murphy, D. J. (1982) The importance of non-planar bilayer regions in photosynthetic membranes and their stabilisation by galactolipids, *FEBS Lett.*, **150**, 19–26.

Newcomb, E. H. and **Frederick, S. E.** (1971) Distribution and structure of plant microbodies (peroxisomes), pp. 442–57 in Hatch, M. D., Osmond, C. B. and Slatyer, R.O. (eds), *Photosynthesis and Photorespiration*, Wiley Interscience, New York.

Nobel, P. S. (1974) *Biophysical Plant Physiology*, W. H. Freeman, San Francisco.

Park, R. B. (1976) The chloroplast, pp. 115–45 in Bonner, J. and Varner, J. E. (eds), *Plant Biochemistry* (3rd edn), Academic Press, New York.

Park, R. B. and **Sane, P. V.** (1971) Distribution of function and structure in chloroplast lamellae, *A. Rev. Plant Physiol.*, **22**, 395–430.

Staehelin, L. A. and **Arntzen, C. J.** (1979) Effects of ions and gravity forces on the supramolecular organisation and excitation energy distribution in chloroplast membranes, pp. 147–75 in *Chlorophyll Organisation and Energy Transfer in Photosynthesis*, Ciba Foundation Symposium 61 (N.S.), Excerpta Medica, Amsterdam.

Staehelin, L. A., Armond, P. A. and **Miller, K. R.** (1977) Chloroplast membrane organization at the supramolecular level and its functional implications, *Brookhaven Symp. Biol.*, **28**, 278–315.

Thornber, J. P. and **Alberte, R. S.** (1977) The organization of chlorophyll *in vivo*, pp. 574–82 in Trebst, A. and Avron, M. (eds), *Encyclopedia of Plant Physiology* (N.S.), Vol. 5, *Photosynthesis I*, Springer-Verlag, Berlin.

Thornber, J. P. (1975) Chlorophyll–proteins: Light harvesting and reaction centre components of plants, *A. Rev. Plant Physiol.*, **26**, 127–58.

CHAPTER 5

Water splitting and the electron transport chain

Photon capture by the photosystem antennae and excitation transfer to PSII and I, provide the energy for oxidation of water and electron movement to acceptors, which donate e^- to biochemical processes, and for passage of protons into the thylakoid lumen, for synthesis of ATP. The electron transport system may be considered in five parts (1) a water-splitting complex, (2) a photosystem II complex, (3) an electron carrier chain, (4) a PSI complex, and (5) a group of e^- carriers which reduce acceptors ($NADP^+$, O_2).

Excitation causes the PSII reaction centre to eject an electron to an acceptor, starting e^- transport along the chain of redox components. Oxidized PSII is reduced by e^- from a water-splitting complex *via* intermediates S and Z (Fig. 5.1). The energized e^- passes, from more to less negative potential, to the primary acceptor pheophytin and then in sequence to the quinone acceptors Q, Q_B and plastoquinone (PQ). Quinones are important carriers of e^- and H^+ in

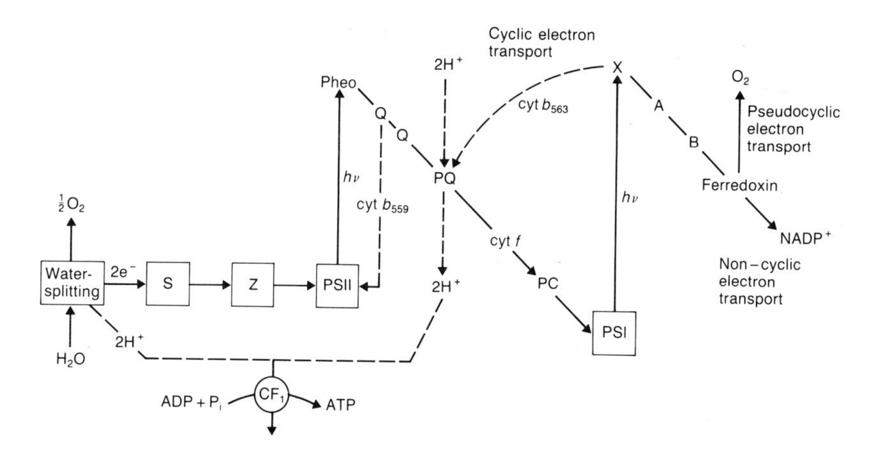

FIG. 5.1 Scheme of electron transport in the thylakoid membrane from water to $NADP^+$ with two light reactions, PSII and I in series; S and Z are intermediates in water splitting, Pheophytin (pheo), quinone acceptors Q, Q_B and PQ (plastoquinone), cytochrome *f* (cyt *f*) and plastocyanin (PC) pass e^-. X, A, B are intermediates in e^- movement to ferredoxin. Proton transport leads to synthesis of ATP. Electrons also cycle from PSI back to the intermediate chain or pass to oxygen.

many biological processes. From PQ the electron passes to cytochrome *f* and plastocyanin before reducing an oxidized PSI reaction centre. Here it is energized again by excitation energy from the chlorophyll matrix and passes *via* intermediates X (or A_2), A and B to oxidized ferredoxin (Fd) and $NADPH^+$, which are reduced and enter into biochemical reactions in the chloroplast stroma.

Electron transport chains bridge the thylakoid membrane, allowing electrons from water inside the thylakoid lumen to pass across the membrane to ferredoxin on the stromal side. Plastoquinone in the membrane is reduced by electrons; the H^+ from the stroma attaches to reduced plastoquinone and is carried to the lumen, where it is released and the plastoquinone oxidized. Thus electron transport is coupled to a plastoquinone cycle which carries ('pumps') H^+ from stroma to the thylakoid lumen in the reverse direction to electron transport, increasing H^+ concentration in the thylakoid lumen and forming the proton concentration gradient, the energy of which drives ATP synthesis (p. 110). Trebst (1974) and Velthuys (1980) have reviewed aspects of electron transport and the books edited by Govindjee (1975) and the texts by Clayton (1971, 1980) and Gregory (1979) provide detailed discussion.

Photosystem II complex

This photosystem consists of chlorophyll pigment antenna, reaction centre complex with a molecule of P680 joined to structural proteins and components linking P680 to the water-splitting enzyme complex and to the electron acceptors of P680. Purified spinach reaction centre complexes have been analysed by methods such as separation of proteins by SDS polyacrylamide gel electrophoresis with urea to denature and dissociate the subunit structure. Although the protein components are still uncertain, there are polypeptides of 43, 47 and 50 kD associated with chl *a* and a 32 kD polypeptide which binds herbicides, such as atrazine. Another polypeptide of similar mass has yet to be assigned a function. A 7–10 kD subunit is the protein linked to cytochrome b_{559}. For each P680 there is a closely linked chl *a* antenna (approximately 50 molecules). The electron acceptor of P680 is probably a single molecule of pheophytin (a chlorophyll without Mg^{2+}) although a chl *a* molecule has also been suggested to precede pheophytin. Pheophytin has been identified from its characteristic ESR and ENDOR signals and is linked to secondary acceptors, 1 or 2 molecules of bound plastoquinone, which pass electrons to the main plastoquinone pool. Pheophytin is associated with β-carotene (10 molecules), an unknown component giving ESR signals, and to a cytochrome (cyt b_{559}), with the Fe of which it interacts magnetically. Cyt b_{559} has different potential forms, a high energy (0.4 mV) form which is oxidized at low temperatures by PSII, may be linked to e^- transport from the water-splitting complex.

A reaction centre will accept or 'quench' excitation if it is reduced (P680), i.e. contains an electron which can be ejected by excitation but will not use excitation when oxidized ($P680^+$). In the latter case excitation may migrate to reduced reaction centres. Ejection of an electron occurs within nanoseconds; the fewer reduced reaction centres the longer the time that excitation dwells in the antenna and the greater chlorophyll fluorescence. If Q is oxidized, fluorescence yield (p. 240) is about 3 per cent but increases to 12 per cent if Q is reduced. Events at the reaction centre are summarized by the following sequence with Z the donor, P680 the reaction centre, pheophytin (pheo) the primary acceptor and Q and Q_B secondary acceptors:

$$\text{Z.P680.pheo.Q} \xrightarrow{h\nu} \text{Z.P680}^{+}\text{.pheo.}^{-}\text{Q} \longrightarrow \text{Z}^{+}\text{.P680.pheo.Q}^{-} \qquad [\mathbf{5.1}]$$

$P680^{+}$ is a very powerful oxidant (+1.2 V or greater) able to remove electrons from water (+0.8 V) and produce a relatively weak reductant (−0.6 V). The state of Z.P680.pheo.Q^{-} is non-quenching, as electrons cannot be transferred even if P680 contains an electron. Possibly recombination of $pheo^{-}$ with $P680^{+}$ gives rise to the high variable fluorescence. Charge separation forms the radical pair $P680^{+}.pheo^{-}$, which can be detected by ESR (see Malkin 1982a and b). For efficient separation, back reactions of the electron with $P680^{+}$ must be limited; this is achieved by loss of energy as heat, delocalization of the electron and formation of a triplet-like state, and most importantly, spatial separation by rapid transfer to the secondary acceptor, Q.

The secondary acceptor, Q, is a quinone bound to protein; it passes electrons to another acceptor Q_B, which is bound to a 32 kD lysine-free polypeptide when oxidized but when reduced it diffuses into the PQ pool (Velthuys 1982). Q_B is a 2 e^{-} transfer molecule. Q_B^{-} is tightly bound to the protein but Q_B^{2-} is not bound. Inhibitors such as the herbicide DCMU, bind to and block the site of Q_B attachment and thereby prevent e^{-} transport. Different herbicides, for example atrazine and DCMU, bind to the same site or closely linked sites. The peptide probably regulates H^{+} movement to Q_B^{-}, Q_B reduction by added reductants and also controls herbicide activity. Trypsin destroys the protein and influences granal stacking. Herbicides such as atrazine which bind to the protein inhibit electron transport, but in many species of plants, mutants have arisen which are insensitive to these herbicides; possibly the binding site has changed so that the herbicides cannot block e^{-} transport.

Understanding of the nature and role of electron acceptors has come from measurement of changes in absorption spectra and ESR as the intermediates are reduced and oxidized, either with light or chemically. Absorption changes in the $P680^{+}$ to P680 reduction occur at 690 nm and 820 nm and there is an ESR signal. An absorption signal at 550 nm is probably from pheophytin, due to changes in its electrical environment caused by the reduction of Q. Another absorption change is called X320, X for the 'unknown' and 320 for the wavelength (nm), thought to be derived from molecules, closely bound to P680, probably the plasto-semiquinone anion of Q. An ESR signal arises from the interaction of Q with iron, probably from cytochrome in the PSII complex. The acceptor is possibly on the outer surface of the thylakoid, shielded by, and linked to, the herbicide binding protein (giving the 'X320-B-enzymatic complex') which stops H^+ joining to the reduced acceptor and slows reduction by added reductants and regulates herbicide activity.

Electron accepting side of PSII

Under physiological conditions $P680^+$ oxidizes a donor, Z (also called D), as the kinetics of Z^+ formation mirror the reduction of $P680^+$ under a wide range of conditions. Z is, in turn, reduced by the water-splitting complex M. Z is poorly characterized, but is possibly PQH_2^+, which has the required redox potential; an ESR signal shows it to be a quinone free-radical located on the e^- donor side of PSII. This signal II, as it is called, has transients with different decay rates which have been designated II_s for 'slow' (1 second), II_f for 'fast' (500 μs) and II_{vf} for 'very fast' (20 μs), the latter is produced by the oxidized state Z^+. The II_{vf} signal is stopped by Tris, which removes Mn^{2+} (and possibly polypeptides of the water-splitting complex) and DCMU also suppresses II_f. Perhaps Z is a form of manganese tightly bound to the reaction centre, or an oxidized plastoquinone molecule.

Water splitting

The water-splitting complex, the structure of which is poorly understood, is on the lumen side of the thylakoid membrane, and is designated as M, L, S or Y (depending on the author!). Its role in the production of 4 H^+, 4 e^- and O_2 from water is well characterized, including the sequence of electron, proton and O_2 release, clearly establishing the mechanisms.

When Allen and Frank gave a sequence of short flashes of bright light, separated by darkness, to algae previously in darkness, the yield of O_2 per flash depended on the position of the flash in the sequence. Joliot and Kok (see Diner and Joliot 1977) showed with refined rapid measurements that O_2 was evolved with characteristic oscillations (Fig. 5.2). The first two flashes evolved

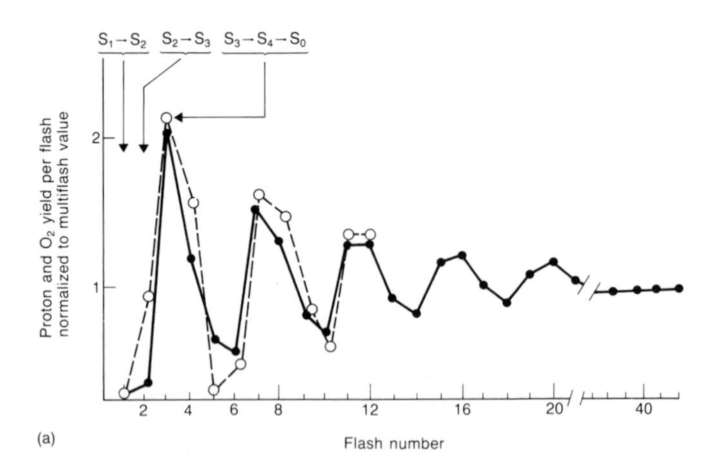

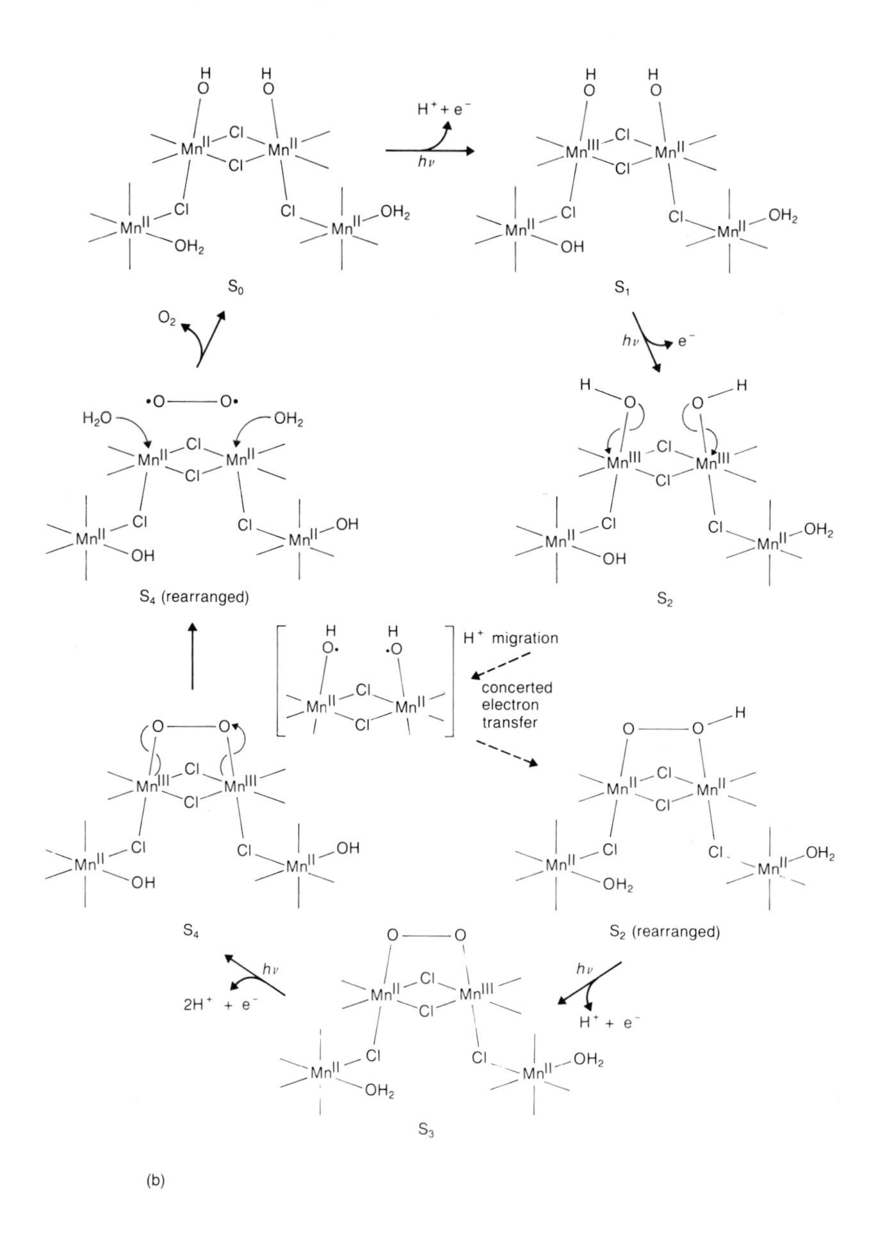

FIG. 5.2(a) Oxygen evolution and proton release by chloroplasts given short (2 μs) intense flashes of light separated by darkness. The number per flash is expressed relative to the production after many flashes. ●——●, O_2 evolution; x---x, H^+ evolution. (After Kok, 1976). (b) An hypothetical model of the arrangement of manganese and chloride at the water-splitting site and the associated proton and electron movement. (After Critchley and Sargeson 1984).

little or no O_2, the third a large 'gush' and the fourth a smaller amount of O_2 than the third but more than the first, that is, a periodicity of four. Oscillations were damped and after some 20 flashes yield per flash was constant. This pattern is characteristic of algae and chloroplasts. Oxygen production is slower just after illumination than with longer illumination and this 'lag' period is inversely proportional to light intensity × time, showing that activation of an intermediate of water splitting is needed for O_2 production. The states induced by light flashes are not stable. If two flashes are separated from a sequence of flashes by a long dark period, the characteristic sequence starts again as if after darkness; the first two flashes activate the process and induce states which decay in darkness. Light captured by one photosystem cannot contribute to another photosystem; O_2 evolution would be proportional to light intensity (i.e. number of photons captured) if co-operation were possible. O_2 evolution uses only light captured by PSII. The characteristics of the system do not change in the flash sequence as the optical cross section (area of pigment available for photon absorption) and the quantum yield are similar at all stages of water splitting.

Formation of intermediate oxidation states of the water-splitting complex

Four photons captured by PSII co-operate to dissociate water, they cause the reaction centre of PSII to eject four electrons. Four oxidation equivalents accumulate on an intermediate, S, before they accept four electrons from two molecules of water, releasing O_2. If S^{4+} is the oxidized component which reacts with water:

$$2\,H_2O + S^{4+} \rightarrow S + 4\,H^+ + O_2 \qquad \textbf{[5.2]}$$

and S is a 'charge'-accumulating chemical device. As photon capture is infrequent in dim light, the intermediate oxidized states must remain stable for sufficient time to enable four positive charges to accumulate, and allow water oxidation.

A model by Kok, Forbush and McGloin (1970) explains the periodicity of O_2 as a cycle of S states. It proposes that S accumulates four oxidizing equivalents solely from P680 and that only O_2 is liberated in a single process. If the water-splitting and PSII complex, written as S.Z.P680.Q, is equivalent to S_0,

	S_0	$h\nu$ →	S_1	$h\nu$ →	S_2	$h\nu$ →	S_3	$h\nu$ →	(S_4)	→	S_0
		e^- H^+		e^-		e^- H^+		e^- $2H^+$		O_2	
Number of atoms of											
Mn^{3+}	4		3		2		1		(0)		4
Mn^{4+}	0		1		2		3		(4)		0
Change in											
Mn charge		+1		−1		+1		(+1)		−3(−4)	
total charge		0		+1		0		−1			
between S states											
Water state	H_2O H_2O		OH^- H_2O		OH^- H_2O		OH^- OH^-		$[O^{2-}\ O^{2-}] \rightarrow O_2$		H_2O H_2O

then S_1 is $S^+.Z.P680.Q^-$ with the reaction centre refilled with an electron and one oxidizing equivalent accumulated on S; further flashes cause the sequence:
Events involving PSII reaction centre and water splitting are therefore:

(1) Activation of the reaction centre chlorophyll, P680 and charge separation.
(2) Reduction of acceptor and rapid donation of an electron from the water-splitting complex, S, *via* Z to $P680^+$.
(3) Repetition until S_4 is formed releasing 4 H^+.
(4) Removal of four electrons from two water molecules by S_4, and liberation of O_2.

To explain why most O_2 is released on the third flash (rather than the fourth as expected), it is assumed that in darkness S is 75 per cent in the state S_1 and 25 per cent in state S_0. The damped oscillations observed experimentally are thought to result from 5 per cent of the light flashes causing double hits and two oxidations and 10 per cent of the photons not hitting a target, so averaging the states (see Clayton 1980). The rate of conversion of S_0 to S_1 and S_2 is less than 100 μs, S_2 to S_3 and S_3 to S_4 require about 0.5 ms and S_4 to S_0 1 ms so the O_2 evolving step is rate limiting.

Decay of S states

In darkness S_2 and S_3 are unstable and are deactivated to S_1 by addition of electrons, probably from reduced compounds *via* the electron transport chain and cyclic electron transport. Electrons flow from PQ *via* Q to P680 then to S_2, but not to S_3. States S_0 and S_i equilibrate in darkness and S_0 goes to S_i if artificially reduced by ferricyanide or DCPIP and ascorbate. Hydroxylamine (NH_2OH) and ammonia bind tightly to the O_2-evolving complex and donate electrons, removing two oxidizing equivalents and increasing the lag in O_2 evolution which cannot be reversed by washing.

The S_2 and S_3 states are rapidly destabilized by a diverse group of very lipid soluble, weak acid anions with —NH and —OH groups, called ADRY agents from Acceleration of the Deactivation Reactions of Y (alternative sign for the water-splitting system), an example is Ant 2p (2(3-chloro-4-trifluoreomethyl) anilino-3,5-dinitrothiophene). ADRY reagents transport electrons from Q over the energy barrier to S.

The mechanism of O_2 evolution involves more than accumulating four oxidizing equivalents before O_2 release. Measurements of pH changes in thylakoids (treated with uncoupling agents (p. 121) to prevent accumulation of H^+ caused by the electron flow through the thylakoid membrane) show a complex release of protons (H^+) in relation to the light flashes. Single protons are released at the S_0 to S_1 and S_2 to S_3 steps, whilst 2 H^+ are released from S_3 to S_0 *via* S_4. Release of H^+ neutralizes the charge accumulation on S, so it is not justified to call it a 'charge accumulating' device as previously stated (see Velthuys 1980).

The structure of the water-splitting complex

The nature of the water-splitting enzyme complex, that is the type, structure and number of the associated electron donors and acceptors and how they bind to H_2O, and to structural components of the complex, is poorly known and a challenging area of research. Manganese is an essential component of the water-splitting complex. In thylakoids two fractions related to O_2 production have been identified, a fraction (2 Mn atoms) tightly bound to PSII and probably involved in charge accumulation and a more loosely bound fraction, (2 Mn per PSII). For very active O_2 production 4 Mn per PSII complex seems essential. The labile fraction is removed by treatment with Tris buffer, chelating agents (EDTA), or by heating, which stop O_2 liberation but not electron transport. Mn^{2+} added back to thylakoids stimulates O_2 evolution only after chemical reduction or illumination. This 'photoreactivation' requires phosphorylation and uses only PSII reactions; there is no interaction between PSII centres. Photoreactivation has two dark reactions, one changes Mn^{2+} to Mn^{3+}, the next Mn^{3+} to Mn^{4+}.

Proteins in the M complex

Although the characteristics of the water-splitting system are established, the nature and structure of the protein components, and of the site of manganese attachment and water splitting are unclear. Controversy surrounds the identification of the Mn–protein complexes and their role in O_2 evolution. However it is now thought (Yocum 1984) that three major polypeptides form the water-splitting complex; a 33 kD polypeptide contains the two tightly bound Mn atoms responsible for O_2 evolution; a 24 kD polypeptide is associated with (and essential for) O_2 evolution and has a regulatory role; an 18 kD polypeptide also has regulatory function. Studies with salt- and urea-extracted proteins, suggest that the 33 kD polypeptide is linked to PSII in the thylakoid by hydrophobic binding on a part of the polypeptide. The 24 kD unit is probably electrostatically bound to the 33 and 18 kD polypeptides. Closely associated with PSII and the M complex is cytochrome b_{559}, a 10 kD protein, with two polypeptides per haem. Cyt b_{559} may donate e^- to P680 and may stimulate O_2 evolution as an acceptor of H^+ from the S states.

Manganese is probably bound to the 33 kD polypeptide in a highly structured arrangement (see Calvin 1981; Govindjee and Whitmarsh 1982) which alters the redox potential of the Mn, aligning the electronic orbitals and providing a stable electronic configuration during charge separation. (One, speculative, model of events at the active site is given in Fig. 5.2b.)

Probably the Mn atoms sit in an 0.4 nm wide and 0.25 nm deep cleft of the 33 kD polypeptide, judged from the size and shape of analogues of water (e.g. NH_2NH_2, NH_2OH etc.) which bind to the enzyme active site and block O_2 release. Two water molecules may bind, 0.15 nm apart, to Mn at the active site. This would allow loss of 4 e^- and release of 2 O as O_2. The Mn in the complex does not produce NMR or EPR signals at room temperature, so changes in Mn associated with water splitting have been difficult to analyse. The lack of signal in S_0 state may be due to coupling of the Mn^{2+} ions in such a way that an antiferromagnetic state is produced. However at low temperatures, EPR signals are detected in the S_1 states, which suggests that a dimer of MnIII and MnII, on the 34 kD polypeptide, loses e^- and H^+ to give MnIII and MnIV states. The stabilized complex may allow potentially reactive chemical states of O_2, such as —OH radical, hydrogen peroxide and superoxide to form but as 'cryptospecies' which are stabilized by interaction with the delocalized orbitals of Mn. Chloride ions are essential for water oxidation and may stabilize the positive charges in the Mn complex and prevent inhibitors blocking O_2 release. Bicarbonate ions are also essential for O_2 evolution; they may act by improving the function of the acceptor of PSII.

Manganese is a period IV transition metal with oxidation states of +2 (most common) up to +7. Two electrons lost from each of two linked Mn atoms would give four oxidizing equivalents (4+). Mn^{3+} is stable in complexes and could act as intermediary oxidant; Mn^{4+} is also stable and water splitting could take place by single electron transfer steps. The reactions and stable state should not be too long-lived. The Mn^{2+} complex with O_2 is unstable, so that O_2 is rapidly released as required for fast water splitting. Oxygen must bind without releasing singlet or other forms of O_2 which could damage the protein. Manganese fulfils these requirements; ESR studies at low temperature on previously heated thylakoids, which release Mn^{2+}, and at low temperature, strongly suggest that manganese is involved in the S-states changes, and may be the physical entity corresponding to intermediate S. The molecular mechanism of H_2O binding to the complex and the role of ions such as chloride and nitrate which act around PSII and are required for O_2 evolution, are unknown.

It is important to know how water splitting is carried out *in vivo* because it may show how to produce H_2 for fuel from an unlimited source – H_2O. Methods of doing this chemically, by reactions using ruthenium atoms in membrane systems for example, are being explored (see Calvin 1981).

Electron transport beyond the primary acceptor of PSII

The energized electron from P680 reduces Q_B (which may be a form of plastoquinone) to the semiquinone and anionic plastoquinone which accepts protons to give hydroquinone. Reduced Q_B, in the Q_B^{2-} form, diffuses from the binding site and enters the plastoquinone pool. Quinones are important in many biological electron transport processes, carrying e^- and H^+. The structure and sequence of reduction for plastoquinone are shown in Fig. 5.3. PQ is found almost exclusively in thylakoid membranes probably between the lipid layers (p. 63) because of its solubility. Several forms exist but PQ B and C are usually less than 20 per cent of the amount of PQ A and their structures are poorly known. PQ is a substituted benzoquinone with two methyl ($—CH_3$) groups

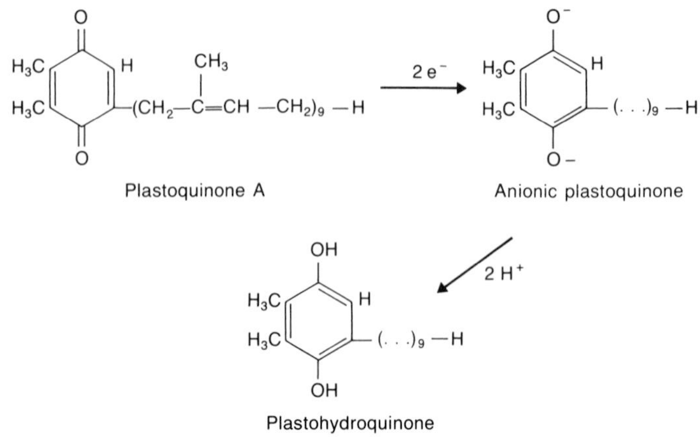

FIG. 5.3 Structure of plastoquinone A, the anionic plastoquinone and plastohydroquinone.

attached to the ring and a long chain of nine isoprene groups; other types of PQ have three and four isoprene units. PQ A is colourless, absorbs light in the ultraviolet (260 nm) and is reduced at +0.1 V *via* plastosemiquinone to plastohydroquinone which absorbs at 290 nm. PQ is implicated as an electron carrier because e^- transport is inhibited when it is extracted with solvents, and restored when it is added back to the membrane, and from spectroscopic measurements of its oxidation and reduction in light and dark. In bright light most, but not all, of the PQ in thylakoids is reduced. The pool of PQ engaged in electron transport in the light is several times larger than that of other carriers. PQ has a molar ratio to chlorophyll of 1:10, with about 40 molecules of PQ per PSII reaction centre; this large pool accepts e^- from several PSII units, acting as a 'buffer' with PSI. PQ is oxidized by cytochrome *f* (cyt *f*) which itself is mainly oxidized under bright light because electrons are more rapidly removed by PSI than they are supplied from PQ, that is, the rate limiting step lies between PQ and cyt *f*. The large pool of PQ and rapid oxidation by cyt *f*, ensure that PQ is not completely reduced in bright light; this enables PSII to function and donate e^-, even though the transport rate from Q to cyt *f* is slow.

PQ also has the important ability to transport protons across the thylakoid membrane (Fig. 5.1). Reduced (anionic)PQ on the stromal side picks up protons and carries them to the lumen, thus increasing the H^+ concentration inside as a consequence of electron transport. The method by which PQ 'turns over' in the membrane is not yet clear. Probably PQH_2, which is hydrophobic, diffuses laterally between PSII and cyt *f* (distances of the order of 100 nm) within the membrane, losing H^+ due to the electrical charge on the membrane. If PSII and PSI are unequally distributed between the granal and stromal thylakoids (p. 76) transport of electrons between them over some distance may be by diffusion of PQ.

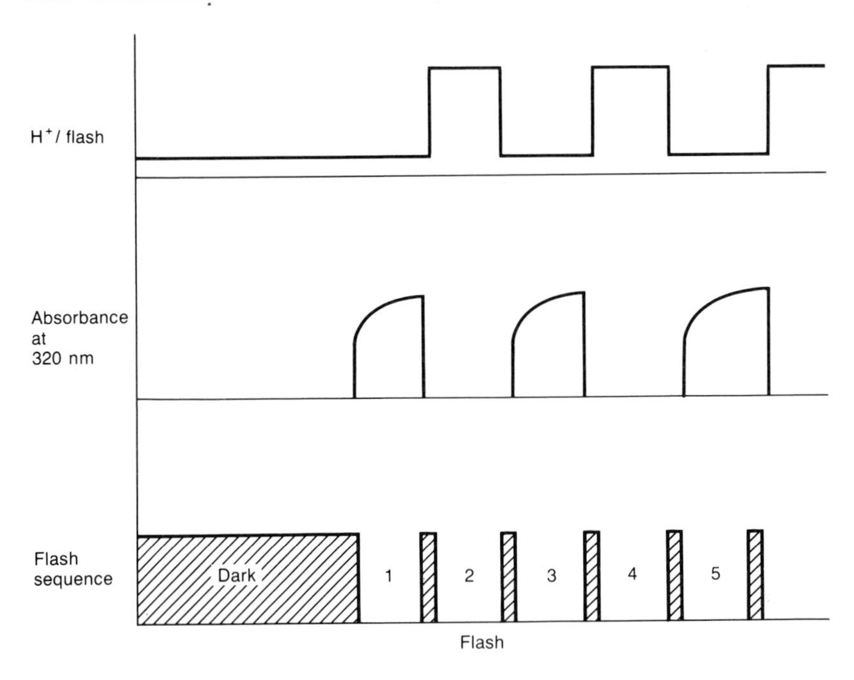

FIG. 5.4 Changes in absorbance at 320 nm, illustrating the conversion of plastoquinone to the plastosemiquinone in relation to H^+ release.

The transfer of e^- and H^+ has been measured from absorbance at 320 nm (where PQ^- but not PQ or PQH_2 absorb) and of H^+ accumulation during a sequence of discrete millisecond light flashes using dark-adapted chloroplasts. PQ^- is formed (Fig. 5.4) by the first flash but H^+ is taken up only every second flash. Regeneration of PQ occurs after the second flash. Some H^+ is taken up with the first, third, fifth flashes etc., as PQ is partially reduced, even in darkness. With time and increasing number of flashes the oscillations are damped. The cycle of PQ reduction and oxidation, which transfers H^+ across the thylakoid, is central to the generation of the pH gradient component of the proton motive force for ATP synthesis (p. 113) and for coupling to e^- transport. Mitchell (1975) suggests that PQH_2 loses e^- to Rieske Fe–S centres and H^+ to the lumen of the thylakoid. The PQH then 'dismutates', giving e^- to cyt b_{563} and to cyt b_{559} low potential form, and losing H^+ to the lumen. The cytochrome cycles e^- back to reduce PQH on the stromal side of the thylakoid and picks up more H^+. This model (referred to as the proton motive 'Q cycle') provides a role for the cytochromes and could also give 2 H^+ per e^- transported. Presently, however, the experimental evidence is equivocal; 1 and 2 H^+ per e^- have been measured.

Electrons pass to cyt b_{563} (called b_6) and cyt *f* in a protein complex which has been isolated from thylakoids. The complex may be associated with the Rieske centres. The protein from cyt *f* is rather hydrophilic and is probably exposed at the lumen surface of the thylakoid. The redox characteristics of cyt *f* are not altered by incorporation in the complex. Electrons pass from PQ to cyt *f via* a Rieske Fe–S centre, with characteristic ESR signal and mid-point potential of +0.2 V, which functions in a one electron transfer,

$$PQH_2 + Fe{-}S_{oxidized} \rightarrow PQH + H^+ + Fe{-}S_{reduced} \quad [5.3]$$

The 2 e^- from PQH_2 go to separate Fe–S molecules as the Rieske centres are single e^- carriers; it has been suggested that they carry H^+ also. Mutants of *Lemna* without the Fe–S centre lack non-cyclic electron transport.

Cytochromes, porphyrin-proteins containing iron bound into the porphyrin as a haem, are important oxidation-reduction electron carriers. Cytochrome *f* (from *folium* = leaf) and two cytochrome *b*'s are involved in thylakoid electron transport. Cytochrome *f* from higher plants has a molecular mass of about 65 kD (although determinations differ) and an absorption band at 554 nm when reduced; it has a potential of +0.37 V. Cytochrome *f* donates single electrons to plastocyanin; it may move towards the outside of the thylakoid on illumination as the membrane changes shape and becomes thinner, increasing its ability to accept electrons from the stromal side during cyclic electron flow. However as mentioned, cyt b_{563}, and cyt b_{559} may link cyt *f* to the stromal side of the membrane. Cyt b_{559} has two forms: a low potential form of about +0.02 V which may have a role in e^- transport around PSII with PQ and cyt b_{563} and the high potential form of +0.4 V which may be involved in water splitting. However, much evidence on cytochromes has accumulated without their role being clarified.

Plastocyanin (PC) a 40 kD protein of four polypeptides, each with one Cu atom co-ordinated by two N atoms from histidine and two S atoms from cysteine, and a redox potential of 0.37 V, is found in thylakoid lamellae of oxygenic tissue, in the ratio 1 molecule to 400 chlorophylls. PC is blue when oxidized, with a major absorption band at 597 nm and is inhibited by treatment with mercury ($HgCl_2$) which blocks the —SH groups, stopping all electron flow to PSI. PC accepts one electron from cyt *f* and may form a pool of electrons which can be passed to PSI (Katoh 1977).

Photosystem I

This photosystem (Fig. 5.5) is organized in chlorophyll–protein complexes, CPI of 130 kD mass and CP_a of 200 kD. It consists of ~200 antenna chl *a* with perhaps an inner core antenna of 30–40 chl *a* per P700 (the reaction centre chlorophyll), 10–20 *β*-carotenes orientated with respect to the chlorophyll iron–sulphur proteins (non-haem-iron), and proteins which bind and structure the complex. The main polypeptide, to which P700 and chl *a* attach, is of 65–70 kD mass (and is specified by the chloroplast DNA). Other polypeptides of 25, 18, 16 and 8 kD have been isolated; some may be exposed to the lumen, others to the stroma. Excitation from the antenna leads to oxidation of P700, the reaction centre chlorophyll. Electron movement around PSI is so fast that

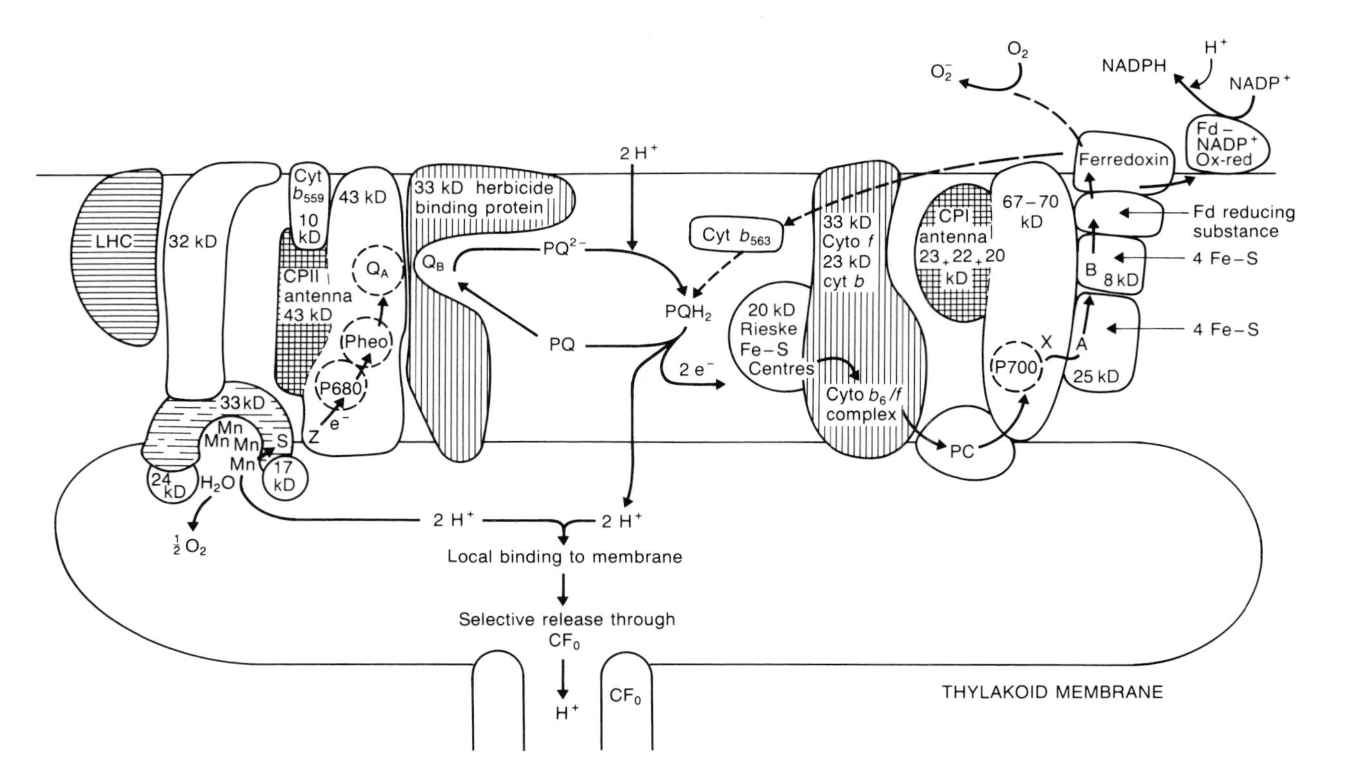

FIG. 5.5 Light harvesting and electron transport systems in thylakoid membranes shown schematically to illustrate the number of, and relationship between, protein complexes and other components. Mass of components is indicated. LHC is light-harvesting chlorophyll-protein, CPII and I are antennae of PSII and I respectively; other components are identified in the text.

there is little fluorescence from the antenna at room temperature, although at the temperature of liquid nitrogen (−196 °C) fluorescence at 730 nm is observed from PSI antenna chlorophyll. The primary acceptor, X, is possibly a pheophytin anion, or a chlorophyll monomer linked to a 16 kD polypeptide in a special environment. A secondary acceptor is found in the form of ESR signals characteristic of 2Fe–2S or, more probably, 4Fe–4S centres of an 18 kD protein; it is called Fe–S centre B (Fe–S.B); another Fe–S centre (Fe–S.A) may take electrons from Fe–S.B. With oxidation/reduction of P700 an optical spectroscopic signal occurs at 430 nm (hence called P430), possibly from the 4Fe–4S centres. The Fe–S centres delocalize electrons, stabilizing them for long enough to enable chemical reactions to take place. The different Fe–S units associated with PSI, may allow e^- to be transported to different processes depending on conditions; if $NADP^+$ is limiting then e^- may pass to the electron chain *via* another centre.

Donation of electrons to ferredoxin

The very negative mid-point potential of the reduced carriers enables ferredoxin (Fd) to be reduced. As P700 is activated cyt *f* is oxidized and Fd reduced, indicating its role as electron acceptor. Ferredoxins are electron-carrying iron–sulphur redox proteins, found in animals and plants, with Fe at the active centre (but not as haem), usually as 2Fe–2S (or sometimes 4Fe–4S) (Fig. 5.6). Plant ferredoxins are of 10–12 kD mass and reddish brown in colour and transfer one electron at each step. They have a redox potential of −0.43 V (soluble ferredoxin), characteristic absorption spectra below 500 nm and ESR spectra which are important for their identification and quantitative estimation.

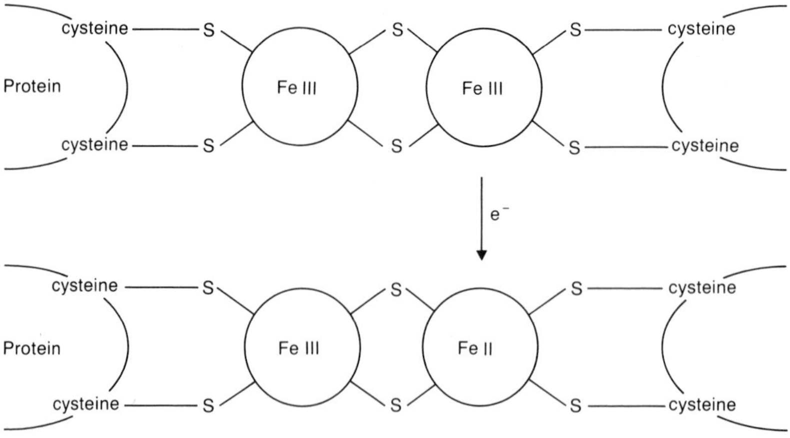

FIG. 5.6 The active Fe–S centre of ferredoxin and the oxidised (Fe III or ferric) and reduced (Fe II or ferrous) form of the iron atom (from Hall and Rao 1977, Figure 2).

Ferredoxins in the thylakoid membrane accept electrons from the secondary acceptors of PSI but soluble ferredoxin is situated on the stromal surface of the thylakoid and may receive electrons from a ferredoxin-reducing substance (FRS), but there is some doubt about the role of this protein.

Ferredoxins from chloroplasts pass electrons to many biological processes, back to PQ in the electron chain for example, giving cyclic electron transport (see p. 116), or to $NADPH^+$, reducing it, the normal route in photosynthesis:

$$2\ \text{ferredoxin}^- + NADP^+ + H^+ \rightarrow NADPH + \text{ferredoxin} \quad [5.4]$$

Ferredoxin reduces $NADP^+$ *via* the flavoprotein enzyme ferredoxin-$NADP^+$ oxido-reductase. The higher plant enzyme is of 40 kD mass, contains one molecule of FAD, an —S—S— bridge and 4 —SH groups, one of which, buried in the molecule and essential for catalytic activity, is located in the stromal side of the thylakoid. It forms a complex with ferredoxin and $NADP^+$ which binds to a lysine amino acid. The reductase also catalyses other reactions with ferredoxin or NADPH; for example NADPH reduces cyt *f* or NAD^+, and NADPH may be oxidized by transferring an electron to ferricyanide, dyes etc. (diaphorase activity). It is therefore an important control point in the electron chain and may redistribute electrons to other substrates if the normal acceptor, $NADP^+$, is in short supply. Ferredoxin also provides electrons to reduce sulphate and nitrate (Ch. 7) and activates some enzymes of the photosynthetic carbon reduction cycle (see p. 138). Reduced ferredoxin reacts with molecular oxygen forming the superoxide radical O_2^- and H_2O_2 in the Mehler reaction. Ferredoxin links electron transport with the chemical reactions and provides flexibility in metabolism.

$NADP^+$ and NADPH, the oxidized and reduced forms of the photosynthetic reductant with characteristic spectra, are water soluble molecules (Fig. 5.7). The structure (which resembles that of ATP, see p. 108) is complex and maintains a stable redox state under cellular conditions and allows specific binding to enzymes, etc. Reduction involves a 2 e^- transfer and one proton 'adds on'. The simplest model of electron flow from PSI to $NADP^+$ is:

$$P700 \rightarrow A_1 \rightarrow (A_2) \rightarrow (A + B) \rightarrow Fd_{sol} \rightarrow NADP^+ \quad [5.5]$$

but models with parallel branches have been suggested (see Clayton 1980) allowing more flexible dispersal of electrons if $NADP^+$ limits electron flow.

Energetics of electron transport

To lift an electron from water (+0.8 V potential) to ferredoxin (−0.42 V) requires two photons acting in series (Ch. 3). Figure 5.8 shows the Hill and Bendall 'Z' scheme (so-called from its appearance) of electron transport. P680 ejects e^- to Q at about 0 V potential and $P680^+$ (potential of +1.2 V) removes e^- from H_2O *via* S + Z. Electrons from Q pass to P700 at +0.4 V; the coupling of electron transport to ATP synthesis is considered in Chapter 6. P700 ejects an electron to ferredoxin at −0.43 V, an accumulated energy of 1.2 V. However

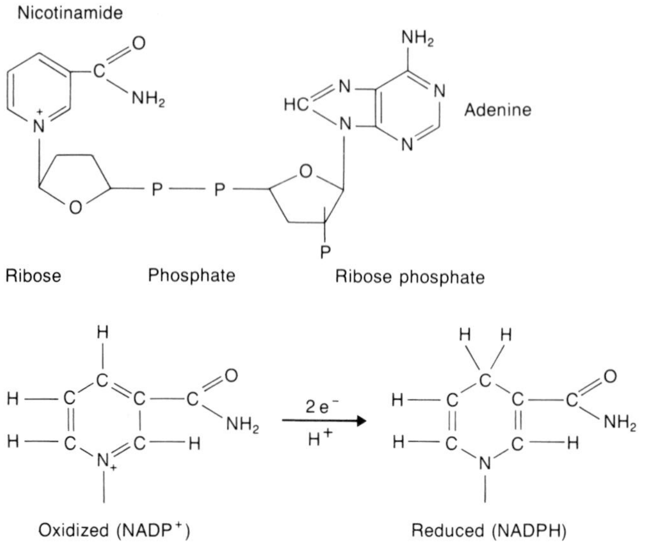

Nicotinamide

Adenine

Ribose   Phosphate   Ribose phosphate

$2e^-$, H^+

Oxidized ($NADP^+$)   Reduced (NADPH)

FIG. 5.7 Structure of NADP, nicotinamide adenine dinucleotide phosphate, and the oxidized and reduced forms.

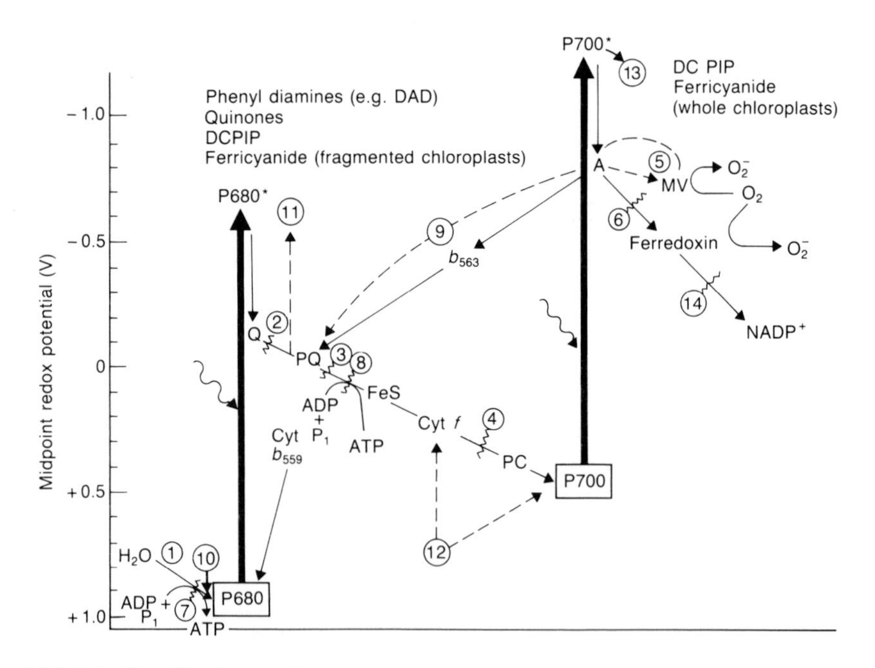

FIG. 5.8 The 'Z' scheme of photosynthetic electron transport and redox potential diagram for the photosystems and components of the electron transport chain. Numbers show the places at which inhibitors act, or at which e^- can be added or removed from the chain. Some compounds acting at these points are listed in Table 5.1.

Table 5.1 Sites of inhibitor activity identified by numbers on the electron transport scheme (Fig. 5.8), the processes affected and examples of compounds affecting them

Site	Process	Compound
1	Water-splitting	NH_2OH; TRIS
2	e^- transport to PQ	DCMU; atrazine
3	e^- and H^+ transport by PQ	DBMIB; DAD
4	e^- transport in Fe–S centres	HCN; mercurichloride; amphotericin B
5	e^- removal from PSI and auto-catalytic production of oxygen radicals	bipyridyl herbicides (Paraquat)
6	Reduction of $NADP^+$	DSPD
7 & 8	Uncoupling ATP synthesis from e^- transport	NH_4^+; FCCP; CCCP; DCCD; Dio-9; DNP; phlorizin
14	$NADP^+$ reduction	2-phosphoadenosine diphosphate ribose
Artificial electron donors and acceptors		
9	e^- carrier from PSI to PQ giving cyclic flow	PMS
10	e^- donor to PSII	Catechol, DPC, ascorbate; phenylenediamine
11	Acceptor from PSII	DCPIP; KFeCN (fragmented chloroplasts); DAD
12	PSI donors	ascorbate; DPC; PMS; DAD; DCPIP
13	Acceptors from PSI	DCPIP; KFeCN (whole chloroplasts); Paraquat

Abbreviations

Atrazine	2-chloro-4-(2-propylamino)-6-ethylamine-5-triazine
NH_2OH	Hydroxylamine
DCMU	3(3,4-dichlorophenyl)-1,1-dimethylurea, (Diuron)
DBMIB	2,5-dibromo-3-methyl-6-isopropyl-*p*-benzoquinone, (Dibromothymoquinone)
DCPIP	2,6-Dichlorophenolindophenol
HCN	Hydrogen cyanide
DAD	2,3,5,6-tetramethyl-*p*-phenylenediamine, (Diaminodurol)
CCCP	Carbonylcyanide-*m*-chlorophenylhydrazone
Dio-9	Antibiotic
DSPD	Disalicylidenepropanediamine
DPC	Diphenylcarbazide
PMS	Phenazinemethosulphate, (5-methylphenazonium-methylsulphate)
FCCP	Carbonylcyanide-*p*-trifluoromethoxyphenylhydrazone
DCCP	Dicyclohexylcarbodiimide
DNP	Dinitrophenol
TRIS	Tris (hydroxymethyl) aminomethane

the total energy in the two photoacts is +0.8 to −0.2 V in PSII and +0.4 to −0.8 V in PSI, a total of 2.2 V. Of the 1 V lost, part is recovered in ATP synthesis. The efficiency of photoactivation and electron transport is $1.4/2.2 \times 100 = 64$ per cent or greater. With a minimum of two photons per e^- transported, eight quanta of red light are needed per oxygen released, an energy of 23.4×10^{-19} J. The energy accumulated by four electrons passing to $NADP^+$ and synthesizing 2.7 ATP is around 8.8×10^{-19} J so maximum efficiency is 38 per cent. However, if the efficiency with respect to the total solar spectrum, not just red light, is calculated the maximum efficiency of the photochemical processes at the reaction centres is about 20 per cent.

Rates of processes in electron transport

Physical reactions within the pigment bed are much faster than electron transport processes. Photon capture and excitation migration to the reaction centre are fast (10^{-15} s and 5×10^{-12} s respectively). Fluorescence from chlorophyll occurs in 10^{-9} s when PSII is reduced but is slower when it is oxidized. Fluorescence from PSI is very limited as e^- transport from the reaction centres occurs in picoseconds. Electron transfer from reaction centres to acceptors requires $\sim 20 \times 10^{-9}$ s. What happens between 10^{-9} and 10^{-12} s is unknown. An electron is transferred from water to $NADP^+$ in 0.02 s. Water splitting (0.2 ms) and electron movement from the PSII acceptor to plastoquinone (400 μs) are fast in comparison with e^- movement from plastoquinone to cyt *f* which requires about 20 ms and is the rate limiting step in the process; the large pool of PQ minimizes this limiting step in e^- transport. Electron transfer from plastocyanin to PSI takes 0.2 ms and from PSI to $NADP^+$ 10 μs. The plastoquinone 'pump' of H^+ across the thylakoid controls the rate of H^+ gradient development, a six-fold difference in rate (16 to 90 ms) occurs when changing from conditions of high to low back pressure of H^+; proton transport is probably slower (60 ms) than PQ to cyt *f* transfer and the development of ΔpH requires 30 s or so. Protons diffuse across the thylakoid in 5 s if Cf_0−CF_1 (see Ch. 7) is not functioning. Proton transport rates across the membrane may be controlled by the herbicide-binding protein of PSII.

Artificial electron donors and acceptors

Many artificial redox compounds donate electrons to or accept them from the electron transport chain. Chemical structures of some are shown in Fig. 5.9. They have been important in analysis of photosynthetic processes since Hill's discovery in 1937 that ferricyanide accepts electrons from water with oxygen evolution (now called the Hill reaction, and the substances Hill reagents):

$$4Fe(CN)_6^{3-} + 2H_2O \xrightarrow[\text{chloroplasts}]{\text{light}} 4Fe(CN)_6^{4-} + 4H^+ + O_2 \qquad [\mathbf{5.6}]$$

Changes in the absorbance spectra of ferricyanide or a dye like DCPIP indicates oxidation–reduction state of the electron acceptors and donors and electron

Ferricyanide

$$[Fe(CN)_6]^{3-} \xrightarrow[h\nu]{1e^-} [Fe(CN)_6]^{4-}$$

Yellow — Colourless

2,6-Dichlorophenolindophenol (DCPIP)

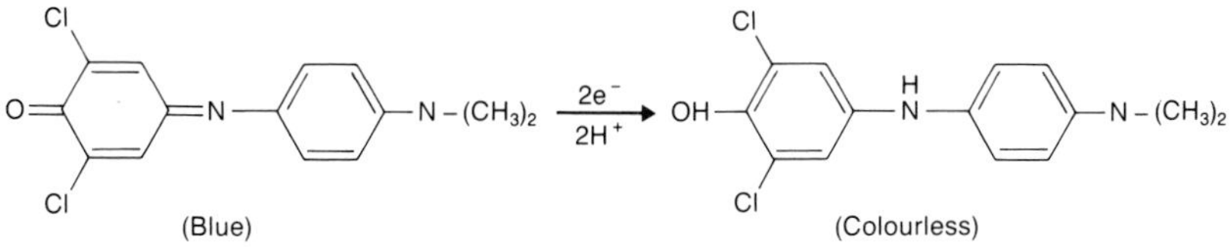

$$\xrightarrow[2H^+]{2e^-}$$

(Blue) — (Colourless)

Dibromothymoquinone (DBMIB)

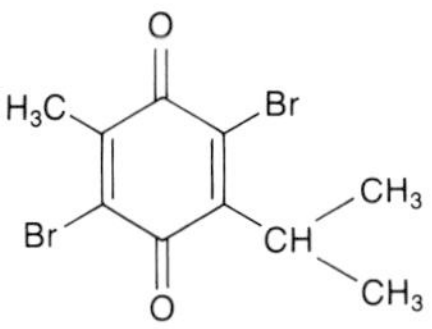

Carbonylcyanide-p-trifluromethoxyphenylhydrazone (FCCP)

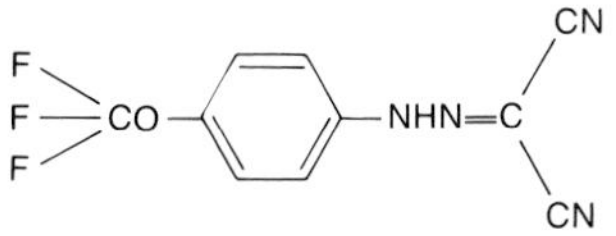

Diaminodurol (DAD)

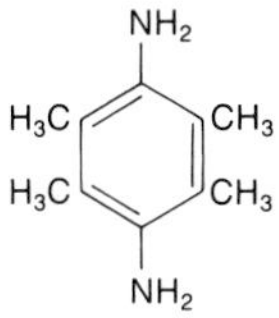

3(3,4-dichlorophenyl)-1,1-dimethyl urea) (DCMU) = Diuron)

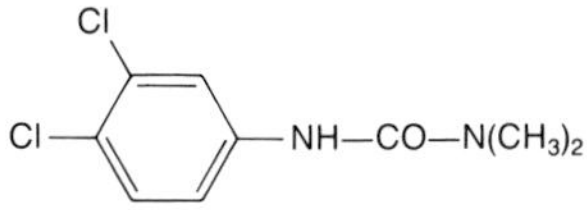

5-methylphenazonium-methylsulphate (PMS)

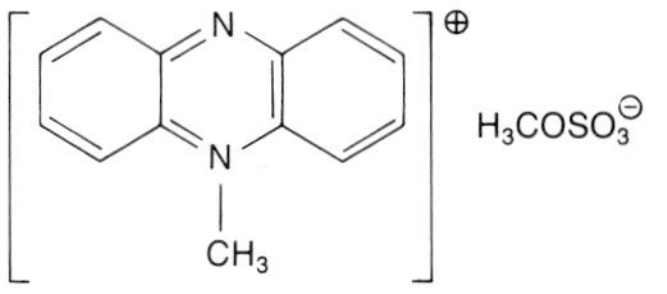

$H_3COSO_3^{\ominus}$

Paraquat = Methylviologen
(N,N-dimethyl-4,4′-dipyridyl-dichloride)

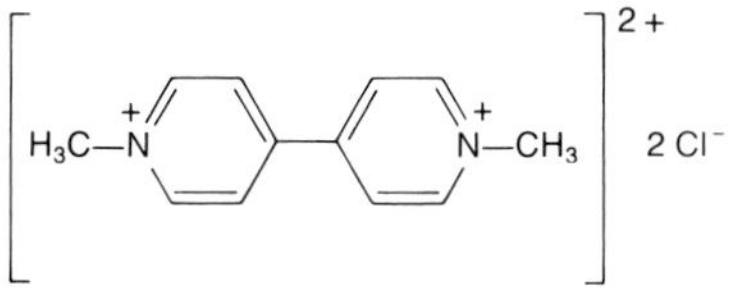

$2\,Cl^-$

Diquat = (N,N, ethylene-2,-dipyridyl-dibromide)

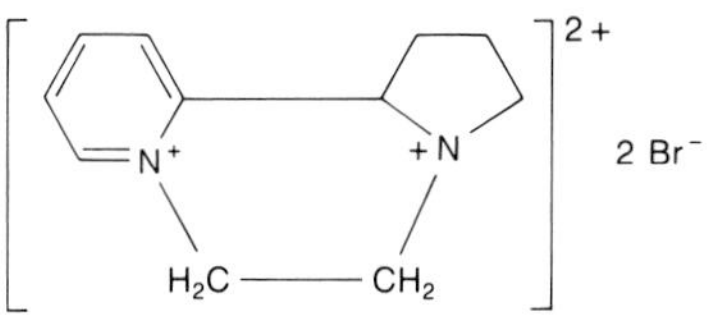

$2\,Br^-$

FIG. 5.9 Structure of some artificial donors and acceptors of electrons or inhibitors of e^- transport used in studies of photosynthesis.

flow. As their redox potentials are known, the potentials of the electron chain components may be determined. Differences in lipid solubility, molecular size, etc. can also be exploited to indicate where a reaction is taking place in the thylakoid or chloroplast.

Electron transport is blocked by compounds which remove electrons from different parts of the chain or are non-functional analogues of compounds in the chain. Hydroxylamine, as mentioned on p. 90, may occupy the position of H_2O in the water-splitting complex and this inhibits water splitting. The viologens, for example methyl viologen (called commercially 'Paraquat') transfer electrons to O_2, forming singlet oxygen, which destroys lipids etc. Viologen recycles, so the process is autocatalytic and destroys tissues rapidly in the light. DCMU is a quinone analogue and blocks the acceptor of PSII or between Q and PQ, thus separating PSII + PSI. One molecule of DCMU per 100 chlorophylls completely stops electron transport. Atrazine stops e^- flow at PQ by inhibiting the binding of Q to the protein complex. DBMIB is a structural analogue of PQ and prevents electron transfer to cyt *f*. An analogue of $NADP^+$, phosphoadenosine diphosphate ribose, blocks ferredoxin $NADP^+$ reductase. The antibiotic DIO-9 inhibits ATP synthesis at CF_1.

Electrons are donated to PSI by phenyldiamine and to cyt *f* by DCPIP plus ascorbate. Using DBMIB to block electron flow, DCPIP (plus ascorbate for reduction) donates electron to PSI and methyl viologen is the acceptor; these reagents provide a test system for PSI and the effects of conditions on it. PSII is measured by electron flow to DCPIP with transport blocked by DBMIB. Open chain electron transport is measured with water as donor and viologen or $NADP^+$ as acceptor. Cyclic electron transport is measured in broken chloroplasts which have lost soluble ferredoxin, by adding PMS which carries electrons from PSI back to PC or cyt *f*. The sites of action of some compounds are shown with approximate redox potentials in Fig. 5.8 and Table 5.1. Some are commercial herbicides such as DCMU, the viologens and atrazine.

Formation of reactive forms of oxygen by photosynthesis

Oxygen in its diatomic form (O_2) which contains two unpaired electrons with parallel spins ('triplet') is not only a product of photosynthesis but is 'assimilated' into different forms as a consequence of it. Light energy (and electrons from other biochemical reactions not considered here) causes the formation of reactive O_2 species (see Elstner 1982):

$$O_2 \xrightarrow{e^-} O_2^{\circ -} \xrightarrow[H^+]{} HO_2^{\circ} \xrightarrow[H^+]{e^-} H_2O_2 \xrightarrow[H^+]{e^-} OH^{\circ} \xrightarrow[H^+]{e^-} 2H_2O \qquad [5.7]$$

superoxide — perhydroxyl radical — hydrogen peroxide — hydroxyl radical

In photosynthesis, excited pigments (e.g. triplet chlorophyll, ^3Chl) donate energy to O_2 giving $^1O_2$ (singlet oxygen):

$$Chl \xrightarrow{h\nu} {}^3Chl; {}^3Chl + O_2 \rightarrow Chl + {}^1O_2 \qquad [5.8]$$

or by electron transfer, producing superoxide:

$$Chl \xrightarrow{h\nu} {}^3Chl + O_2 \rightarrow Chl^+ + O_2^{\circ -} \qquad [5.9]$$

Photosystems may pass e^- to O_2 directly or *via* intermediates, paraquat for example, which reduces O_2 to superoxide and recycles to carry more e^-, or from physiological intermediates such as ferredoxin which produces superoxide or, in the Mehler reaction, H_2O_2:

$$Fd_{red} + O_2 \rightarrow Fd_{ox} + O_2^{\circ -}$$
$$Fd_{red} + O_2^- + 2H^+ \rightarrow H_2O_2 + Fd_{ox} \qquad [5.10]$$

Hydrogen peroxide and O_2^- react to give $^\circ OH$ in a reaction catalysed by metal salts:

$$H_2O_2 + O_2^{\circ -} \xrightarrow{Fe_{salt}} O_2 + {}^\circ OH + OH^- \qquad [5.11]$$

Superoxide reacts with H^+ in the presence of the enzyme superoxide dismutase (SOD), in alkaline conditions, to give H_2O_2 which is destroyed by catalase:

$$O_2^{\circ -} + O_2^{\circ -} + 2\,H^+ \xrightarrow{SOD} H_2O_2 + O_2$$
$$2\,H_2O_2 \xrightarrow{catalase} 2\,H_2O + O_2 \qquad [5.12]$$

SOD is found throughout the plant cell including chloroplasts, and some catalase is in the chloroplast, probably bound to PSI, not soluble. These two enzymes (together with the carotenoid quenching of reactive forms of oxygen) are an important defence against highly reactive oxygen states. Superoxide reacts with unsaturated fatty acids causing lipid peroxidation and thereby destroying membranes, and with chlorophyll causing photobleaching. Accumulation of malondialdehyde indicates destruction of lipids. Superoxide also oxidizes sulphur compounds, NADPH and ascorbic acid, or reduces cyt *c* and metal ions; such reactive O_2 species must be rapidly removed. SOD is a fast enzyme which, together with catalase, protects the thylakoid. By coupling oxidation/ reduction of the thiol (—SH) groups of the tripeptide glutathione (which is at high concentration in chloroplasts) with the reduction state of ascorbic acid and NADP, H_2O_2 and superoxide content is regulated:

$$H_2O_2 + \text{ascorbate} \xrightarrow[\text{peroxidase}]{\text{ascorbate}} \text{dehydroascorbate} + H_2O$$
$$\text{dehydroascorbate} + \text{glutathione}_{red} \rightarrow \text{glutathione}_{ox} + \text{ascorbate}$$
$$\text{glutathione}_{ox} + NADPH \xrightarrow[\text{reductase}]{\text{glutathione}} \text{glutathione}_{red} + NADP^+ \qquad [5.13]$$

Bright light damages mutant plants lacking SOD, catalase or other enzymes which control the oxidation/reduction state of the chloroplast, or when adverse conditions cause stomatal closure and restrict the consumption of NADPH allowing high energy states and reducing compounds to accumulate in the light-harvesting apparatus.

References and Further Reading

Åkerlund, H-E. (1983) Peptides involved in photosynthetic oxygen evolution with special emphasis on a 23-kilodalton protein, pp. 201–8 in Inoue, J., Murata, N., Crofts, A. R., Renger, G., Govindjee and Satoh, K. (eds), *The Oxygen Evolving System of Photosynthesis*, Academic Press, Tokyo.

Amesz, J. (1977) Plastoquinone, pp. 238–46 in Trebst, A. and Avron, M. (eds), *Encyclopedia of Plant Physiology* (N.S.), Vol. 5, *Photosynthesis I*, Springer-Verlag, Berlin.

Avron, M. (1975) The electron transport chain in chloroplasts, pp. 374–86 in Govindjee (ed.), *Bioenergetics of Photosynthesis*, Academic Press, New York.

Blankenship, R. E. and **Parson, W. W.** (1978) The photochemical electron transfer reactions of photosynthetic bacteria and plants, *A. Rev. Biochem.*, **47**, 635–53.

Bonaventura, C. and **Myers, J.** (1969) Fluorescence and oxygen evolution from *Chlorella pyrenoidosa*, *Biochim. Biophys. Acta*, **189**, 366–83.

Calvin, M. (1981) Photoactivation of oxygen and the evolution of O_2. *Special Publication, Roy. Soc. Chem. No. 39*, pp. 45–67, The Royal Society of Chemistry, London.

Clayton, R. K. (1971) *Light and Living Matter: A Guide to the Study of Photobiology.* Vol. 1. *The Biological Part*, McGraw-Hill, New York. Reprinted in 1977 by R. E. Krieger, Huntington, New York.

Clayton, R. K. (1980) *Photosynthesis: Physical Mechanisms and Chemical Patterns*, I.U.P.A.B. Biophysics Series, Cambridge University Press.

Cogdell, R. J. (1978) Carotenoids in photosynthesis, *Phil. Trans. R. Soc. Lond. B*, **284**, 569–79.

Cox, R. and **Olsen, L. F.** (1982) The organisation of the electron transport chain in the thylakoid membrane, pp. 49–79 in Barber, J. (ed.), *Electron Transport and Photophosphorylation, Photosynthesis 4*, Elsevier Biomedical Press, Amsterdam.

Cramer, W. A. and **Crofts, A. R.** (1982) Electron and proton transport, pp. 387–467 in Govindjee (ed.), *Photosynthesis*, vol. 1, *Energy Conversion by Plants and Bacteria*, Academic Press, New York.

Cramer, W. A. (1977) Cytochromes, pp. 227–37 in Trebst, A. and Avron M. (eds), *Encyclopedia of Plant Physiology* (N.S.), vol. 5, *Photosynthesis I*, Springer-Verlag, Berlin.

Critchley, C. and **Sargeson, A. M.** (1984) A manganese-chloride cluster as the functional centre of the O_2 evolving enzyme in photosynthetic systems. *FEBS Lett.*, **177**, 2–5.

Diner, B. A. and **Joliot, P.** (1977) Oxygen evolution and manganese, pp. 187–205 in Trebst, A. and Avron, M. (eds), *Encyclopedia of Plant Physiology* (N.S.), vol. 5, *Photosynthesis I*, Springer-Verlag, Berlin.

Elstner, E. F. (1979) Oxygen activation and superoxide dismutase in chloroplasts, pp. 410–15 in Gibbs, M. and Latzko, E. (eds), *Encyclopedia of Plant Physiology* (N.S.), vol. 6, *Photosynthesis II*, Springer-Verlag, Berlin.

Elstner, E. F. (1982) Oxygen activation and oxygen toxicity, *A. Rev. Plant Physiol.*, **33**, 73–96.

Ferguson, S. J. and **Sorgato, M. C.** (1982) Proton electrochemical gradients and energy transduction processes, *A. Rev. Biochem.*, **51**, 185–217.

Forti, G. (1977) Flavoproteins, pp. 222–6 in Trebst, A. and Avron, M. (eds), *Encyclopedia of Plant Physiology* (N.S.), vol. 5, *Photosynthesis I*, Springer-Verlag, Berlin.

Govindjee (ed.) (1975) *Bioenergetics of Photosynthesis*, Academic Press, London.

Govindjee and **Whitmarsh, J.** (1982) Introduction to photosynthesis: Energy conversion by plants and bacteria, pp. 1–18 in Govindjee (ed.), *Photosynthesis*, vol. 1, *Energy Conversion by Plants and Bacteria*, Academic Press, New York.

Gregory, R. P. F. (1979) *Photochemistry of Photosynthesis*, Academic Press, New York.

Hall, D. O. and **Rao, K. K.** (1977) Ferredoxin, pp. 206–16 in Trebst, A. and Avron, M. (eds), *Encyclopedia of Plant Physiology* (N.S.), vol. 5, *Photosynthesis I*, Springer-Verlag, Berlin.

Halliwell, B. (1978) Biochemical mechanisms accounting for the toxic action of oxygen on living organisms – the key role of superoxide dismutase, *Cell Biol. Internat. Repts.*, **2**, 113–28.

Hauska, G. (1977) Artificial acceptors and donors, pp. 253–65 in Trebst, A. and Avron, M. (eds), *Encyclopedia of Plant Physiology* (N.S.), vol. 5, *Photosynthesis I*, Springer-Verlag, Berlin.

Hauska, G. (1978) Vectorial redox reactions of quinoid compounds and the topography of photosynthetic membranes, pp. 185–96 in Hall, D. O., Coombs, J. and Goodwin, T. W. (eds), *Proceedings of the 4th International Congress on Photosynthesis*, The Biochemical Society, London.

Hoch, G. E. (1977) P700, pp. 136–48 in Trebst, A. and Avron, M. (eds), *Encyclopedia of Plant Physiology* (N.S.), vol. 5, *Photosynthesis I*, Springer-Verlag, Berlin.

Izawa, S. (1977) Inhibitors of electron transport, pp. 266–82 in Trebst, A. and Avron, M. (eds), *Encyclopedia of Plant Physiology* (N.S.), vol. 5, *Photosynthesis I*, Springer-Verlag, Berlin.

Izawa, S. (1980) Acceptors and donors for chloroplast electron transport, pp. 413–675 in San Pietro, A. (ed.), *Methods in Enzymology*, vol. 69, *Photosynthesis and Nitrogen Fixation*. Part C. Academic Press, London.

Joliot, P. and **Kok, B.** (1975) Oxygen evolution in photosynthesis, pp. 388–412 in Govindjee (ed.), *Bioenergetics of Photosynthesis*, Academic Press, New York.

Junge, W. (1977) Physical aspects of light harvesting, electron transport and electrochemical potential generation in photosynthesis of green plants, pp. 59–93 in Trebst, A. and Avron, M. (eds), *Encyclopedia of Plant Physiology* (N.S.), vol. 5, *Photosynthesis I*, Springer-Verlag, Berlin.

Junge, W. and **Auslander, W.** (1978) Proton release during photosynthetic water oxidation: kinetics under flashing light, pp. 213–28 in Metzner, H. (ed.), *Photosynthetic Oxygen Evolution*, Academic Press, London.

Katoh, S. (1977) Plastocyanin, pp. 247–52 in Trebst, A. and Avron, M. (eds), *Encyclopedia of Plant Physiology* (N.S.), vol. 5, *Photosynthesis I*, Springer-Verlag, Berlin.

Kok, B. (1976) Photosynthesis: the path of energy, pp. 845–85 in Bonner, J. and Varner, J. E. (eds), *Plant Biochemistry* (3rd edn), Academic Press, New York.

Kok, B., Forbush, B. and **McGloin, M.** (1970) Co-operation of charges in photosynthetic oxygen evolution. 1. A linear four step mechanism, *Photochem. Photobiol.*, **11**, 457–75.

Malkin, R. (1982a) Photosystem 1, *A. Rev. Plant Physiol.*, **33**, 455–79.

Malkin, R. (1982b) Redox properties and functional aspects of electron carriers in chloroplast photosynthesis, pp. 1–47 in Barber, J. (ed.), *Electron Transport and Photophosphorylation, Photosynthesis 4*, Elsevier Biomedical Press, Amsterdam.

Mitchell, P. (1975) The protonmotive Q cycle: a general formulation, *FEBS Lett.*, **59**, 137–9.

Moreland, D. E. (1980) Mechanism of action of herbicides, *A. Rev. Plant Physiol.*, **31**, 597–638.

Radmer, R. (1983) Studies of oxygen evolution using water analogs and mass spectrometry, pp. 135–44 in Inoue, J., Murata, N., Crofts, A. R., Renger, G., Govindjee and Satoh, K. (eds), *The Oxygen Evolving System of Photosynthesis*, Academic Press, Tokyo.

Radmer, R. J. and **Ollinger, O.** (1984) Studies of oxygen evolution using substituted hydroxylamine and hydrazine compounds as water analogs, pp. 269–72 in Sybesma, C. (ed.), *Advances in Photosynthesis Research*, vol. 1, Martinus Nijhoff/Dr W. Junk Publishers, The Hague.

Renger, G. (1983) Photosynthesis, pp. 515–42 in Hoppe, W. *et al.* (eds), *Biophysics*, Springer-Verlag, Berlin.

Renger, G., Echert, H.-J. and **Weiss, W.** (1983) Studies on the mechanism of photosynthetic oxygen formation, pp. 73–82 in Inoue, J., Murata, N., Crofts, A. R., Renger, G., Govindjee and Satoh, K. (eds), *The Oxygen Evolving System of Photosynthesis*, Academic Press, Tokyo.

Rottenberg, H. (1977) Proton and ion transport across the thylakoid membranes, pp. 338–49 in Trebst, A. and Avron, M. (eds), *Encyclopedia of Plant Physiology* (N.S.), vol. 5, *Photosynthesis I*, Springer-Verlag, Berlin.

Trebst, A. (1974) Energy conservation in photosynthetic electron transport of chloroplasts, *A. Rev. Plant Physiol.*, **75**, 423–58.

Trebst, A. (1978) Plastoquinones in photosynthesis, *Phil. Trans. R. Soc. Lond. B*, **284**, 591–99.

Trebst, A. (1980) Inhibitors in electron flow: tools for the functional and structural localization of carriers and energy conservation sites, pp. 675–715 in San Pietro, A. (ed.), *Methods in Enzymology*, vol. 69, *Photosynthesis and Nitrogen Fixation.* Part C. Academic Press, London.

Trumpower, B. L. (1982, ed.) *Function of Quinones in Energy Conserving Systems*, Academic Press, New York.

Velthuys, B. R. (1980) Mechanisms of electron flow in photosystem II and towards photosystem I, *A. Rev. Plant Physiol.*, **31**, 545–67.

Weaver, E. C. and **Corker, G. A.** (1977) Electron paramagnetic resonance spectroscopy, pp. 168–78 in Trebst, A. and Avron, M. (eds), *Encyclopedia of Plant Physiology* (N.S.), vol. 5, *Photosynthesis I*, Springer-Verlag, Berlin.

Wydrzynski, T. J. (1982) Oxygen evolution in photosynthesis, pp. 469–506 in Govindjee (ed.), *Photosynthesis*, vol. 1, *Energy Conversion by Plants and Bacteria*, Academic Press, New York.

Yocum, C. F. (1984) Photosynthetic oxygen evolution: an overview, pp. 239–42 in Sybesma, C. (ed.), *Advances in Photosynthesis Research*, vol. 1, Martinus Nijhoff/Dr W. Junk Publishers, The Hague.

CHAPTER 6

Synthesis of ATP: photophosphorylation

Metabolic processes, such as the assimilation of carbon dioxide, protein synthesis and ion pumping, require ATP. Ruben recognized in 1943, that 'assimilatory power', now equated with ATP, as well as reduced pyridine nucleotides, was essential for photosynthesis. In 1954 Arnon demonstrated ATP synthesis in higher plant chloroplasts and Frenkel discovered it in photosynthetic bacteria. In both systems the photoreactions drive the synthesis of ATP, hence the term photophosphorylation. Mitochondria and aerobic bacteria synthesize ATP by respiration of preformed substrates. Despite the different sources of energy for photo- and respiratory ('oxidative') phosphorylation, both involve electron flow along a chain of redox components coupled to ATP synthesis; this chapter considers the mechanism of phosphorylation in chloroplasts.

The metabolic role of ATP

Before considering the mechanism of ATP synthesis it is important to outline the nature of ATP and to emphasize its essential role in metabolism, including photosynthesis. ATP has two anhydride (pyrophosphate) bonds which are hydrolysed according to eqns 6.1a and 6.1b at positions I and II (Fig. 6.1) respectively.

$$\text{ATP} + H_2O \longleftrightarrow \text{ADP} + P_i + \text{energy}\,(-31\ \text{kJ mol}^{-1}) \qquad [\mathbf{6.1a}]$$

$$\text{ADP} + H_2O \longleftrightarrow \text{AMP} + P_i + \text{energy} \qquad [\mathbf{6.1b}]$$

Under cellular conditions the energy values of the ATP/ADP + P_i reaction may be 50–60 kJ mol^{-1}. The energy of the reaction comes from the electrostatic repulsion of negative charges on the phosphate groups, and to resonance stabilization and the large enthalpies of solvation of the reaction products. The Gibbs free energy released in the reactions is 'coupled' by biochemical mechanisms to do work, for example the ion ATPases couple the hydrolysis of ATP with the transport of ions. In many metabolic reactions hydrolysis of ATP is stoichiometrically linked to the chemical transformations. ATP is used in most cellular reactions, directly or *via* other phosphorylated nucleotides, for example guanosine and uridine nucleotides (e.g. GTP and UTP), to which ATP transfers P_i groups.

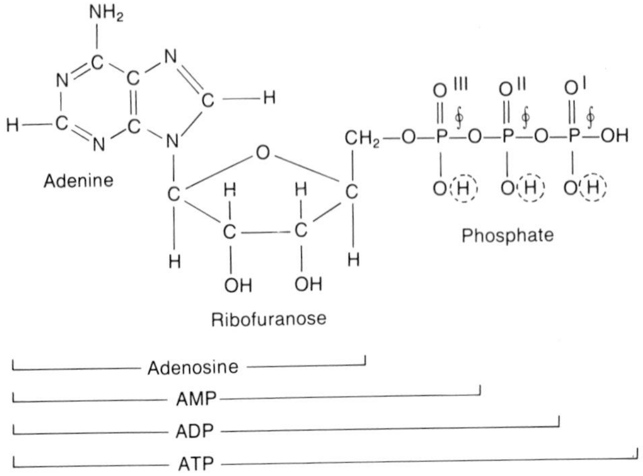

FIG. 6.1 The structure of adenosine triphosphate, ATP, the principal phosphate group transfer compound in biochemical reactions. (H) is dissociable hydrogen. ATP bonds (I, II, III) are cleaved enzymatically at $\oint_i$, releasing P_i and energy.

$$\text{ATP} + \text{GDP} \xrightarrow{\text{nucleotide diphosphokinase}} \text{ADP} + \text{GTP} \qquad [6.2]$$

ATP is required in metabolic pathways where phosphorylated intermediates are interconverted, for example the photosynthetic carbon reduction cycle. Cleavage of ATP at position I allows the P_i group to be transferred to water (the enzymes are called ATPases) according to eqn 6.1a or to other compounds in the presence of suitable enzymes. There are three different mechanisms; (1) the enzyme may itself be phosphorylated (e.g. ion ATPase), (2) no covalent bond forms between enzyme and ATP (e.g. adenylate kinase, eqn 6.4) and (3) the enzyme forms phosphorylated intermediates in the course of interchanging groups (e.g. glutamine synthetase, which uses ATP in transferring NH_3 to glutamate to form glutamine). When ATP donates phosphate groups to compounds, it increases their reactivity, for example glucose is phosphorylated to glucose-6-phosphate by ATP and hexokinase before consumption in glycolysis and respiration.

$$\text{glucose} + \text{ATP} \xrightarrow{\text{hexokinase}} \text{glucose-6-phosphate} + \text{ADP} \qquad [6.3]$$

The synthesis of ATP is central to any discussion of photosynthetic processes. The anhydride bonds of ATP are not, as often said, 'richer in energy' or of 'higher energy' than those of many other compounds and do not therefore 'drive' metabolism simply by providing energy. The ability of a chemical reaction to do work is related to the state of equilibrium of the reaction, the further from equilibrium the more energy available. The ATP reaction is important because under the conditions in the cell it is displaced from equilibrium and the Gibbs free energy change, ΔG, is favourable for doing metabolic work. ATP has an intermediate phosphate group transfer potential, as defined by the free energy of hydrolysis, and can function as a phosphate group carrier. In an analogous way NADP (p. 97) acts as an electron and H^+ carrier in metabolism. ATP is, however, very stable at normal temperatures and near neutral pH in the cell, unless involved in enzyme reactions, when it is an almost universal donor and acceptor of phosphate groups to other molecules, activating them in biochemical reactions.

Different cell compartments contain ATP, which because of differences in rates of reactions, will be in different equilibrium states. Exchange of phosphorylated compounds (often not ATP directly) takes place between compartments where reactions consume or produce ATP, so regulating the ATP pools in different parts of the cell. Cell metabolism is balanced with respect to the energy available for ATP synthesis and supply of substrates. Turnover of ATP in cells is rapid, the total 'pool' in leaf cells may be broken down and resynthesized within 50 ms; therefore metabolism responds quickly to the supply of – and demand for – ATP. If synthesis slows, metabolism also slows and as different pathways require different amounts of ATP or are differentially regulated by the concentration of ATP or ADP, so the response of the system is modified. The proportions of ATP, ADP and AMP in cellular compartments, under different metabolic conditions which change the requirement for nucleotide, is controlled by adenylate kinase:

$$\text{ATP} + \text{AMP} \xrightleftharpoons{\text{adenylate kinase}} 2\,\text{ADP} \qquad [\mathbf{6.4}]$$

This means that the forms of adenylate are regulated close to an optimum for the many processes involved in metabolism. The phosphorylation state in tissues is expressed by the energy charge (EC) (Atkinson 1977):

$$\text{EC} = \frac{[\text{ATP}] + \frac{1}{2}[\text{ADP}]}{[\text{AMP}] + [\text{ADP}] + [\text{ATP}]} \qquad [\mathbf{6.5}]$$

An EC of 1 would be a condition of all ATP, and an EC of 0, all AMP. Atkinson has generalized the response of enzymes to EC by suggesting that the ATP-regenerating enzymes have minimum velocity at large EC and maximum at small, whereas enzymes consuming ATP act in reverse. So in cells with rapid synthesis of ATP compared to demand for ATP, EC is large and ATP synthesis is slow. Conversely with little ATP synthesis and large demand, EC is small and ATP synthesis is rapid, given the required conditions. Equilibrium is attained between supply and demand because of the response of enzyme systems to EC. However, the rate of reactions is not linear with EC but changes most rapidly as EC decreases from approximately 0.9 to 0.6 and only slowly with a further decrease. Metabolism is therefore very sensitive to small change in EC and is closely regulated by the supply of, and demand for, ATP.

Control of enzyme reactions is not only by EC or the availability of ATP, ADP or AMP as substrate but by these molecules acting as allosteric effectors of enzyme reactions. The effectors bind away from the reaction site and change the catalytic behaviour of the enzyme. An important photosynthetic example is the effect of ATP on 3-phosphoglycerate kinase of chloroplasts (see p. 137); the enzyme uses ATP in formation of 1,3-diphosphoglycerate and is stimulated at high EC. It also generates ATP in the reverse reaction, which is inhibited by ATP and high EC (0.9 to 1.0) and stimulated by low EC (0.7). Both forward and reverse reactions are inhibited by AMP (see Preiss and Kosuge (1976) for details). Such complex control based on phosphorylated adenylates provides for a very subtle balance between processes and is an essential feature of cellular metabolism and maintenance of cellular homeostasis.

Measurement of ATP

Because of the rapid turnover of ATP, tissues or cells must be killed quickly (milliseconds) if the state of the system is not to change during extraction of ATP; this is done by plunging tissues into very cold solvents, for example pentane at −20 °C, or by clamping between the jaws of metal tongs at liquid nitrogen temperature (−196 °C). Adenylates are extracted by solvents, which also denature the enzymes. The concentration of ATP may be measured by one of several methods, for example by chromatographic separation from other nucleotides on ion exchange columns and detection of ATP with ultraviolet light after elution. Enzymatic methods are frequently used to measure the ATP in extracts, for example hexokinase converts glucose to glucose-6-phosphate using ATP, the glucose-6-P is oxidized by $NADP^+$ and the resultant NADPH is detected spectrophotometrically, as 1 mol ATP consumed produces 1 mol NADPH. Another sensitive method for ATP, measures the bioluminescence produced when an extract from the light organs (lanterns) of fireflies, containing luciferin and the enzyme luciferase, reacts with ATP:

$$\text{luciferin} + \text{ATP} \xrightarrow[\text{luciferase}]{Mg^{2+}} \text{luciferase–AMP} + \text{pyrophosphate} \qquad [\mathbf{6.6a}]$$

$$\text{luciferin–AMP} + \tfrac{1}{2}O_2 \xrightarrow{\text{luciferase}} \text{oxyluciferin} + AMP^{2-} + \text{light} \qquad [\mathbf{6.6b}]$$

The emitted photons are measured with a sensitive photometer.

Energy transducing mechanisms in phosphorylation – the chemiosmotic theory

ATP is synthesized by reversal of eqn 6.1a; an ATP synthetase enzyme is required and an energy source. The enzyme is found in all energy transducing membranes and is reversible, the direction in which the reaction proceeds depending on the concentration of ATP and on electron flow in the membrane. In mitochondria the enzyme is called F1 and is on the inner (matrix) surface of the inner membrane; in photosynthetic bacteria it is on the inside of the

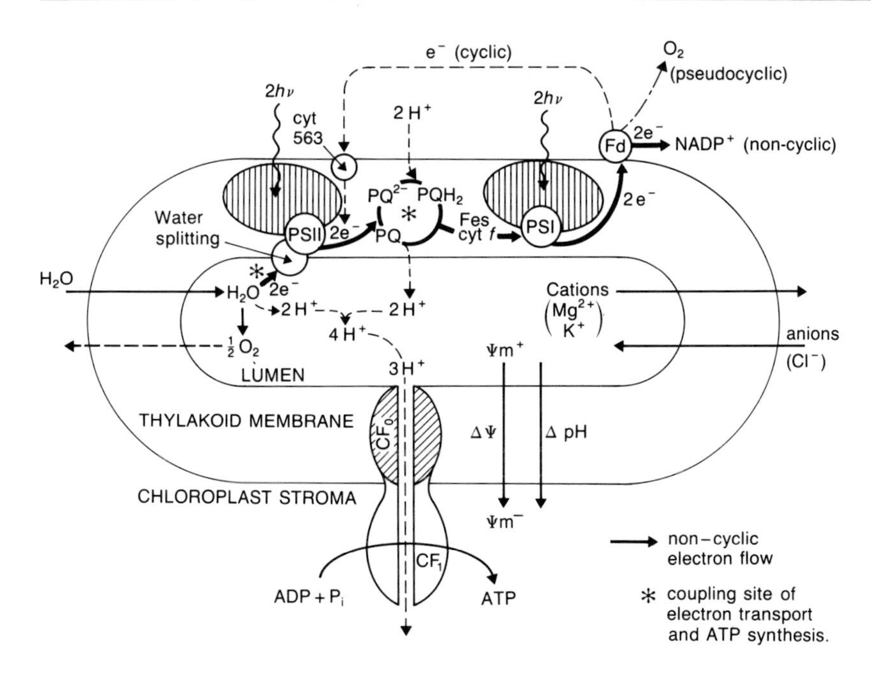

FIG. 6.2 Schematic relationship between electron transport driven by light, proton accumulation in the thylakoid lumen and ATP synthesis by the enzyme complex CF_0-CF_1. $\Delta\psi$ and ΔpH are, respectively, the electrical potential difference and H^+ gradient providing the proton motive force for ATP synthesis.

membrane. The enzyme of chloroplasts is called coupling factor; it is composed of a basal part, CF_0, in the lipid membrane and a head part, CF_1 attached to CF_0, and projecting from the thylakoid membrane into the stroma (Fig. 6.2). Despite their different topographies each ATP synthesizing system is essentially a lipid-membrane enclosed vesicle with the ATP synthetase projecting from one side of the compartment to the other. The inside of the bacterial membrane is equivalent to the inside of the mitochondrion and to the outside of the chloroplast thylakoid. In chloroplasts the 57 kJ of energy required for synthesis of a mole of ATP is provided by the energy lost by photoenergized electrons as they are transported in the thylakoid membrane along a chain of redox carriers, as discussed in Ch. 5.

The mechanism by which the redox potential energy of the electron in the membrane is coupled to the synthesis of the anhydride bond of ATP, was a matter for heated debate for many years. Several hypotheses were advanced to account for the observations that electron transport was coupled at three sites (in mitochondria) or two (in chloroplasts) to ATP synthesis and that the demand for ATP and the supply of ADP and P_i could regulate electron transport. The relationship between ATP synthesis and electron transport was measured by determining ATP production or P_i consumption in relation to O_2

consumption in mitochondria or O_2 evolution in chloroplasts. Any hypothesis also had to account for ion accumulation (e.g. calcium in mitochondria) which was known to be related to the electron transport processes and could be driven by ATP hydrolysis in the absence of electron transport. Another phenomenon requiring explanation was how artificial compounds of very diverse type, applied to mitochondria or chloroplasts, could uncouple ATP synthesis from electron transport.

Electron transport was thought to produce a high energy form of an energy-transducing intermediate (often called ~ 'squiggle') which could provide energy for ATP synthesis and related processes. Uncouplers, it was suggested, prevented the formation of ~ or destroyed it. If a component in the membrane, electron carrier or protein for example, altered its configuration as the redox potential changed, allowing ADP and P_i to bind and form the anhydride bond, it should have been detectable spectroscopically or by other means. Despite much experimentation no 'high energy' chemical intermediate has been identified and direct coupling between the electron chain and ATP synthesis is not now thought to be the mechanism of energy transduction.

The chemiosmotic hypothesis, developed by Mitchell, has provided a general mechanism for coupling the energy in electron transport to phosphorylation. It combines into a coherent scheme many experimental observations related to ATP synthesis, for example, the need for intact organelles (e.g. mitochondria) or membrane vesicles (e.g. the thylakoids in chloroplasts), the requirement for coupling factor attached to the membrane and the need for electron transport. It also accounts for the observed increase in alkalinity of the chloroplast stroma in the light, in addition to the other features mentioned previously. The hypothesis is that photophosphorylation in chloroplasts and in photosynthetic bacteria and respiratory chain phosphorylation in mitochondria employ the same basic energy transducing mechanism, driven by a proton flux through the ATP synthetase enzyme. Energy from electron transport is conserved as a high concentration of protons on one side of the vesicle membrane and a low concentration on the other. The high concentration is outside the cell membrane in the photosynthetic bacteria, between the inner and outer mitochondrial membranes, and inside the thylakoid lumen in chloroplasts. In photosynthesis, the energy of electrons, activated by light reactions is conserved by coupling electron transport in thylakoid membranes (Fig. 6.2) to H^+ transport across the membrane into the thylakoid. Capture of four photons leads to accumulation of 2 H^+ from oxidation of one molecule of water and 2 H^+ are transported from the stroma (which becomes alkaline) by the plastoquinone pump (p. 92).

The difference in H^+ concentration across the membrane is equivalent to a gradient in pH, ΔpH. Electron transport and proton pumping also cause an electrical potential difference, $\Delta\psi$, to develop across the membrane, which in chloroplasts is positive inside the thylakoid lumen and in mitochondria, negative in the matrix. Under the influence of $\Delta\psi$ and of ΔpH acting together, protons in the thylakoid lumen will tend to move from the lumen, across the membrane, to the stroma to preserve electrical neutrality and decrease the gradient of H^+. This force is called the proton motive force, pmf, (also called the proton electrochemical potential, $\Delta\mu_{H^+}$). Thus:

$$\text{pmf} = \Delta\mu_{H^+} = \Delta\psi + \Delta\text{pH} \qquad [6.7]$$

The Gibbs energy change for the transfer of one mol of H^+ down a gradient of H^+ between the inside, i, and outside, o, of the membrane vesicle in the absence of an electrical potential, is:

$$\Delta G = 2.3\,RT\log_{10}\frac{[H^+]_o}{[H^+]_i} \qquad [6.8]$$

The Gibbs energy change for the transfer of one mole of ions down the electrical potential gradient $\Delta\psi$ (in millivolts) is:

$$\Delta G = -mF\Delta\psi \qquad [6.9]$$

where m is the number of charges (for protons, $m = 1$), and F is the Faraday constant (9.65×10^4 Coulombs mol^{-1}).

As Gibbs energy differences are additive, the total Gibbs energy difference or pmf resulting from the transfer of one mol of ions down an electrical potential gradient of $\Delta\psi$ (mV) and an H^+ concentration gradient is:

$$\Delta G = -mF\Delta\psi + 2.3\,RT\log_{10}\frac{[H^+]_o}{[H^+]_i} \qquad [6.10]$$

In units of electrical potential the combined energy for the two sources of proton flow, with $m = 1$, is:

$$\Delta\mu_{H^+} = \text{pmf} = \Delta\psi - \frac{2.3\,RT}{F}\Delta\text{pH} \qquad [6.11]$$

which at 30 °C becomes, with units of millivolts:

$$\Delta\mu_{H^+} = \Delta\psi - 60\Delta\text{pH} \qquad [6.12]$$

Lipid membranes have very low permeability to H^+. The protons, however, can flow in a controlled manner through the enzyme CF_0-CF_1 complex, producing the change in enzyme configuration required for ATP synthesis. Possibly a certain 'pressure' of H^+ is needed to change the enzyme configuration. The flow of protons is called protonicity, by analogy with electricity and it is a property of the flow of H^+ in CF_1 which synthesizes ATP. At present the chemiosmotic hypothesis does not provide a full description of the molecular mechanism of energy transduction at the level of enzyme sites (enzymological studies may provide that), but rather a mechanism of coupling ATP production with electron transport.

The magnitudes of $\Delta\psi$ and ΔpH have been measured by several techniques. Synthetic lipophilic ions (e.g. tetraphenyl-phosphonium) called Skulachev ions after the discoverer, penetrate lipid membranes, even though charged, due to the extensive π orbital system (p. 24). Their absorption spectra change with conditions so that, with careful calibration, an estimate of $\Delta\psi$ can be made optically. The ΔpH has been measured from fluorescence quenching of 9-aminoacridine in chloroplasts. Micro-electrodes have also been employed to measure pH and ion concentrations. The difference in concentration of ions between the medium and organelle has been measured using isotopes, after separation of the organelles and medium, following experimental changes. The ionic concentrations have been used to calculate the concentration gradient of the ion and to estimate $\Delta\psi$. Rapid centrifugation through silicone oil is employed to separate the organelle from the incubation medium. Allowance must be made for the contamination of the space in the organelle freely available to the ion. All techniques suffer to a smaller or greater degree from changing the gradient they are intended to measure (e.g. by introducing compounds into the vesicle) and uncertainty about their reaction within the vesicle; Nicholls (1982) considers the techniques in some detail. Carotenoids, bound in the thylakoid (and other membranes) are important indicators of $\Delta\psi$. They respond to the large electrical field (3×10^5 V cm^{-1}) which develops, for example, on illumination, by a rapid (nanosecond) change in absorption towards longer wavelengths. This electrochromic shift is readily measured without disturbing the conditions. The electrochromic shift may be calibrated to provide a measure of $\Delta\psi$, however it reflects the $\Delta\psi$ only in the immediate vicinity of the carotenoid in the membrane, not necessarily at the membrane surface.

Under steady illumination $\Delta\psi$ is 10–50 mV across the thylakoid membrane, whereas ΔpH is 3 units, equivalent to 180 mV, and therefore the most important component of pmf in thylakoids; in mitochondria $\Delta\psi$ is the most important component. However, in suddenly illuminated chloroplasts (Fig. 6.3) $\Delta\psi$ may also be more important than ΔpH in developing pmf; within 10^{-8} s $\Delta\psi$ develops due to electron transport but ion transport is much slower. The positive charge in the thylakoid lumen causes anions to move in from the stroma, to balance the electrical charge. Chloride ions (particularly *in vitro*) enter to join the H^+ in the lumen; in Jagendorf's instructive phrase 'thylakoids pump hydrochloric acid into themselves in the light'. Cations, Mg^{2+} and K^+, move into the stroma and with time $\Delta\psi$ decreases as H^+ accumulates. Up to 1 milliequivalent of H^+ accumulates per mg of chlorophyll and would increase the membrane potential and, as the lumen is small (10^{-21} m^3, p. 56) a small change in H^+ (0.001 equivalents) would greatly alter pH and could damage the membrane. However most of the protons (99%) are buffered by proteins in the lumen. Large $\Delta\psi$ is, possibly, important with the onset of illumination and during fluctuations in intensity, allowing a pmf to develop before H^+ accumulates appreciably, so that ATP synthesis may start quickly.

ATP synthesis can be completely separated from electron transport if the ΔpH is caused by subjecting intact thylakoids to an acid–base transition. Thylakoids are incubated in the dark, with a buffer solution, for example succinic acid at pH 4, which diffuses into the lumen providing a controlled

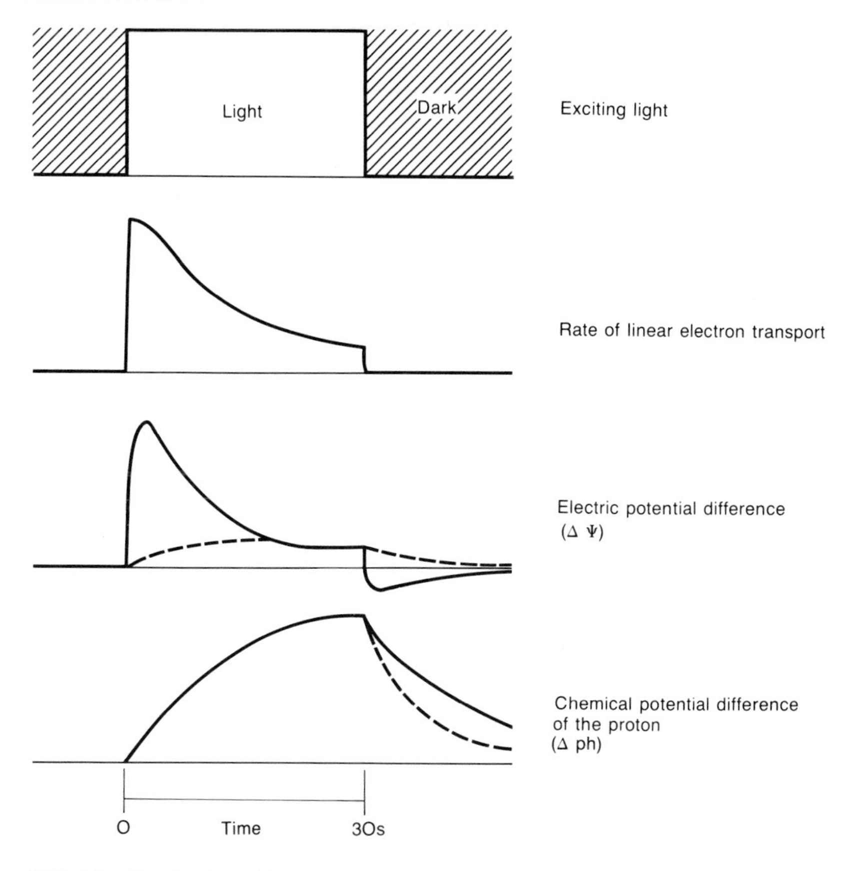

FIG. 6.3 Illustration of the relative roles of the two components of proton motive force, ΔpH and Δψ, as a consequence of electron transport upon illumination following darkness. (Modified from Junge 1977, Figure 8).

internal concentration of H^+; then the external pH is raised quickly. With a gradient of four pH units ATP is synthesized, but a gradient smaller than two pH units is ineffective. Even with electron transport inhibited by DCMU ATP is synthesized, showing the pH gradient to be the driving force, not electron flow. A general relationship between ATP synthesis and the ΔpH is shown in Fig. 6.4a. ATP synthesis is related to the cube of the H^+ concentration (Fig. 6.4b), evidence that synthesis of one ATP requires 3 H^+ to flow through coupling factor. However there is still uncertainty about the 'true' value, because there is a basal rate (non-phosphorylating) of H^+ flux, so that the phosphorylation per H^+ depends on ΔpH; values of 2.4 H^+ per ATP have been suggested. For chloroplasts the accepted $H^+/2\ e^-$ ratio is 4, for non-cyclic electron transport. With 3 H^+ required per ATP synthesized, the ATP/2 e^- ratio is 1.33; with 2.4 H^+/ATP the ATP/2 e^- ratio is 1.7.

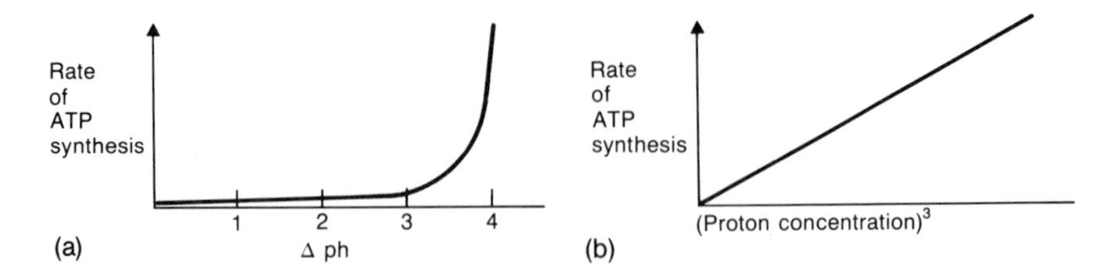

FIG. 6.4(a) ATP synthesis related to differences in pH across the thylakoid membrane. (b) Rate of ATP synthesis by thylakoids plotted against the cube of the $[H^+]$ gradient, showing the relationship to be 1 ATP formed per 3 H^+.

Paths of electron flow and coupling to phosphorylation

Three paths of photosynthetic electron flow linked to ATP synthesis have been recognized and are shown in Fig. 6.2. ATP synthesis coupled to a 'linear' flow of electrons from water to $NADP^+$ is called non-cyclic photophosphorylation because the electrons pass on to an acceptor, which passes them to metabolic reactions, and do not return to the electron transport chain. In cyclic photophosphorylation, electrons are cycled from PSI back to the electron transport chain. Pseudo-cyclic photophosphorylation involves electron flow to O_2, and to H^+ and water rather than to $NADP^+$ and is a variant of non-cyclic electron transport, but with a different electron acceptor. It is, of course, the electron movement which is cyclic or non-cyclic not phosphorylation. However, the inexact but historical term remains.

Non-cyclic photophosphorylation

With adequate substrates to oxidize NADPH, primarily CO_2, linear (i.e. 'non-cyclic') flow of electrons is coupled to ATP synthesis. Both PSII and PSI are needed. The action spectra for both CO_2 assimilation and ATP synthesis are very similar and they saturate at similar light intensities. Measured ATP/2 e^- ratios for non-cyclic photophosphorylation are between 1.5 and 2; higher ratios may be obtained if allowance for the basal rate of electron transport is made. However, it is not established if the basal rate occurs when the processes are coupled or only when uncoupled. It is important to know the *in vivo* rates of ATP formation per H^+ because at least 1.5 ATP must be synthesized for each 2 e^- if most of the ATP required for CO_2 fixation is produced non-cyclically, as is probable.

Cyclic photophosphorylation

When non-cyclic electron flow is prevented, electrons from PSI or more probably ferredoxin (shown by the greater sensitivity to antimycin, an Fe–S inhibitor), pass back to plastoquinone *via* cytochrome 563, a *b* type cytochrome also called cyt b_6 (Fig. 6.2). Cytochrome involvement is shown by spectral changes. The ratio of cytochromes to PSI is 2:1, so two cytochromes take one electron each from ferredoxin fed by two different PSI centres and reduce plastoquinone. Cyt b_{563} has a potential of −0.18 V and donates e^- to plastoquinone at about zero potential. Coupling with ATP synthesis is associated with the transfer of H^+ from plastoquinone to the thylakoid lumen. Only energy of PSI is used, that is, it is driven by light above 680 nm. With no net transport of e^-, water is not split and no O_2 evolved. However when the acceptors of PSI are fully reduced the e^- flow cannot start; a slow flux of e^- from PSII to PSI maintains the correct redox potentials ('poises' the system) so that electrons can move. DCMU, which stops electron flow from PSII to PSI, inhibits cyclic photophosphorylation by interfering with 'poising'.

The importance of cyclic photophosphorylation *in vivo* is not clear; it may be most important in physiologically intact tissues when non-cyclic electron transport is slowed by lack of CO_2 and O_2, and in dim light (see Gimmler, 1977). Cyclic and non-cyclic photophosphorylation may co-operate, the former poises the system and the latter generates ATP. Reduced $NADP^+$ and ferredoxin (when nearly all reduced) also regulate electron flow in cyclic photophosphorylation.

Pseudo-cyclic photophosphorylation

This requires both photosystems, like non-cyclic photophosphorylation but with ferredoxin reducing an 'oxygen reducing factor' which passes electrons to molecular oxygen as the terminal electron acceptor. The two-step O_2 reduction forms the superoxide radical O_2^- and then, by the action of superoxide dismutase (p. 102), hydrogen peroxide. Electrons can also be donated from the reduced acceptors of PSI to H^+ giving H_2, the reaction is catalysed by hydrogenase. Pseudo-cyclic photophosphorylation is measured by the uptake of O_2 caused by light and by the effect of ADP and P_i on it. However, in photosynthesizing tissues H_2O_2 is destroyed by catalase, O_2 is released and there is no net exchange of O_2. Electrons go from water to O_2 back to water so the process is not cyclic as the same electrons are not recycled. The rate is greater at high O_2 concentration than low and when CO_2 fixation is slow; it saturates at higher light intensity than cyclic photophosphorylation. It has the same P/2 e^- ratio as non-cyclic photophosphorylation as the same sites of coupling are employed.

Enzyme mechanism of phosphorylation

In chloroplasts the thylakoid lumen is a separate compartment from the stroma (Ch. 4) with the 'coupling factor' complex, CF_1, projecting into the stroma and its base portion, CF_0, spanning the thylakoid membrane (Fig. 6.5). CF_1 was first detected by Avron in 1963, as a protein which when removed from the

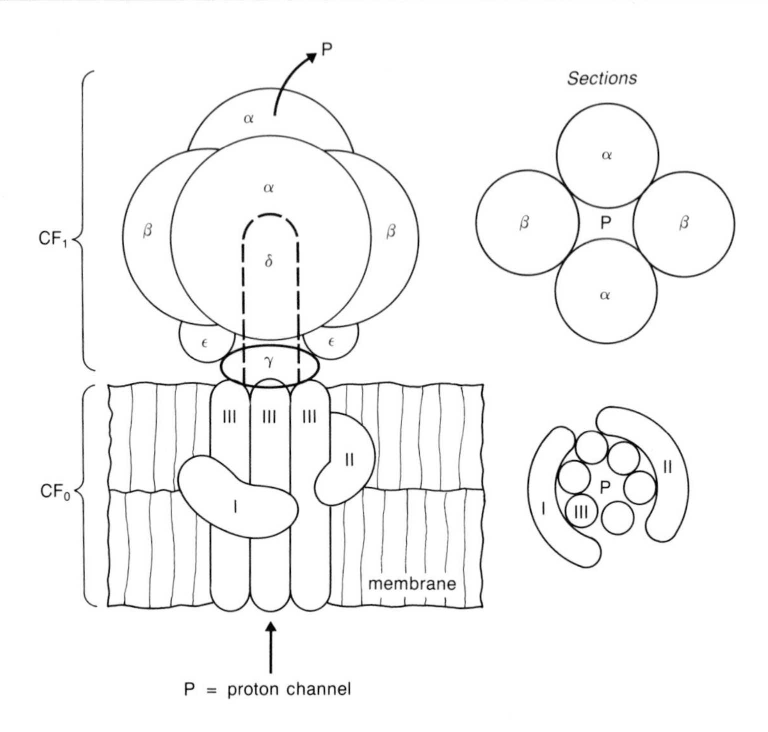

FIG. 6.5 Idealized structure of chloroplast coupling factor, the CF_1 (with polypeptide subunits α to ε) CF_O (with subunits I, II and III) enzyme complex responsible for synthesis of ATP.

thylakoid, uncoupled photophosphorylation but when replaced, restored ATP synthesis. CF_1 is only loosely attached and is readily removed if the cation concentration (particularly Mg^{2+}) is low, for example after treatment with EDTA, a cation chelating chemical. CF_1 is thought to move around on the membrane under the influence of ionic and electrical charges.

CF_1 has a molecular weight of 325 kD and may be separated into five individual types of polypeptide subunits of different masses (Fig. 6.5) by compounds such as urea, which break hydrogen bonds, followed by electrophoresis in gels (polyacrylamide plus the detergent sodium dodecyl sulphate (SDS)). There are two spherical (10 nm diameter) α subunits (58–62 kD, the masses differing slightly according to authors) which probably regulate proton permeability and therefore enzyme activity, two spherical (5.8 nm diameter) β subunits (55–57 kD) with the active sites for synthesis of ATP and binding of nucleotides on the α or β units. Subunit ε (14 kD) is elongated and inhibits ATPase enzyme activity, as its removal allows CF_1 to break down ATP; ε may be bound to γ, a 37–38 kD subunit which allows protons to enter the CF_1 complex. This is an $\alpha_2\beta_2\gamma\delta\varepsilon_2$ CF_1. More recent determinations suggest an $\alpha_3\beta_3\gamma\delta\varepsilon_2$ (or ε_1) CF_1 complex of about 400 kD mass. These polypeptides are shown, by X-ray analysis, to form a hollow, water-filled sphere of about 10 nm diameter; some 10 000 molecules of water are bound per molecule of CF_1. The single δ subunit (20–21 kD and elongated in shape, 2.5 nm diameter by 10 nm long), attaches the other subunits to CF_0, a hydrophobic protein of 93 kD mass made up of three polypeptides (one each of I and II and 6 of III) of different molecular masses embedded in the membrane. Their masses are 15–18, 13–16 and 8 kD respectively and I and II link to ε of CF_1. Protons leak through the proton conducting channel, formed from the six units of III, if CF_1 is removed, thus destroying the pH gradient without ATP synthesis (see section on chemiosmotic coupling) and uncoupling electron transport from phosphorylation. One molecule of dicyclohexylcarbodiimide (DCCD) binds to the III subunits and inhibits proton translocation.

CF_1 plus CF_0 constitutes a controlled leakage path for proton flow across the membrane. For ATP synthesis ADP + P_i bind firmly to CF_1 (a process inhibited by arsenate which is a phosphate analogue and competes with phosphate) but the binding and mechanism by which passage of protons through the complex produces the anhydride bond of ATP is not known. Changes in conformation, which are important for catalysis and involve the γ subunit and probably the α and β units also, were elegantly shown by incubating CF_1 with tritiated water (Jagendorf 1975). CF_1 was removed, denatured in urea, and incorporation of radioactive tritium measured. Illumination or a pH gradient, which energize the complex, caused tritium from the medium to exchange with hydrogen on the protein. Under de-energizing conditions some 100 H atoms per CF_1 were not exchanged showing them to be hidden in the de-energized state, associated with conformational changes. Proton exchange was essential for ATP synthesis. ADP + P_i bound to CF_1 and altered the tritium exchange, demonstrating that molecular arrangement of subunits changes with conditions and is involved in catalysis. Sulphydryl groups on the CF_1 polypeptides (measured with N-ethyl maleimide, which reacts with —SH groups on the γ subunit) are hidden within the complex in the dark but become exposed in the light. Possibly four —SH groups on γ subunits control H^+ flux, like a 'plug' in the 'flow' system. Reagents which bind to —SH groups (e.g. dithiothreitol) increase ATPase, suggesting that sulphydryl groups regulate ATP synthesis. Light activates CF_0–CF_1, not only *via* ΔpH but by the thioredoxin system which regulates other chloroplast enzymes (p. 138) changing sulphydryl groups on subunits. In the unactivated state a much greater ΔpH is needed to drive photophosphorylation.

The structure of the site for ATP synthesis on CF_1 and the mechanism by which the terminal anhydride bond is formed, are not understood and most models are speculative. Enzyme conformation may change with proton concentration or energy state of the membrane; light increases subunit reactivity and may alter the position of subunits, opening up a channel for protons. ADP and P_i bind tightly to CF_1 at two sites, about 90 μm of binding sites per chloroplast. A CF_1-ADP and P_i complex may be formed, which is inhibited by arsenate. Studies with ^{18}O labelled ATP, P_i and H_2O show that as protons move through CF_0 they remove an oxygen atom on the phosphate group,

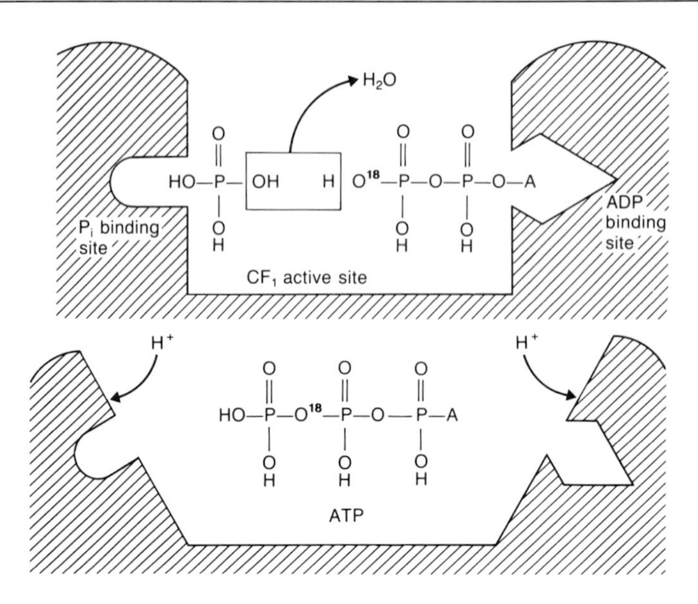

FIG. 6.6 A possible sequence for the synthesis of ATP on CF_1. P_i and ADP bind to the enzyme and in a spontaneous reaction H^+ is removed from ADP terminal phosphate (shown by incorporation of ^{18}O from ADP into ATP) and the OH from P_i. ATP is bound to the complex. When H^+ passes through CF_1, conformational changes release ATP and allow P_i and ADP to bind again, thus 'resetting' the complex. A = adenosyl residue.

forming water (Fig. 6.6). The electron on the phosphate group moves to the ADP terminal phosphate forming the anhydride bond. This is thought to require little change in energy. Protons in CF_1 alter the conformation of the peptide chains and change the binding energy between ATP and enzyme complex, allowing ATP to escape into the stroma. ADP + P_i then bind again to the enzyme reaction site in its energized conformation, which then returns to its original form, producing the anhydride bond.

CF_1 activity is regulated by several factors, including H^+ and the concentration of substrates, ADP and P_i. ATP inhibits ATP synthesis and ADP and P_i inhibit ATP hydrolysis so that the complex is allosterically controlled. Nucleotides bind to multiple sites, not active in ATP synthesis, depending on the energized state of CF_1 (indicating conformational changes in CF_1 proteins) but their role in catalysis is not understood. Energy from the proton gradient is essential for releasing ATP by changing conformation of CF_1; this stage may be the most energy demanding.

In illuminated thylakoids, ATP synthesis is the main function of CF_0–CF_1 but this complex also catalyses hydrolysis of ATP. Treatment of isolated CF_1 with the protein-digesting enzyme trypsin, heat, or sulphydryl reagents (e.g. dithiothreitol) stimulates ATPase activity which requires Ca^{2+}; the ε subunit controls ATPase activity. This may have a physiological significance, allowing ATP to drive proton accumulation, providing control over the ionic balance of thylakoids, for example in darkness, when regulation of the state of the membranes and ionic concentration creates the conditions needed for rapid synthesis of ATP on illumination.

Relationship between ATP formed and e^- and H^+ flow

Mitchell's hypothesis was that 2 H^+ were needed for 1 ATP but, in chloroplasts, the stoichiometry is close to 3, from several lines of evidence. With 2 coupling sites in the electron chain, 8 photons give 8 H^+ for 4 e^- transported to $NADP^+$. Thus the ATP/2 e^- ratio is 1.33. The energy available to drive synthesis is calculated from eqn 6.12 with ΔpH of 3 or 180 mV and $\Delta\psi$ of 20 mV, a total pmf of 200 mV.

A mole of ATP requires about 30 kJ for synthesis under equilibrium conditions which when converted to redox potential difference, $\Delta E'$, by

$$\Delta G = -nF\Delta E'$$

where n is the number of reducing equivalents and F the Faraday constant, gives about 150 mV. Under cellular conditions the low ATP concentration may require more energy, 57 kJ mol^{-1}, about 230 mV, more than the system can provide. Possibly the ΔpH is 3.5, which would suffice to generate ATP with 3 H^+, as the pmf is 210 mV and $\Delta\psi$ of 30 mV would give about -70 kJ mol^{-1}, sufficient to synthesize ATP under unfavourable equilibrium conditions. Energetically the chemiosmotic hypothesis is feasible and is supported by experimental evidence of many types. It is now the accepted model for energy transduction in phosphorylation.

Uncouplers of ATP synthesis

An intact enclosed vesicle, as in the thylakoid system, is needed for H^+ accumulation with a bounding membrane impermeable to H^+; the lipid membrane is an effective barrier. Illumination of intact thylakoids (Fig. 6.7) without ADP or P_i, causes H^+ accumulation from water splitting and a 'back pressure' on the PQ pump, slowing electron transport to a basal rate, corresponding to leakage. With the substrates ATP and P_i, H^+ flows through CF_1, ATP is synthesized, the back pressure drops and electron transport rate increases until equilibrium is attained. Treatments causing loss of ΔpH or $\Delta\psi$ uncouple electron transport and ATP synthesis.

Different uncoupling mechanisms are known; if only some CF_1 complex is removed by EDTA, which chelates Mg^{2+}, then H^+ leaks through CF_0 and no ATP is made. Inhibition of ATP synthesis by arsenate or thiophosphate allows H^+ flow through CF_1 and destroys the ΔpH. Phlorizin (a glucoside from roots) blocks ATP synthesis but prevents H^+ flow. Detergents disrupt the membrane preventing ΔpH formation. Membranes are made 'leaky' by lipid-soluble

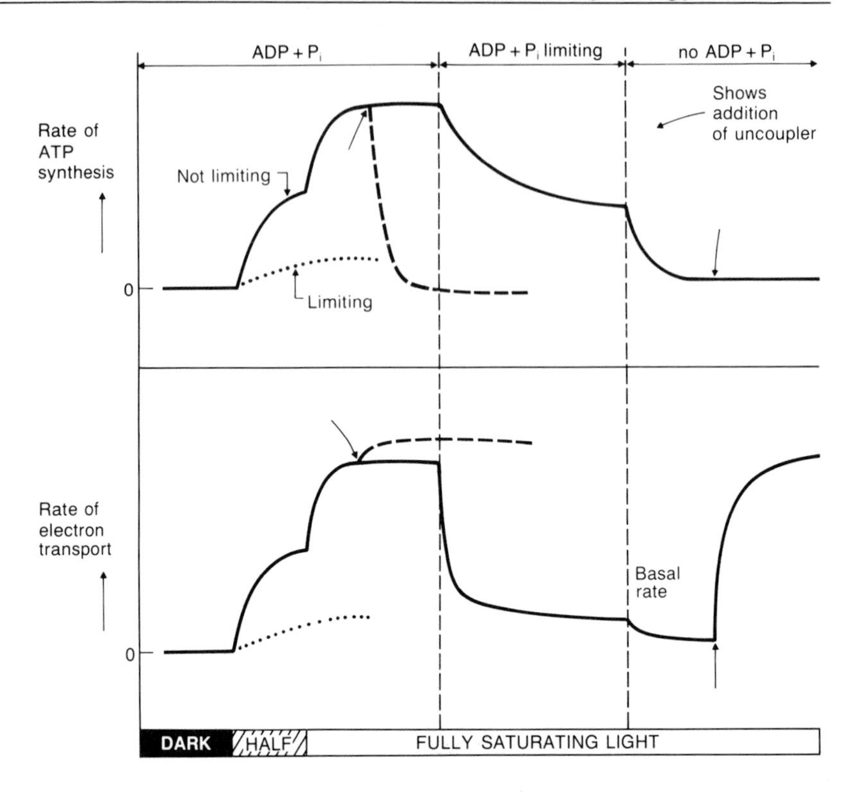

FIG. 6.7 Relationship between photophosphorylation by chloroplasts and the rate of electron transport in darkness and light with limiting or non-limiting ADP + P_i and the effect of an 'uncoupler' such as nigericin (see text) which uncouples ATP synthesis from electron transport.

proton ionophores – H^+ transporting compounds, such as carbonylcyanide *p*-trifluoromethoxyphenyl hydrazone (FCCP). The negatively charged molecule moves along the $\Delta\psi$ gradient into the lumen, where it is protonated. It moves back to the medium, is deprotonated and recycles, collapsing ΔpH. Dinitrophenol (DNP) carries anions and cations (H^+) in response to $\Delta\psi$, and is a very effective uncoupler in mitochondria but not in thylakoids. Other ionophores carry ions; for example valinomycin (Fig. 6.8), a depsipeptide (with alternating hydroxy and amino acids) antibiotic from bacteria, dissolves in the membrane and carries K^+ out of the lumen, changes $\Delta\psi$ and uncouples. Together valinomycin and DNP are very effective uncouplers in chloroplasts, carrying H^+ and K^+ from the lumen and collapsing ΔpH and $\Delta\psi$. Nigericin (Fig. 6.8) binds H^+ and K^+ reversibly and transports them across the membrane and changes the pmf. Gramicidin is a peptide which forms a very efficient channel (carrying 10^7 ions s^{-1}) for monovalent ions, including H^+; only one molecule per thylakoid uncouples ATP synthesis completely.

Ammonia is an uncoupler. On entering the lumen (Fig. 6.8) it is protonated

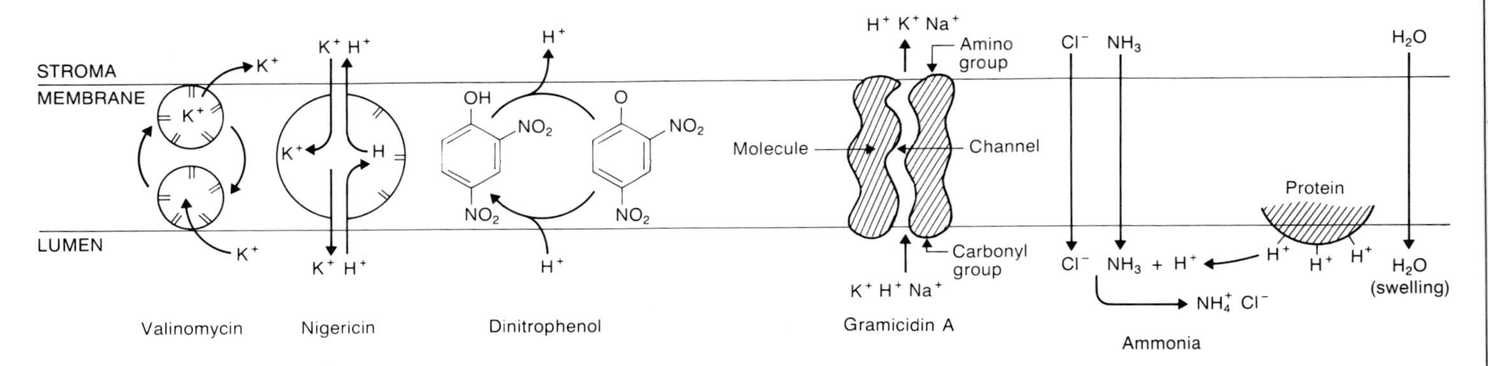

FIG. 6.8 Substances which uncouple ATP synthesis from electron transport by dissipating the H^+ gradient across the thylakoid membrane or consuming H^+ inside the lumen.

to NH_4^+ and destroys ΔpH. Anions enter to restore neutrality, osmotic potential increases and water enters, causing swelling. Ammonia only uncouples above 10^{-3} M concentration as H^+ is an effective buffer. As ammonia is formed under normal physiological conditions in chloroplasts it must be rapidly metabolized to prevent damage.

Photophosphorylation and physiological control

Photosynthesis is regulated, in ways not well understood, by the balance between ATP and NADPH formation and the supply of CO_2, P_i etc. In dim light non-cyclic electron transport is slow and ATP synthesis limits the rate of metabolism if $NADP^+$ is available as electron acceptor. With increased light intensity electron flow will be non-cyclic if $NADP^+$ is available (with adequate CO_2 or other substrates) and ATP synthesis is adequate for metabolism. However if the rate of ATP synthesis is limiting, $NADP^+$ is reduced and therefore non-cyclic e^- flow is slowed and cyclic electron flow may increase the rate of ATP synthesis. The same may apply when light is abundant but CO_2 limiting. Pseudo-cyclic electron flow may also permit ATP synthesis when $NADP^+$ is not available, perhaps in very intense light. Low demand for ATP, which slows the rate of supply of ADP to CF_1 or inadequate P_i supply, inhibits phosphorylation. Photosynthesis by C3 plants (p. 129) requires an ATP/2 e^- ratio of at least 1.5 for CO_2 reduction; as other processes also consume ATP, particularly in the light, either the ratio *in vivo* is greater than *in vitro* or additional ATP is synthesized in other ways. Regulation of the ATP/NADPH ratio is important for photosynthesis but poorly understood. Affinity of components of the e^- transport chain for substrates may be expected to determine the relative flux of e^- into parts of metabolism. Ferredoxin-$NADP^+$ reductase has a much greater affinity for $NADP^+$ than NAD^+ (a K_m of 10^{-5} M compared to 3×10^{-3} M) but under unphysiological conditions it catalyses H^+ transfer between NADPH and NAD^+ and also oxidation of NADPH by several electron acceptors. Also, the e^- flux will depend on environmental conditions (e.g. light) which determine the saturation of the e^- transport chain, relative to CO_2, O_2 and NO_3^- supply. Conditions during growth may also affect the relative amounts of different components. C3 and shade plants for example, may have many grana and grow in dim light but are relatively inefficient in bright light. Some C4 plants (e.g. *Zea mays*) are agranal and inefficient in dim light, but are very efficient in bright light. Rate of ATP synthesis may limit CO_2 assimilation in bright light in C3 plants and in dim light in C4. Granal number and size may regulate the area of membrane not only for light harvesting and electron flow between photosystems (p. 80) but H^+ flux through CF_1 under dim light conditions.

References and Further Reading

Atkinson, D. E. (1977) *Cellular Energy Metabolism and its Regulation*, Academic Press, New York.

Berzborn, R. J., Müller, D., Roos, P. and **Andersson, B.** (1981) Significance of different quantitative determinations of photosynthetic ATP-synthase CF_1 for heterogeneous CF_1 distribution and grana formation, pp. 107–20 in Akoyunoglou, G. (ed.), *Photosynthesis III. Structure and Molecular Organisation of the Photosynthetic Apparatus*, Balaban International Science Services, Philadelphia.

Chappell, J. B. (1977) *ATP*. Carolina Biology Readers. Carolina Biological Supply Co., Burlington, NC.

Cross, R. L. (1981) The mechanism and regulation of ATP synthesis by F_1-ATPases, *A. Rev. Biochem.*, **50**, 681–714.

Denton, R. M. and **Pogson, C. I.** (1976) *Metabolic Regulation*, Chapman and Hall, London. Holsted Press, John Wiley & Sons Inc., New York.

Gimmler, H. (1977) Photophosphorylation *in vivo*, pp. 448–72 in Trebst, A. and Avron, M. (eds), *Encyclopedia of Plant Physiology* (N.S.), vol. 5, *Photosynthesis I*, Springer-Verlag, Berlin.

Hinkle, P. C. and **McCarty, R. E.** (1978) How cells make ATP, *Sci. Amer.*, **238**, 104–23.

Jagendorf, A. T. (1975) Mechanisms of photophosphorylation, pp. 414–92 in Govindjee (ed.), *Bioenergetics of Photosynthesis*, Academic Press, New York.

Jagendorf, A. T. (1977) Photophosphorylation, pp. 307–37 in Trebst, A. and Avron, M. (eds), *Encyclopedia of Plant Physiology* (N.S.), vol. 5, *Photosynthesis I*, Springer-Verlag, Berlin.

Junge, W. (1977) Membrane potentials in photosynthesis, *A. Rev. Plant Physiol.*, **28**, 503–36.

Junge, W. and **Jackson, J. B.** (1982) The development of electrochemical potential gradient across photosynthetic membranes, pp. 589–646 in Govindjee (ed.), *Photosynthesis*, vol. 1, *Energy Conversion by Plants and Bacteria*, Academic Press, New York.

McCarty, R. E. (1979) Roles of a coupling factor for photophosphorylation in chloroplasts, *A. Rev. Plant Physiol.*, **30**, 79–104.

Mitchell, P. (1966) Chemiosmotic coupling in oxidative and photosynthetic phosphorylation, *Biol. Rev.*, **41**, 445–502.

Nelson, N. (1977) Chloroplast coupling factor, pp. 393–404 in Trebst, A. and Avron, M. (eds), *Encyclopedia of Plant Physiology* (N.S.), vol. 5, *Photosynthesis I*, Springer-Verlag, Berlin.

Nelson, N. (1982) Structure and function of the higher plant coupling factor, pp. 81–104 in Barber, J. (ed.), *Electron transport and photophosphorylation (Topics in Photosynthesis 4)*, Elsevier Biomedical Press, Amsterdam.

Nelson, N., Nelson, H. and **Schatz, G.** (1980) Biosynthesis and assembly of the proton translocating ATPase complex from chloroplasts, *Proc. Natl. Acad. Sci. USA*, **77**, 1361–64.

Nicholls, D. G. (1982) *Bioenergetics. An Introduction to the Chemiosmotic Theory*, Academic Press, London.

Ort, D. R. and **Melandri, B. A.** (1982) Mechanism of ATP synthesis, pp. 537–87 in Govindjee (ed.), *Photosynthesis*, vol. 1, *Energy Conversion by Plants and Bacteria*, Academic Press, New York.

Pradet, A. and **Raymond, P.** (1983) Adenine nucleotide ratios and adenylate energy charge in energy metabolism, *A. Rev. Plant Physiol.*, **34**, 199–224.

Preiss, J. and **Kosuge, T.** (1976) Regulation of enzyme activity in metabolic pathways, pp. 278–336 in Bonner, J. and Varner, J. E. (eds), *Plant Biochemistry* (3rd edn), Academic Press, New York.

Roth, R. and **Nelson, N.** (1984) Conservation and organisation of subunits on the chloroplast proton ATPase complex, pp. 501–10 in Sybesma, C. (ed.), *Advances in Photosynthesis Research*, vol. II, Martinus Nijhoff/Dr W. Junk Publishers, The Hague.

Schlodder, E., Gräber, P. and **Witt, H. T.** (1982) Mechanism of photophosphorylation in chloroplasts, pp. 105–75 in Barber, J. (ed.), *Electron transport and photophosphorylation (Topics in Photosynthesis, 4)*, Elsevier Biomedical Press, Amsterdam.

Shavit, N. (1977) Bound nucleotides and conformational changes in photophosphorylation, pp. 350–68 in Trebst, A. and Avron, M. (eds), *Encyclopedia of Plant Physiology* (N.S.), vol. 5, *Photosynthesis I*, Springer-Verlag, Berlin.

Shavit, N. (1980) Energy transduction in chloroplasts: structure and function of the ATPase complex, *A. Rev. Biohem.*, **49**, 111–38.

Simonis, W. and **Urbach, W.** (1973) Photophosphorylation *in vivo*, *A. Rev. Plant Physiol.*, **24**, 89–114.

CHAPTER 7

The chemistry of photosynthesis

Carbon dioxide reduction is the major energy consuming process in photosynthesis; nitrate and sulphate reduction (p. 149) use less than 5 per cent of total energy. Enzymes of CO_2 assimilation are in the chloroplast stroma, and are affected by conditions there, for example ion and NADPH concentrations or products of enzyme reactions, which act as enzyme effectors. Enzyme activity determines the rate of CO_2 fixation and the balance between processes, such as starch synthesis in the chloroplast or carbon export from it, and may regulate respiration and photosynthesis, which have common intermediates and enzymes.

Carbon dioxide assimilation

The mechanism was speculative (see Rabinowitch 1945) until radioactive carbon isotopes became available; ^{14}C with a half-life of 5760 years is eminently suited as tracer for carbon and Calvin, Benson and Bassham (amongst others) exploited it to analyse CO_2 assimilation. Briefly, as this major achievement has been frequently recounted, Calvin's group grew algae (mainly *Chlorella*; other plants were also studied) under constant conditions in $^{14}CO_2$ for different periods. The algae were rapidly killed and the assimilates extracted in solvents. The radioactive compounds were separated by chromatography, and the radioactivity detected by autoradiography. Products were identified chemically and by co-chromatography with authentic compounds. Analysis of the kinetics of ^{14}C incorporation into assimilates, their specific radioactivity (SA = $^{14}C/(^{12}C+^{14}C)$) and ^{14}C distribution in the molecules established the sequence of reactions.

CO_2 assimilation is a cyclic, autocatalytic, process (Fig. 7.1), variously called the Calvin cycle, reductive pentose phosphate pathway, photosynthetic cycle or photosynthetic carbon reduction cycle (PCR cycle). PCR cycle emphasizes the carbon, reduction and cyclic aspects. The PCR cycle is the fundamental CO_2 assimilatory process in all photosynthetic organisms, including prokaryotes; it appears to have developed early in evolution and to have retained its characteristics. Additional processes for accumulating CO_2 have arisen which do not replace the PCR cycle, but rather add to it (C4 and CAM mechanisms, see Ch. 9).

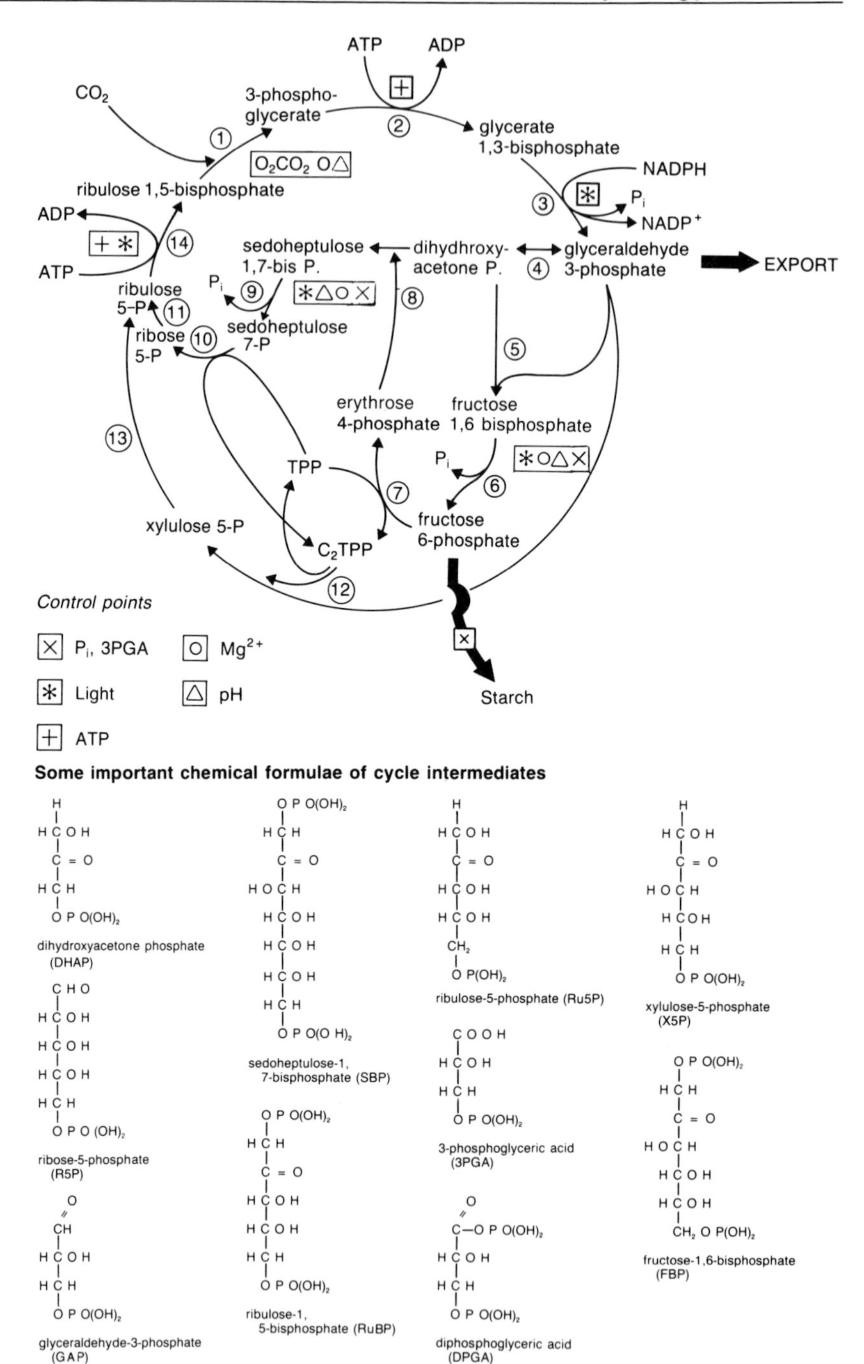

FIG. 7.1 The photosynthetic carbon reduction cycle, with numbered reactions corresponding to the enzymes listed in Table 7.1. The upper part of the figure includes the carboxylation and reduction steps, the lower part shows the regeneration of the CO_2 acceptor ribulose-1,5-bisphosphate (RuBP).

Mechanism of the PCR cycle

An acceptor molecule, ribulose bisphosphate (RuBP), combines with CO_2 in a carboxylation reaction, in the presence of the enzyme RuBP carboxylase, giving a product 3-phosphoglyceric acid (3PGA). The acceptor is regenerated from 3PGA in reactions consuming NADPH and ATP. A scheme of the cycle is given in Fig. 7.1 and reactions are identified by numbers which correspond to the enzymes listed in Table 7.1. If more carbon is fixed than is used to regenerate RuBP, carbon is exported from the cycle as triosephosphates or used in starch synthesis in the chloroplast. Control of export is essential to prevent depletion of components of the cycle and to maintain the autocatalytic process. Rate of cycle turnover in the steady state depends on the primary reactions of electron transport and ATP synthesis and on enzyme activity.

Assimilation of CO_2 is described by:

$$3\,CO_2 + 9\,ATP + 6\,NADPH + 5\,H^+ \rightarrow C_3H_5O_3P + 9\,ADP + 8\,P_i + 6\,NADP^+ + 3\,H_2O + 468\,kJ\,mol^{-1} \qquad [7.1]$$

with triosephosphate as the product; for each CO_2 assimilated a minimum of 3 ATP and 2 NADPH + H are needed. The light reactions generate sufficient NADPH but may not produce the required ATP by non-cyclic electron transport (p. 116).

An arbitrary starting point in the PCR cycle is carboxylation of RuBP giving two molecules of 3PGA (reaction (1)), a reaction unique to the PCR cycle and catalysed by RuBP carboxylase: details of enzyme and reaction are given later. Reaction (2) uses ATP to phosphorylate 3PGA to a more reactive state in 1,3-diphosphoglyceric acid with two acid anhydride bonds. In reaction (3) NADP glyceraldehyde-3-phosphate dehydrogenase substitutes H^+ for the phosphate group in 1,3-diphosphoglycerate. The enzyme is $NADP^+$ dependent, in contrast to the respiratory enzyme which requires NAD^+. This is the only reduction in the PCR cycle and is of the greatest importance. Glyceraldehyde-3-phosphate (GAP) is converted (4) to dihydroxyacetone phosphate (DHAP); these two compounds are used in (8) and (12) where three-carbon units are converted to five carbons in the regeneration of RuBP. Triosephosphates are condensed (5) to the 6-carbon compound fructose bisphosphate; the aldolase has maximum activity in alkaline conditions. Dephosphorylation of FBP by fructose bisphosphatase, an enzyme unique to the PCR cycle, gives fructose-6-phosphate (F6P). The reaction has a free energy change of $-25\ kJ\ mol^{-1}$, so is not reversible and is a control point in the cycle.

Regeneration of RuBP is achieved by interconversion of 3-, 4-, 5- and 6-carbon compounds. Transketolase removes 2-carbon fragments (glycoaldehyde) from F6P and sedoheptulose-7-phosphate (S7P), attached to the thiamine pyrophosphate (TPP) co-factor of the enzyme. Erythrose-4-phosphate (E4P) reacts with DHAP (8) giving sedoheptulose-1,7-bisphosphate. This is dephosphorylated by the sedoheptulose bisphosphatase, unique to the PCR cycle and an important control point. Ribose-5-phosphate (Ru5P) is made from S7P (10),

Table 7.1 Enzymes of the photosynthetic carbon reduction cycle (reactions are numbered as in Fig. 7.1) with their approximate mass, specific activity (SA = μmol min^{-1} mg chlorophyll^{-1}) and Michaelis constants (K_m). The free energy change ΔG^s, at the steady state physiological (i.e. stromal) concentrations of substrates is for *Chlorella* in 40 Pa CO_2 and 21 kPa O_2. Large negative ΔG^s indicates a probable control reaction

Reaction number	Enzyme	SA	Mass (kD)	ΔG^s	Increased by	Decreased by	K_m
1	Ribulose bisphosphate carboxylase/oxygenase	10	550	−41	high pH, CO_2 Mg^{2+}, FBP	gluconate, SBP?	CO_2 12 μM, O_2 250 μM, RuBP 40 μM
2	Phosphoglycerate kinase	900	47	+16	3PGA, ATP Mg^{2+}	DGPA, ADP	3PGA 0.5 μM, ATP 0.1 μM
3	Triosephosphate dehydrogenase	100	140	−6.7	light	P_i?	DPGA 1 μM, NADPH 4 μM
4	Triosephosphate isomerase		53	−7.5		PEP, RuBP, glycolate P	DHAP 1 mM, GAP 0.4 mM
5,8	Aldolase	100	140	−1.6	high GAP/DHAP ratio		FBP 20 μM, GAP 0.3 mM DHAP 0.4 mM
6	Fructose bisphosphatase	100	140	−27	high pH, Mg^{2+}, ATP, reductant, light	P_i	FBP 0.2 mM
7,10,12	Transketolase	150	140	−5.9	Mg^{2+}, high pH		
9	Sedoheptulose bisphosphatase	0.5	70	−29.7	light, pH, Mg^{2+}	P_i	SBP 13μM
11	Ribose-5-phosphate isomerase		54	−0.5	freely reversible		R5P 0.2 mM, ATP 0.1 mM
13	Ribulose-5-phosphate 3 epimerase		46	−0.6	freely reversible		X5P 0.5 mM
14	Ribulose-5-phosphate kinase	320	54	−15.9	reductant, ATP, energy charge?	ADP	R5P 0.2 mM, ATP 0.6 mM

Data from Bassham (1965) and Robinson and Walker (1981).

and converted to ribulose-5-phosphate (11). Another 5-carbon sugar, xylulose-5-phosphate, is formed (12) from glycoaldehyde and glyceraldehyde-3-phosphate and is converted to Ru5P (13). The 'final step' (14) is the phosphorylation of Ru5P to the more reactive RuBP by Ru5P kinase, another enzyme unique to the PCR cycle, with a free energy change of -15 kJ mol^{-1}, the fourth most negative in the cycle.

Radioactive labelling patterns

When *Chlorella* was exposed to $^{14}CO_2$ in 5 per cent CO_2, 70 per cent of the ^{14}C assimilated was in 3PGA after 2 seconds, and some in alanine and other compounds. Longer exposure gave ^{14}C in other amino acids and sugars. When extrapolated back to zero time all ^{14}C was in 3PGA and the proportion decreased with time as label entered other compounds, suggesting synthesis from 3PGA. The rapid labelling of alanine suggested that it had a primary role in assimilation but its saturation kinetics indicated that it was a secondary product. Sucrose was labelled later, evidence that it was a secondary product of assimilation.

After 5 seconds of photosynthesis, chemical analysis showed 95 per cent of the ^{14}C to be in carbon 1 of 3PGA and 2.5 per cent in C2 and C3; with time they become uniformly labelled. This indicated that 3PGA was synthesized from products containing ^{14}C in a cyclic process. Adding CO_2 to a 2-carbon acceptor was expected to produce the three carbons of 3PGA, and a suitable compound was sought but none found. An early assimilation product, F6P, was labelled in C3 and C4 as if two 3-carbon compounds had joined and indeed triosephosphates were later implicated; condensation of DHAP and GAP gave the 6-carbon bisphosphate. Analysis of S7P showed the label in C3, C4 and C5 as if 3- and 4-carbon compounds were joined. A suitable 4-carbon compound (erythrose-4-phosphate) had the required labelling pattern and discovery of the enzymes substantiated the sequence. Inability to find a 2-carbon acceptor suggested that a 5-carbon compound plus CO_2 would give two molecules of 3PGA, and ribulose-1,5-bisphosphate was finally identified as the CO_2 acceptor. It was labelled chiefly in C3, as expected if it were derived from S7P *via* R5P. Reaction of the glycoaldehyde with GAP would give rise to xylulose-5-phosphate with ^{14}C in C3; subsequent conversion to ribulose-5-phosphate produced the observed ^{14}C distribution pattern. Further turns of the cycle would give uniformly labelled 3PGA. Within 5 minutes all cycle components would be saturated with ^{14}C and the specific activity of secondary products would increase.

Enzymes of the PCR cycle

Many enzymes of the cycle were discovered first in non-photosynthetic tissues; some were not detected in early studies or only at low activity, but improved methods of extraction and assay have substantiated the mechanisms.

Ribulose-1,5-bisphosphate carboxylase/oxygenase (RuBPc/o)

The carboxylating enzyme (earlier called fraction 1 protein) occurs in all photosynthetic organisms; it comprises 50 per cent of the soluble protein in leaves (6 mg per mg chl) and has been called the most abundant protein on earth. In higher plants the enzyme is made up of eight large and eight small subunits, of respectively 55 and 15 kD, a total mass of 550 kD; these may be joined as in Fig. 7.2. RuBPc/o enzymes from photosynthetic bacteria are more variable than those of higher organisms, with different numbers of subunits, *Rhodospirillum rubrum*, for example, has only two large subunits. Large subunits from higher plants are closely homologous, that is, they have very similar amino acid compositions, but the small units are not. The complex structure may be necessary for catalysis. Large subunits are synthesized in the chloroplast and coded for by chloroplast DNA, but the small subunit is made outside the chloroplast and coded by nuclear DNA. This may reflect the symbiotic origins of the chloroplast. There is one catalytic site for CO_2 fixation on each large subunit, that is eight per molecule for the higher plant enzymes; the reaction is understood in some detail and there is much research in progress on the catalytic mechanism.

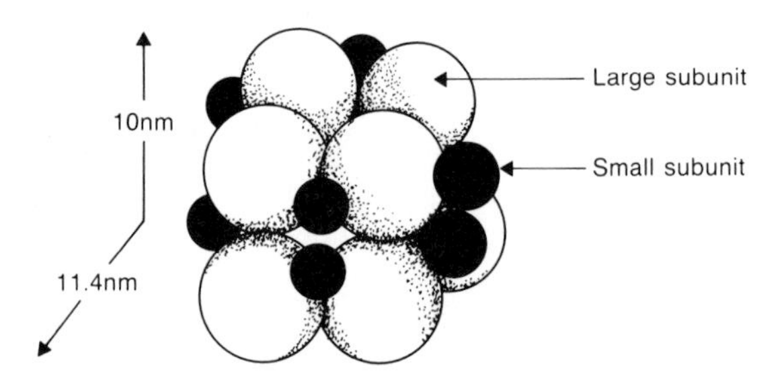

FIG. 7.2 Illustration of the subunit structure of ribulose bisphosphate carboxylase-oxygenase enzyme.

Enzyme activation

In early preparations, the activity of RuBPc/o was too small to account for the measured rates of photosynthesis; the K_m (Michaelis constant) for CO_2 was much greater than the CO_2 concentration in the chloroplast. Addition of CO_2 (as bicarbonate ions) and magnesium activates the extracted enzyme and greatly increases the rate of 3PGA synthesis. Incubation with CO_2 is required before Mg^{2+} is co-ordinated (it is not covalently bound) at the activation site (Fig. 7.3) to complete the activation step. Carbon dioxide binding is slow compared with the Mg^{2+} activation step. Activating CO_2 joins to (carbamylates) a lysine amino acid residue within the activation site, allowing the Mg^{2+} to co-ordinate. Activation may alter the geometry (electronic configuration) of chemical groups of the active site to which RuBP binds, or of the substrate CO_2 which can then

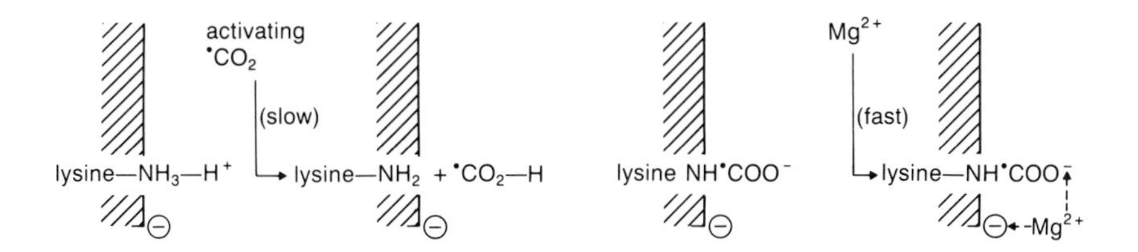

FIG. 7.3 Activation of ribulose bisphosphate carboxylase-oxygenase by a slow reaction with CO_2 and a subsequent fast reaction with Mg^{2+}, which increases the affinity of the enzyme for substrate CO_2.

react with RuBP. Magnesium also plays an important role in regulating the electronic state in other photosynthetic components such as chlorophyll. The K_m of activated RuBP carboxylase is 1–20 μm for substrate CO_2, adequate for the assimilation of CO_2 in leaves in low atmospheric CO_2 concentrations. *In vivo* the enzyme is probably activated even in darkness when stromal Mg^{2+} concentration is low.

Reaction mechanism for CO_2

RuBP binds to the catalytic site by its phosphate groups (Fig. 7.4) to basic, possibly lysine, residues and a tautomeric change in electronic configuration occurs. The C2 of RuBP is normally slightly positive (electron attracting) because bonding electrons are 'pulled away' towards the carboxyl O in the molecule. However, loss of a proton from C3 and formation of a keto-enol equilibrium produces a nucleophilic (proton attracting) enediol allowing the CO_2 to react at the negatively charged C2 of RuBP. As the CO_2 molecule is polarized (Fig. 7.5), the change in RuBP electronic structure allows the CO_2 to join to the C2 of RuBP. The 6-carbon intermediate formed is 2-carboxy-3-ketoarabinitol-1,5-bisphosphate. Water donates OH^- to the C3 of RuBP, cleaving the intermediate to two molecules of 3PGA. Two molecules of the D-stereoisomer of 3PGA are formed in the reactions. However if hydrolysis of the 6-carbon intermediate proceeds non-enzymatically both the L and D isomers are formed; the enzyme controls the stereochemistry of the reaction. Analogues of the 6-carbon intermediate, such as 2-carboxyarabinitol-1,5-bisphosphate (CAB) are effective inhibitors because they bind to, and block the catalytic site.

RuBPc/o oxygenase activity

Ribulose bisphosphate carboxylase is a bifunctional enzyme, catalysing the reaction of molecular oxygen with RuBP at the same catalytic site as the carboxylation.

The oxygenation reaction is:

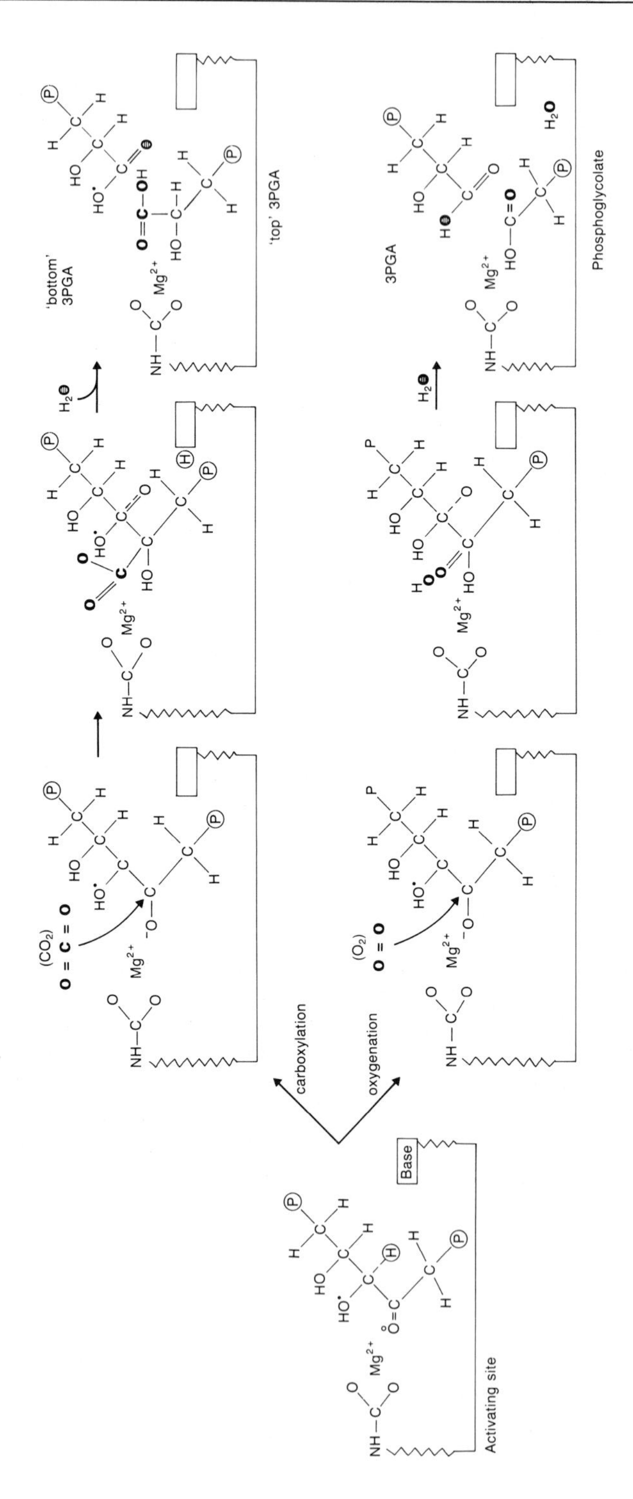

FIG. 7.4 Schematic of the reaction site of ribulose bisphosphate carboxylase and the oxygenase showing molecular events in carboxylation and oxygenation of RuBP to give 3-phosphoglyceric acid (3PGA) or phosphoglycolate and 3PGA respectively. (After Gutteridge and Keys, unpublished).

$$RuBP + O_2 \rightarrow \text{2-phosphoglycolate} + 3PGA \quad [7.2]$$

Phosphoglycolate is a 2-carbon phosphorylated compound. The mechanism of the oxygenation reaction catalysed by the enzyme is still not understood. Molecular oxygen is relatively stable, is uncharged and is not polarized (Fig. 7.5). Other types of oxygenase enzyme contain a transition metal (e.g. Fe) or a redox active prosthetic group, which transfers electrons; but RuBPc/o has no such group and there is no comparable mono-oxygenase enzyme. It is not clear how the oxygen is activated. It may be co-ordinated to the Mg^{2+} in the presence of the enediol form of RuBP on the enzyme. This leads to formation of a hydroperoxide intermediate (Fig. 7.4) which is subsequently hydrolysed by water bound to the metal. If the RuBP oxygenase reaction is performed in the presence of $^{18}O_2$, one ^{18}O joins to C2 of RuBP to produce phosphoglycolate labelled in the carboxyl group, and the other is released as $H_2{}^{18}O$. An O from H_2O reacts with C3 of RuBP to give 3PGA. Possibly the enediol form of RuBP can attract the electrons in C of CO_2 or of O_2; the required electron changes in RuBP are produced by the co-ordinated Mg^{2+} and bound activating CO_2.

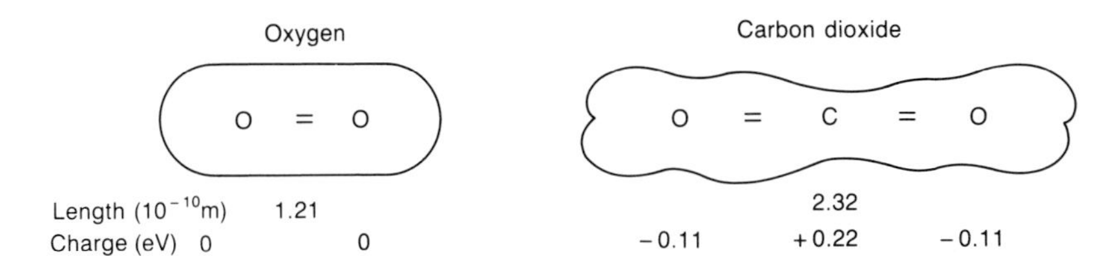

FIG. 7.5 Molecular structures of oxygen and carbon dioxide. (Modified from Ogren 1978, Figure 2).

That oxygenation and carboxylation occur at the same site is shown by inhibition of both functions by CAB, other intermediates of the PCR cycle and pyridoxal phosphate. Also, both functions are activated to the same degree by CO_2 and Mg^{2+}. However the two reactions are differentially affected by pH, and by temperature which changes the O_2/CO_2 solubility ratios in solution. If manganese ions replace magnesium ions in the activation of the isolated enzyme, the ratio of oxygenase to carboxylase activity increases.

Competition between RuBP carboxylase and oxygenase reactions

Oxygen and CO_2 compete for RuBP at the catalytic site; they are mutually competitive inhibitors. The rate, V_c, of the carboxylation of RuBP in the presence of competitive inhibition by O_2 with saturating RuBP is:

$$V_c = \frac{V_{cmax} \times C}{C + K_C (1 + O/K_O)} \quad [7.3]$$

where V_{cmax} is the maximum velocity, C and O are the partial pressures of CO_2 and O_2 in equilibrium with dissolved gases in the chloroplast stroma and K_C and K_O are the Michaelis-Menten constants for CO_2 and O_2 respectively. The rate, V_o, of the oxygenase reaction with saturating RuBP is:

$$V_o = \frac{V_{omax} \times O}{O + K_O\,(1 + C/K_C)} \qquad [7.4]$$

where V_{omax} is the maximum rate of oxygenation.

The K_m for CO_2 in the oxygenation reaction equals the CO_2 inhibition constant for the oxygenase and O_2 has the same K_m in the oxygenase reactions as K_i in the carboxylase reaction. Oxygen competes inefficiently with CO_2 for RuBP at the catalytic site and only at high molar ratio of O_2 to CO_2 is the oxygenase reaction significant.

The two reactions may be expressed as a ratio of oxygenase to carboxylase, α, (Fig. 7.6) which increases as the ratio of O_2 to CO_2 increases:

$$\alpha = \frac{V_o}{V_c} = \frac{V_{omax}}{V_{cmax}} \times \frac{OK_C}{CK_O} \qquad [7.5]$$

The oxygenase activity of RuBP carboxylase is important because the PG formed is oxidized in leaf cells by the glycolate pathway (p. 163), with the

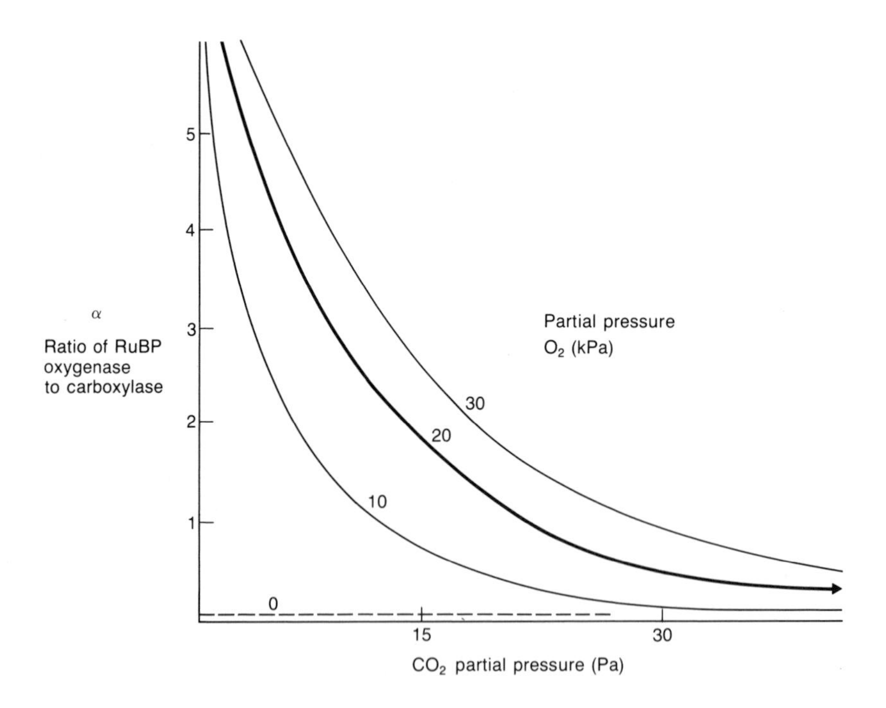

FIG. 7.6 Ratio of RuBP oxygenase to RuBP carboxylase activity, α, in relation to CO_2 and O_2 concentrations for extracted enzyme.

release of CO_2. The CO_2 produced is a form of respiration, which occurs in the light as photosynthesis proceeds, and is therefore called photorespiration. It offsets the CO_2 assimilation by the PCR cycle and thus decreases the efficiency of CO_2 assimilation. When the ratio of O_2 to CO_2 in the chloroplast stroma is very small, α approaches zero and photorespiration is negligible. However in air, with the partial pressure of O_2 a thousand-fold greater than CO_2 (p. 207), α is about 0.4; photorespiration is large and 20–30 per cent of the fixed carbon is lost. When α is 2, carbon lost by photorespiration equals gross photosynthesis (see p. 224) and there is no net gain of carbon. Some types of plants have developed mechanisms to avoid the loss of photorespired carbon, *viz.* C4 and CAM plants; those mechanisms are considered in Chapter 9. Crop plants with low oxygenase to carboxylase ratio would be more efficient at CO_2 assimilation but attempts to select this desirable trait or to modify the enzyme chemically or by genetic manipulation have failed so far; it remains one of the major challenges in plant research.

Control of PCR cycle enzymes

RuBPc/o is activated by large increases in Mg^{2+} and pH (from 6 to 13 mM and 6 to 8 units respectively) which occur in the chloroplast stroma with change from darkness to light; also NADPH and metabolite (e.g. F6P, R5P and E4P) concentrations increase the activity of the carboxylase. In darkness these compounds decrease, ensuring that the PCR cycle stops, thus avoiding depletion of intermediates and CO_2 acceptor. Fructose bisphosphate inhibits RuBPc/o activity at higher concentrations which may occur when triosephosphate export from the chloroplast is slowed by phosphate shortage. RuBPc/o is then slowed, preventing the accumulation of PCR cycle products and depletion of RuBP.

Fructose and sedoheptulose bisphosphatases

Fructose bisphosphatase (FBPase) achieves maximum activity at alkaline pH and with high Mg^{2+}, as expected of a chloroplast enzyme. It is also activated by light, as described on p. 138. FBPase activity is small with low concentrations of FBP but increases greatly above a threshold value of FBP concentration. This characteristic provides control of the PCR cycle and of processes leading to and from FBP. Synthesis of FBP is dependent on the production of triose-phosphate; when this is rapid FBP accumulates, stimulating FBPase. If FBP decreases below the threshold the reaction slows, allowing the cycle to attain a new equilibrium. Regulation of SBPase is similar to that of FBPase.

3-Phosphoglycerate kinase and triosephosphate dehydrogenase

3-Phosphoglycerate kinase catalyses a reaction with large positive free energy change and is therefore controlled by the end products, the only PCR cycle enzyme so regulated. ADP and a low ATP/ADP ratio slow the reaction as does accumulation of glyceraldehyde-3-phosphate, regulating the cycle in relation to ATP synthesis and consumption (p. 124). Triosephosphate dehydrogenase is stimulated by reduced ferredoxin and the thioredoxin system. Control of two steps in 3-phosphoglycerate metabolism by products of the light reactions and the PCR cycle provides co-ordination of cycle function in light and dark, preventing large fluctuations in intermediates or, more importantly, their depletion.

Phosphoribulokinase

ATP stimulates this enzyme, which is regulated by energy charge and light *via* reduced thylakoid proteins.

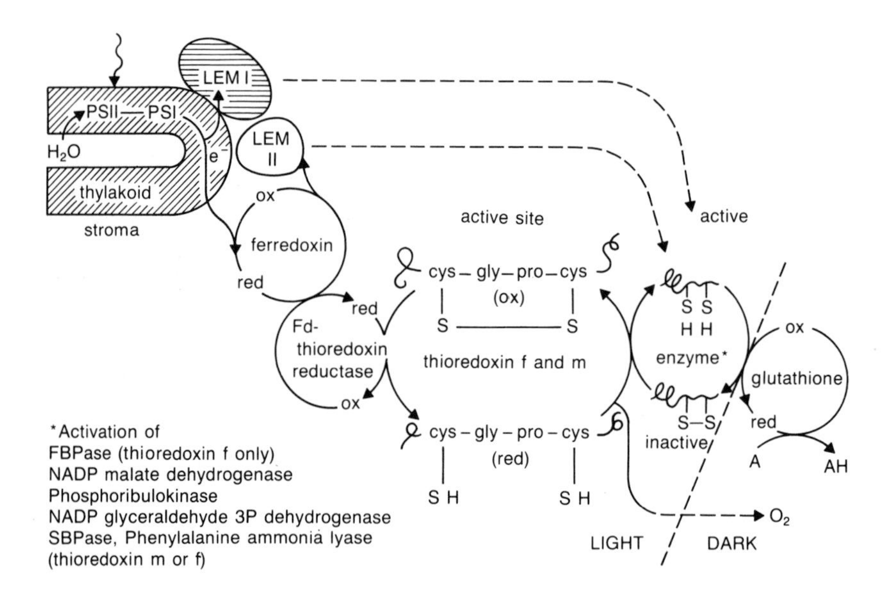

FIG. 7.7 Simplified scheme of the regulation of enzyme structure and activity by the reduction of disulphide bridges with electrons from the thylakoid *via* thioredoxin and the mechanism of deactivation.

Light activation of enzymes

Several PCR cycle enzymes are controlled by reduced components of the electron transport chain (Fig. 7.7). FBPase, SBPase, $NADP^+$-glyceraldehyde phosphate dehydrogenase and phosphoribulokinase are activated in the light by ferredoxin or proteins linked to it or to the reduced acceptor of PSII. The proteins, a water-soluble low mass thioredoxin and ferredoxin-thioredoxin reductase, contain sulphydryl groups which are reduced to —SH, affecting the conformation of the enzyme. Other proteins with similar functions are bound to the thylakoids, and are called 'light effect mediators' (LEMs); they donate electrons to enzymes and regulate catalysis. How the thioredoxin system is deactivated in darkness is not well understood; possibly oxidized glutathione or ascorbate slowly deactivate proteins. Probably a continuous flow of e^- to O_2 oxidizes the thioredoxin so that inhibition of electron supply is accompanied by enzyme inactivation. Such light effectors provide several points at which light can start and regulate carbon flux in the PCR cycle although their function in intact systems, such as leaves, is still unclear.

Induction and control of the PCR cycle

There is a delay or 'lag' of several minutes between illuminating isolated chloroplasts, after a long period of darkness, and attainment of a rapid, constant rate of photosynthesis. Intact systems such as protoplasts and leaves exhibit a shorter lag, although closed stromata may open slowly and prolong the lag period for photosynthesis in leaves. Rate of photosynthesis depends on light, CO_2 supply, nutrition, species etc. The maximum rate of photosynthesis and how quickly it is attained determines the plant's productivity and efficiency. What limits the rate of photosynthesis during induction and at the steady state? Rapid synthesis of RuBP is required and even darkened chloroplasts contain some RuBP, which may enable the cycle to start quickly. Photochemistry and electron transport are very fast following illumination and NADPH is synthesized within a few seconds but ATP synthesis is slow. Reduction of thioredoxin is fast but activation of enzymes is probably slower, requiring minutes, comparable to the enzyme changes induced by alterations in Mg^{2+} or pH in the stroma. Over several minutes ATP concentration and the ATP/ADP ratio increase. Synthesis of inhibitors (e.g. 6-phosphogluconate) stops and as their concentration decreases so the PCR cycle increases. As RuBP is synthesized faster than RuBPc/o activity increases, there is a transient increase in RuBP in tissues in the lag phase followed by a decrease in the steady state. Fructose bisphosphate and sedoheptulose bisphosphate also increase to exceed the steady state values in the early lag phase as they are rapidly formed from 3PGA, *via* triosephosphate, but their conversion to F6P and S7P is slow due to the slow activation (30 s) of the respective bisphosphatases.

If rapid synthesis of 3PGA and the formation of triosephosphates cause depletion of ATP, accumulation of ADP inhibits further activity, thus slowing the rate of R5P and RuBP formation. During induction, as concentration of intermediates rises and falls and enzymes in the cycle are activated at different rates, CO_2 assimilation and the concentration of intermediates oscillate as the system 'hunts' (to use an engineering term) until it approaches dynamic equilibrium, where the rate is determined by a 'limiting factor', either on the rate of supply of light, CO_2 etc. from the environment or in the control mechanisms operating within the system. For example, in bright light CO_2 may limit the rate of assimilation or, with abundant CO_2, it may be regulated by the rate of a particular enzyme reaction. Oscillation in CO_2 fixation and amounts of intermediates are damped because the several pools act as capacitances in the system, smoothing the demand/supply imbalance and giving only small fluctuations in the generally increasing rate of photosynthesis.

In the induction phase every 5 CO_2 assimilated produces an extra RuBP, if no carbon is removed from the PCR cycle, so that the rate of CO_2 assimilation increases with each turn of the cycle. Starting with 1 μmol RuBP per m^2 leaf turning over in unit time, after five turns of the cycle there will be 2 μmol RuBP and 5 μmol of CO_2 fixed, so the CO_2 fixation rate is 1 μmol per unit time, after two turns there will be 4 μmol RuBP and 15 μmol will have been assimilated, with the rate of CO_2 fixation 2 μmol per unit time. A further turn of the cycle gives 8 μmol RuBP and 4 μmol CO_2 per unit time. This is an exponential increase in activity. However, the increase is limited by CO_2, the amount of light or internal processes such as NADPH and ATP synthesis so the rate of CO_2 assimilation becomes constant with time. Thus the system exhibits a lag phase, the duration of which is determined by the amount of RuBP (or materials for its formation) and the turnover rate of the RuBP regeneration and carboxylation system. Isolated chloroplasts lose intermediates of the PCR cycle and therefore contain little RuBP and generate it slowly; the lag period is shortened by addition of intermediates, 3PGA for example, particularly if the ATP/ADP ratio is small, enabling RuBP to be made quickly. However, adding ribulose-5-phosphate increases the lag, because it stimulates the synthesis of RuBP but consumes ATP. Increased ADP then inhibits phosphoglycerate kinase and thereby assimilation.

During rapid photosynthesis the rate of turnover of substrates depends on the size of the pools and the activity and amount of the enzymes. The slowest reaction limits RuBP regeneration, but at present it is not clear which is the limiting factor, although RuBPc/o is the most likely contender followed by FBPase, SBPase and phosphoribulokinase. Supply of organic phosphate to the cytosol may limit assimilation, slowing export of triosephosphate from the chloroplast; formation of large pools of phosphorylated intermediates 'locks up' P_i. Large 3PGA and small P_i concentration stimulate starch synthesis (p. 158), freeing P_i but as it is a slow reaction, the rate of assimilation with deficient P_i is smaller than the rate when P_i is freely available. Rate of ATP synthesis may limit assimilation under some conditions, for example in leaves grown in dim light, which have smaller capacity to form ATP than those grown in bright light. In C3 plants photorespiration limits assimilation by competitive inhibition of carboxylation by O_2 and because RuBP is consumed in the RuBP oxygenase reaction in addition to the CO_2 release. Phosphoglycolate produced by RuBP oxygenase is not completely recycled to form RuBP so the amount of acceptor limits assimilation. Without O_2, CO_2 assimilation increases more than expected from preventing photorespiratory CO_2 release, as extra RuBP becomes available for carboxylation.

During rapid photosynthesis, enzymes turn over quickly, possibly at rates approaching the maximum, and therefore control the rate of assimilation, even if light, CO_2 and nutrients are in excess; this is discussed for leaves in Chapter 11. Control of the PCR cycle is complex; interactions between conditions in the stroma and thylakoid membranes, amount of substrates and of control molecules, regulate enzyme activation and reaction rates which determine the overall function of the system and its rate in relation to demand for assimilates.

Metabolite concentrations in the chloroplast stroma

Under steady state conditions the rate of each step in the PCR cycle is constant and the pool size of each metabolite also; some approximate values are given in Table 7.2. In some cases (RuBP) the concentration exceeds the enzyme K_m, in others (DHAP and GAP) K_m and concentration are similar. Concentration of substrates and the K_m of a reaction determine the physiological rate and, because reactions are interconnected, the rate of CO_2 fixation.

When concentrations of CO_2 or O_2 change or light alters or demand for products changes, then the equilibrium conditions and fluxes of material are no longer constant. A finite time (many seconds) is required for transport of material from one pool to another and the rate of reactions is not constant, so fluctuations in the size of metabolite pools are observed. A step decrease in CO_2 concentration to a photosynthesizing leaf (Fig. 7.8) results in an immediate decrease in 3PGA synthesis and in the amount of 3PGA in the tissue. RuBP is not consumed so the pool increases very shortly after 3PGA falls; triose is not used for RuBP formation so it also rises. As the diminishing pool of 3PGA feeds through to RuBP, transient changes in the amount are observed. Glycolate synthesis increases with the low CO_2/O_2 ratio. In low O_2 concentration,

Table 7.2 Concentrations of photosynthetic carbon reduction cycle intermediates in the chloroplast stroma

Intermediate	mM
RuBP	0.2–0.6
3PGA	3–5
ATP + ADP + AMP	1–4
DPGA	10^{-3} (?)
NADPH + $NADP^+$	0.1–0.5
DHAP + GAP	0.3–0.4
FBP	0.1–0.3
F6P	0.6–1.5
SBP	0.2–1.0
X5P	0.5
R5P	0.1
Ferredoxin	1

Data from Robinson and Walker (1981).

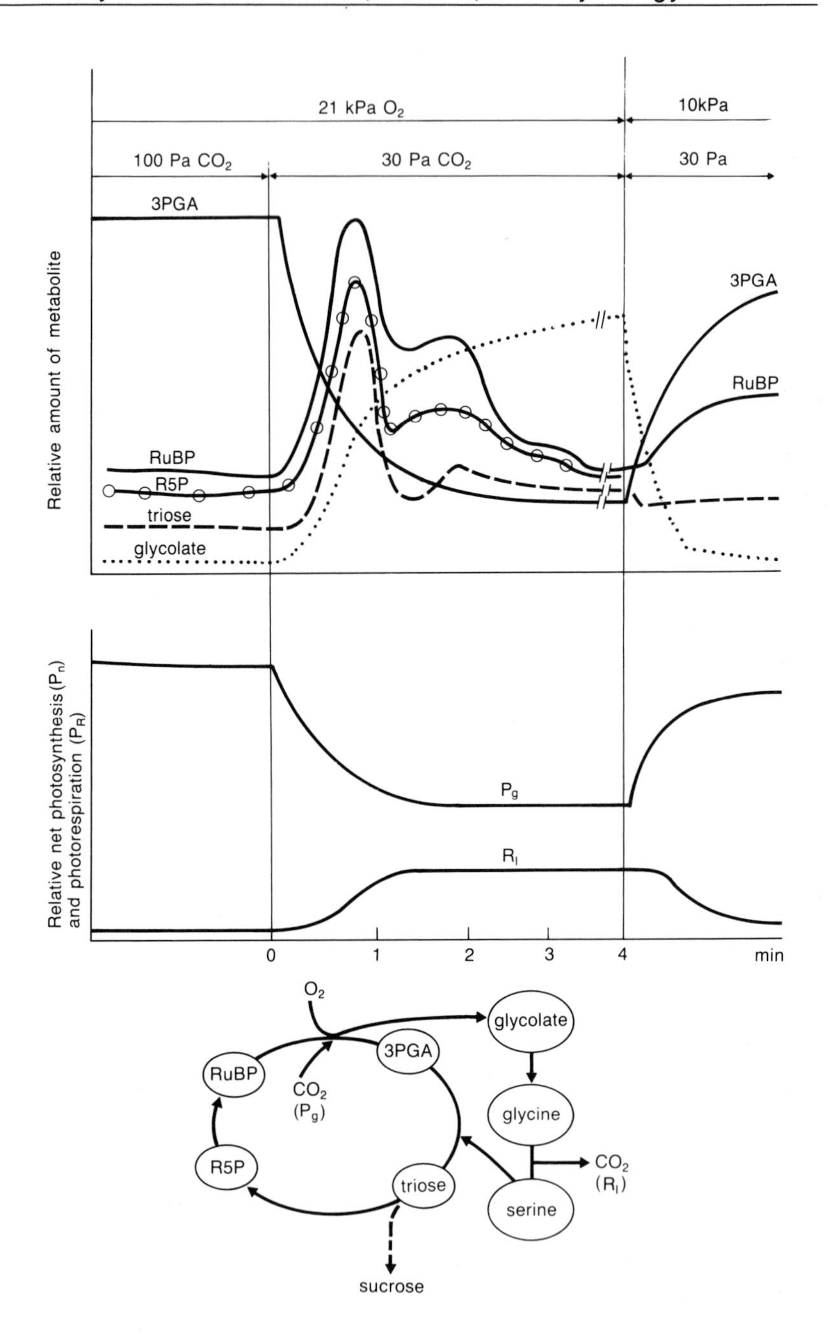

FIG. 7.8 Changes in the metabolite pools of photosynthesizing cells following perturbations in CO_2 and O_2 concentration related to net photosynthesis and photorespiration (semi-schematic). Note the phase shifts between fluctuations in metabolites and CO_2 exchange due to position in the PCR cycle and the different rates at which metabolites are interconverted.

glycolate synthesis slows and 3PGA increases, permitting more RuBP to be made and faster photosynthesis. In a complex, cyclic system the size of the fluctuations, their rate and phasing between metabolites depends on pool sizes, reaction rates and the magnitude of the change perturbing the system. This was appreciated by Wilson and Calvin (1955), who, after describing such fluctuations, wrote: '. . . perhaps the most important result of this work is the general insight it gives into the complicated interrelated system of chemical reactions which occur in living systems.'

Exchange of photosynthate between chloroplast and cytosol

In mature leaves products of the PCR cycle and nitrate assimilation do not accumulate in the chloroplast stroma (with the exception of starch), but are transported across the envelope to the cytosol. The fluxes of material and energy, their control and interaction with chloroplast reactions are studied on isolated chloroplasts, with intact envelopes, free (or relatively so) of the rest of the cell substance and physiologically undamaged as far as can be established. Distribution of assimilates between these chloroplasts and defined media is measured under different conditions of light, CO_2, etc. and with added metabolites, inhibitors etc. Permeability of the envelope is determined from the distribution of metabolites between the medium and the chloroplasts by, for example, using radioactively labelled substrates. How added compounds affect the rate of photosynthesis and their interaction with the individual chloroplast reactions is measured by CO_2 and O_2 exchange. The latter is rapidly determined by the polarographic oxygen electrode (Fig. 7.9). A suspension of chloroplasts is placed in a small chamber with an O_2-sensitive electrode, which produces a voltage proportional to O_2 concentration, allowing this to be measured continuously. Vigorous stirring and small volume increase the speed of the instrument's response. Inhibitors of photosynthesis or substrates may be added without significantly disturbing conditions.

Methods have been developed for determining the compounds exchanged between cell compartments. Fast, efficient separation of organelles is essential for such metabolic studies. Chloroplasts are rapidly separated from the medium by placing the suspension in a centrifuge tube above a killing solution (e.g. perchloric acid) but separated from it by a layer of inert silicone oil. After the reaction has taken place, the tube is centrifuged rapidly, the chloroplasts penetrate the oil and reactions are stopped by the acid within seconds. Such treatment may alter the behaviour of chloroplasts because conditions differ from those in the intact cell, yet the methods provide much information on processes.

Other techniques have been used to analyse chloroplast and cytosol interaction *in vivo*. Distribution of assimilates, radioactive label etc., have been measured in separated organelles. Isolation in aqueous solutions allows redistribution of water-soluble substances between parts of the cell and causes 'smearing' of the distribution pattern. Therefore non-aqueous media are used. Chloroplasts are

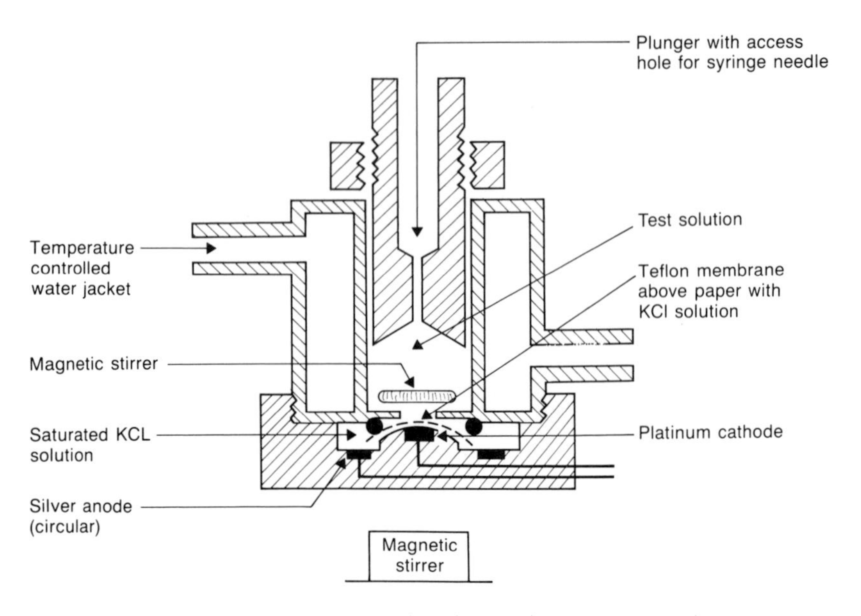

FIG. 7.9 An oxygen electrode apparatus for measurement of O_2 exchange by photosynthesizing or respiring tissues or extracted organelles.

isolated in organic solvents such as carbon tetrachloride, which prevent water-soluble metabolites mixing during isolation; the method provides information on chloroplast interaction with the cytosol. Techniques and detailed discussion of the results are given by Heber (1974) and by Heldt (1979).

The outer membrane of the chloroplast envelope allows many substances to pass but the inner is very selective. This is shown by measuring the volume of chloroplasts which can be penetrated by low molecular mass, neutral substances such as sorbitol or sucrose, or by larger polymeric sugars. The space between the envelope membranes is accessible to sorbitol but the stroma and thylakoids are not (indicating that there is no direct connection between the inner membrane and thylakoid compartments). Changes of osmotic concentration in the medium cause chloroplasts to swell or shrink, altering the proportion of volume accessible to solutes and indicating that the outer membrane is freely permeable but the inner is not.

Movement of compounds across the envelope is not by simple diffusion; CO_2, O_2 and H_2O are major exceptions. CO_2 is soluble in lipid membranes and diffuses rapidly aided by carbonic anhydrase. Membranes have low permeability to cations, many large sugars and phosphorylated intermediates of the PCR cycle (Fig. 7.10). Carbon dioxide, Cl^-, dihydroxyacetone phosphate (DHAP) and P_i penetrate most rapidly; sucrose does not permeate nor does $NADP^+$ or NADPH; ATP and ADP only enter slowly. 3PGA, which might be expected to enter readily, has only limited penetration.

When plants or algae assimilate $^{14}CO_2$ the cytosol contains labelled DHAP,

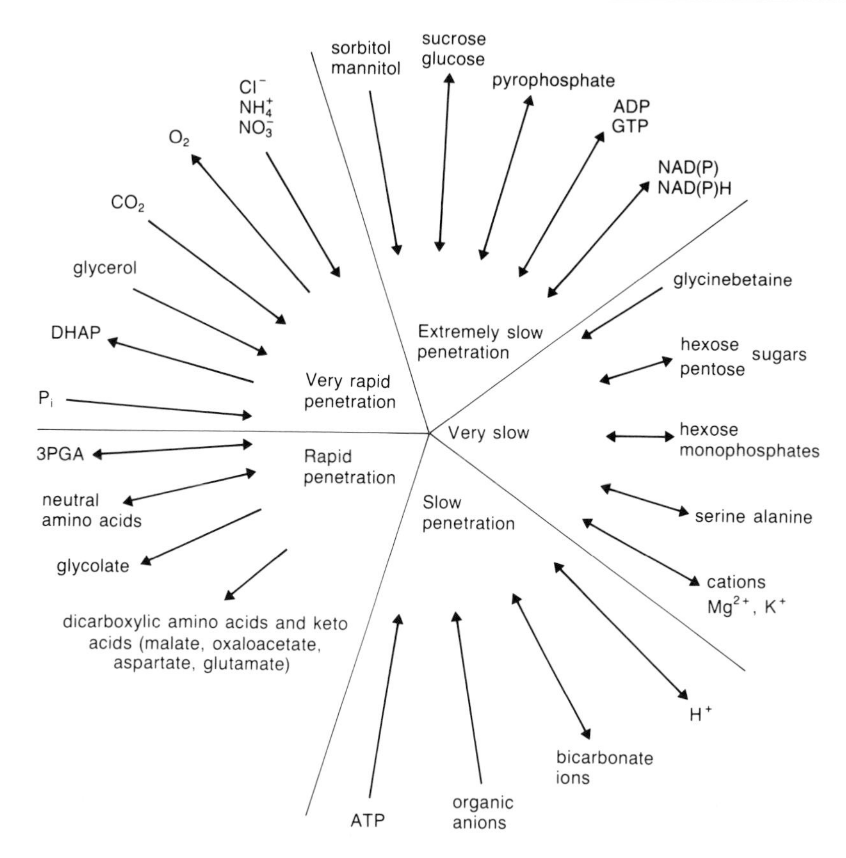

FIG. 7.10 Permeability of the intact chloroplast envelope to substances produced or consumed in photosynthesis by C3 cells, based on information in the references (see particularly Heber 1974).

3PGA, fructose- and glucose-6-phosphates, fructose bisphosphate and the nucleotide sugar uridine diphosphoglucose before labelled sucrose; the delay ('lag') is up to a minute. With longer time intervals, several amino acids are labelled. However, not all are made in the chloroplast and transferred unaltered to the cytosol, most are synthesized from triosephosphate (DHAP), which rapidly passes across the envelope. Fructose bisphosphate, for example, quickly appears outside the chloroplast but is synthesized from triosephosphate. Sucrose, the major translocated photosynthetic product, is made in the cytosol from DHAP (Ch. 8).

The phosphate and dicarboxylate translocators

DHAP is transported rapidly (16 μmol (mg chlorophyll)$^{-1}$ h^{-1}) and 3PGA at half this rate, by a phosphate translocator which exchanges P_i from the cytosol with DHAP from the chloroplast in a strict 1:1 stoichiometry; absence of one of the translocation pair stops all movement. Substrate transport is so fast that it probably does not limit assimilation. If P_i is removed from the chloroplast suspending medium, no DHAP or 3PGA is exported, but addition of P_i increases export. This mechanism prevents depletion of the phosphate and intermediates of the PCR cycle, for triosephosphate export also depletes P_i, slowing the PCR cycle and ATP synthesis. Carbon dioxide assimilation by isolated chloroplasts is increased if the translocator is blocked (for example by pyrophosphate) because depletion of PCR cycle intermediates is stopped, whereas P_i increases export of triosephosphate, and therefore depletion. Thus the translocator regulates interactions between the two compartments. The phosphate translocator is a divalent anion exchange protein, probably two 29 kD particles on the inner membrane. DHAP is divalent but 3PGA trivalent at neutral or alkaline pH; trivalent 3PGA only slowly equilibrates with the divalent form so in illuminated chloroplasts 3PGA is retained and DHAP exported and the 3PGA/triosephosphate ratio is 10 times greater in the light than in the dark. This enables 3PGA to be reduced at low NADPH/$NADP^+$ ratio and low phosphorylation potential. Active sites on the translocator are probably sulphydryl groups and lysine and arginine residues (since activity is blocked by chloromercuribenzoate and benzene sulphonates respectively). Possibly DHAP binds to the protein, changing the conformation and allowing it to accept P_i and opening up a transfer 'channel'.

Dicarboxylic acids, oxaloacetate, malate and α-ketoglutarate and the amino acids aspartate and glutamate are also translocated by a single molecule dicarboxylate transporter, located on the inner envelope, which does not work by counter-exchange as does the phosphate translocator system, so less rigid metabolic control is obtained. Little is known of the carrier, or if several forms exist, specific to a given acid.

Translocators in biological membranes are identified by saturation of the rate of transport with increasing concentration of the compound being transported. From the velocity of movement *versus* concentration, the K_m and V_{max} of the translocator are obtained. Addition of different substances shows the nature of the material transported and if there is competition for the carrier. Analogues of the substrate with related chemical structure compete for the translocator sites and enable the reaction to be analysed. All these characteristics show that an active process is involved, rather than diffusion.

Chloroplast envelopes also regulate the passage of H^+ into the medium. The pH of darkened chloroplast stroma is smaller than that of the medium due to Donnan equilibrium of H^+ with proteins but in the light the pH of the stroma increases as H^+ is pumped into the thylakoid lumen. With an acidic medium or cytosol, H^+ would enter the stroma faster than ions (for example Mg^{2+}) in the light and decrease the pH gradient required for ATP synthesis. However, the envelope has an ATPase and H^+ is pumped (slowly compared to the thylakoid pump) out of the stroma into the medium in exchange for K^+. Anything which penetrates the envelope and transports H^+ into the stroma inhibits photosynthesis, for example the nitrite anion (NO_2^-) at pH 7 gives nitrous acid (HNO_2) which enters and dissociates; H^+ remains in the stroma destroying the pH gradient and NO_2^- diffuses back to the medium and recycles H^+. Other weak acid anions, such as glycolate, decrease photosynthesis in the same way.

Shuttles of assimilate, reducing power and energy across the chloroplast envelope

Combination of the phosphate and dicarboxylate translocators and the H^+ pump provides the mechanism for distributing assimilates and energy between compartments. Pyridine nucleotides ($NADP^+$ and NADPH in the chloroplast, NAD^+ and NADH in the cytosol) cannot pass the intact envelope. ATP and other adenylates move by a translocator, but very slowly (2 μmol ATP mg chlorophyll^{-1} h^{-1}) in older leaves although faster in younger. Synthesis of sucrose from triosephosphate requires a rapid ATP supply but respiration, which may be inhibited in the light, is not a likely source and ATP is probably synthesized in the cytosol *via* a shuttle of DHAP (Fig. 7.11). A large cytosolic DHAP/PGA ratio drives the synthesis of ATP and reducing power in the

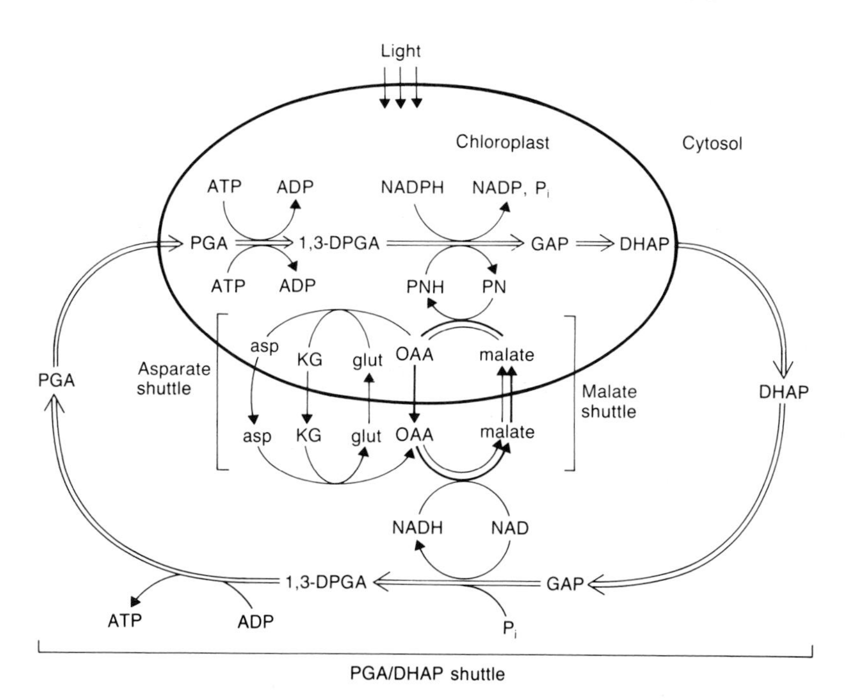

FIG. 7.11 Shuttles of dihydroxyacetone phosphate and 3-phosphoglyceric acid and of dicarboxylic acids between chloroplast and cytosol to generate ATP and NAD(P)H outside the chloroplast in the light or to provide ATP and NAD(P)H to the chloroplast in darkness. (From Heber 1975).

cytosol by triosephosphate oxidation. In the stroma, 3PGA is metabolized to DHAP and in the cytosol DHAP is converted back to 3PGA generating ATP and NADH. The reaction is the reverse of that in the stroma:

$$\text{DHAP} + \text{ADP} + \text{P}_i + \text{NAD(P)}^+ \rightleftharpoons \text{3PGA} + \text{ATP} + \text{NAD(P)H} + \text{H}^+ \qquad [7.6]$$

Exchange of ATP and pyridine nucleotides (PN) is thus directly linked to assimilate transport. Glyceraldehyde phosphate formed in the cytosol may be oxidized to 1,3-diphosphoglyceric acid without synthesis of ATP, and reduce $NADP^+$ in the cytosol before NAD^+, thus favouring reactions using NADPH. Glyceraldehyde phosphate oxidation by this mechanism proceeds before the glycolytic enzyme is activated because the latter requires much larger concentration of substrate. These mechanisms may provide flexibility in the supply of ATP and NADPH, and a mechanism to balance demands in the cytosol, for example in protein synthesis, ion exchange etc., particularly if mitochondria are inhibited.

At equilibrium the balance between stromal (s) and cytosol (c) assimilates and energy across the envelope is given by:

$$\frac{[\text{H}^+]_c}{[\text{H}^+]_s} = \frac{[\text{PGA}]_s}{[\text{PGA}]_c} \times \frac{[\text{DHAP}]_c}{[\text{DHAP}]_s} \times \left\{ \left[\frac{[\text{ATP}]}{[\text{ADP}][\text{P}_i]} \right]_s \Big/ \left[\frac{[\text{ATP}]}{[\text{ADP}][\text{P}_i]} \right]_c \right\} \times \left\{ \left[\frac{[\text{PNH}]}{[\text{PN}^+]} \right]_s \Big/ \left[\frac{[\text{NADH}]}{[\text{NAD}^+]} \right]_c \right\} \qquad [7.7]$$

Thus illumination controls the balance of the redox components by the H^+ gradient across the envelope. Decreasing stromal pH decreases the 3PGA/DHAP ratio and increases the ATP/ADP ratio. If demand for any component of the shuttle changes then re-equilibration takes place. In darkness the DHAP shuttle reverses and supplies the chloroplast with ATP and NADH, starch is metabolized to DHAP which enters the chloroplast and is converted to 3PGA thus forming ATP and NADPH.

Dicarboxylate exchange is also regulated by the proton gradient. Malate dehydrogenase (which catalyses malate oxidation and oxaloacetate (OAA) formation) occurs on both sides of the envelope and the redox potential is determined by:

$$\frac{[\text{H}^+]_c}{[\text{H}^+]_s} = \frac{[\text{OAA}]_s}{[\text{OAA}]_c} \times \frac{[\text{malate}]_c}{[\text{malate}]_s} \left[\frac{[\text{PNH}]}{[\text{PN}^+]} \right]_s \Big/ \left[\frac{[\text{NADH}]}{[\text{NAD}^+]} \right]_c \qquad [7.8]$$

The gradients of malate and oxaloacetate at equilibrium are balanced so that the ratio of:

$$\frac{[\text{OAA}]_s}{[\text{OAA}]_c} \times \frac{[\text{malate}]_c}{[\text{malate}]_s} = 1$$

and therefore:

$$\frac{[\text{H}^+]_c}{[\text{H}^+]_s} = \left[\frac{[\text{PNH}]}{[\text{PN}^+]} \right]_s \Big/ \left[\frac{\text{NADPH}}{[\text{NAD}^+]} \right]_c \qquad [7.9]$$

Coupling the H^+ gradient with the distribution of substrate and pyridine nucleotides enables electrons to flow between compartments against a gradient of reduction potential. Malate concentration is 10^3 times greater than oxaloacetate in tissues, and this inhibits transport of OAA. However transaminations with glutamate give aspartate and α-ketoglutarate so that the stromal NADP and cytosolic NAD systems are linked. The malate/aspartate ratio is larger in the light than in the dark in tissues. Together the DHAP and dicarboxylate shuttles regulate the pyridine nucleotides in both compartments. Chloroplast stroma is more reduced than the cytosol in the light and a gradient of phosphorylation potential develops from cytosol to stroma, that is, there is more ATP outside the envelope than inside, as found experimentally by non-aqueous fractionation. Of course in the dark ATP in the stroma is at low concentration but even in the light there is relatively more ATP in the cytosol than in the chloroplast, due to the action of the shuttles, despite ATP synthesis in the stroma.

C4 plants (Ch. 9) have massive flux of assimilate between the cell types during CO_2 fixation, as well as in translocation. Bundle sheath chloroplasts form 3PGA and DHAP; probably the latter is transported in exchange for P_i as in C3 plants. Large flux of malate or aspartate occurs across the cell membrane but does not involve the chloroplast envelope. Glycolate transport in C4 plants is probably very small. Mesophyll cell chloroplasts of C4 species import pyruvate and 3PGA from oxaloacetate, malate and phosphoenol pyruvate (PEP) which are exported across the envelope to the cytosol. Thus the type of compounds formed and transported are similar to C3 plants but the proportion and size of the fluxes probably differ greatly. In addition there is considerable difference in the energy (ATP) and reductant (NADPH) requirements and movement in C4 plants compared to C3, but control is obscure.

The translocators provide a ‘valve’ for regulating the flows of assimilate and energy. By coupling reactions in one compartment with those in another acting in reverse, net transport of energy and reductant is achieved without physical movement of ATP or NADPH and fine control of metabolism is possible. Also specific translocators control the fluxes of triosephosphate and P_i, and organic acids, enabling close coupling of all cellular factors. Thus transitions between dark and light are controlled; for example starch synthesis occurs in the stroma if P_i is in short supply, or in darkness starch may supply the stroma with ATP. Over-reduction of one compartment is balanced by exchange with another. However, conditions which drive metabolism too far from equilibrium cannot then be balanced and metabolism is damaged. Many interactions and transport processes in the cell are poorly understood and remain to be quantitatively evaluated.

Photosynthetic assimilation of nitrogen and sulphur

An adequate supply of reduced nitrogen compounds is required for plant growth. Light energy is used for reducing nitrate to NH_3 which is used in

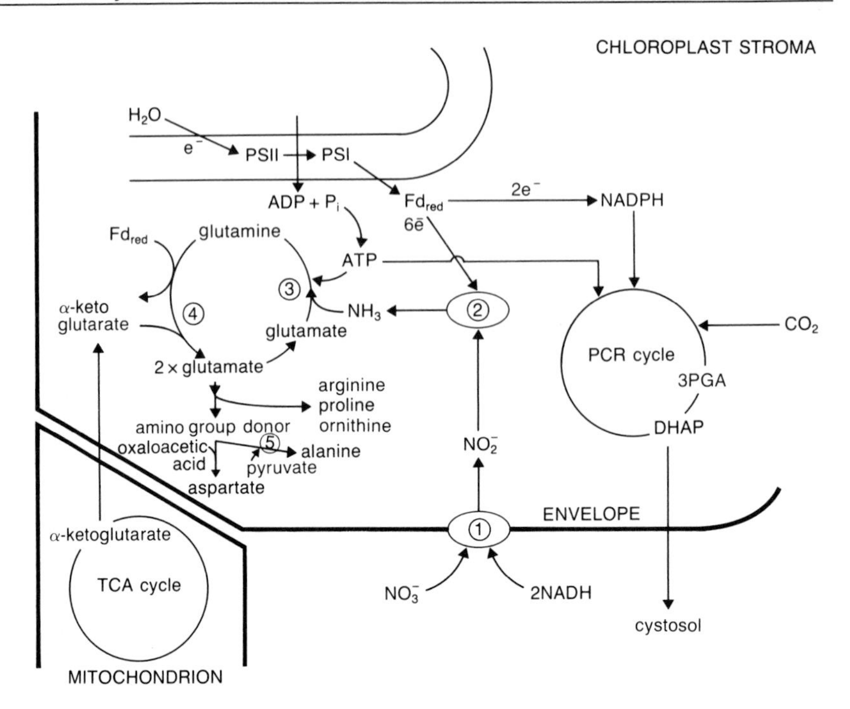

FIG. 7.12 Nitrate reduction in chloroplasts shown schematically with the glutamine/glutamate cycle of ammonia assimilation, but omitting the photorespiratory nitrogen cycle. Enzymes identified in the figure are: (1) nitrate reductase; (2) nitrite reductase; (3) glutamine synthetase; (4) GOGAT; (5) pyruvate aminotransferase.

synthesis of amino acids and proteins. Nitrate reduction in roots, and in leaves in darkness uses energy and reductant derived *via* respiration from carbohydrates and is therefore only indirectly dependent upon photosynthesis. Higher plants absorb nitrogen from their environment as the nitrate ion (NO_3^-), or to a limited extent as ammonia (NH_3) or ammonium (NH_4^+), but before nitrate is utilized for amino acid synthesis it is first reduced to nitrite (NO_2^-) and then NH_3 (Fig. 7.12):

$$NO_3^- + 2e^- + 2H^+ \xrightarrow[\text{reductase}]{\text{nitrate}} NO_2^- + H_2O \qquad [\mathbf{7.10}]$$

$$NO_2^- + 6e^- + 7H^+ \xrightarrow[\text{reductase}]{\text{nitrite}} NH_3 + 2H_2O \qquad [\mathbf{7.11}]$$

Reduction of NO_3^- to NH_3 requires 8 e^-. Nitrate reductase is a cytoplasmic enzyme and uses electrons from NADH (rather than NADPH) passing *via* FAD, cytochrome b_{557} and molybdenum. The electrons are probably supplied to NAD^+ in the cytosol from NADPH in the chloroplast by shuttle systems.

Nitrate reductase is a 200 kD molybdo-flavoprotein enzyme loosely attached to the chloroplast envelope. It is a rapidly turned-over enzyme with a half-life of some four hours. Nitrate reductase activity is controlled by the concentration of nitrate and by light. Concentration of ammonia in leaves is usually so small that it does not control activity.

Nitrite is a toxic compound and therefore must be rapidly removed when formed. Reaction [7.11] proceeds with electrons supplied from ferredoxin, not NAD(P)H, and all the intermediates are bound to the enzyme. The nitrite reductase enzyme (60 kD) is in the stroma and contains sirohaem and two additional atoms of iron and two labile sulphides per molecule. The location of nitrite reductase close to ferredoxin would ensure that nitrite reduction obtains available electrons in preference to nitrate reductase, so that toxic concentrations of NO_2^- would not be reached. In darkness the supply of electrons from respiration probably controls conversion of NO_3^- to NO_2^- and NH_3 reduction is slower than in the light. Assimilation of NO_3^- by isolated intact chloroplasts increases the OH^- ions and changes the pH of the cell; exchange of organic acids balances this.

Ammonia, the product of the nitrite reductase reaction is toxic, uncoupling ATP synthesis from electron transport in the thylakoids; it is therefore rapidly assimilated and the concentration in the leaves is at μM values. Assimilation of ammonia involves formation of the amide glutamine from glutamate:

$$\text{glutamate} + NH_3 + \text{ATP} \rightarrow \text{glutamine} + \text{ADP} + P_i + H_2O \qquad [\mathbf{7.12}]$$

The reaction is catalysed by glutamine synthetase, a protein of 360 kD probably with eight subunits, a pH optimum of 8 and requiring Mg^{2+} for activation; approximately half the enzyme activity in higher plant leaves is found in the chloroplast. Glutamate dehydrogenase (located in the mitochondria and chloroplasts) may operate at very high ammonia concentrations:

$$\alpha\text{-ketoglutarate} + NH_3 + \text{NAD(P)H} + H^+ \rightarrow \text{glutamate} + \text{NAD(P)}^+ + H_2O \qquad [\mathbf{7.13}]$$

Much evidence points to the glutamine synthetase reaction as the primary route for ammonia assimilation; for example isotopic nitrogen (^{15}N or ^{13}N) from NO_3^- accumulates first in glutamine then glutamate. When $^{14}CO_2$ is given to photosynthesizing leaves containing an analogue of glutamate to block further glutamate metabolism, ^{14}C accumulates in glutamine. Glutamine synthetase is found in many plants; it is a very efficient enzyme with a high affinity (low K_m) for NH_3 (10^{-5} M), the concentration of which is therefore kept at rather low physiological levels.

Glutamate is synthesized from glutamine, regenerating glutamate as an acceptor of NH_3, by glutamate synthetase (also called GOGAT, the acronym for the earlier name of the enzyme (glutamine (amide):2-oxoglutarate aminotransferase (oxido reductase NADP)) found in chloroplasts:

$$\text{glutamine} + \alpha\text{-ketoglutarate} + Fd_{red} + H^+ \rightarrow 2\ \text{glutamate} + Fd_{ox} \qquad [\mathbf{7.14}]$$

The dicarboxylic acid α-ketoglutarate is the 'carbon skeleton' for amino acid synthesis. Probably dicarboxylic acids are not formed within the chloroplast but are produced in the mitochondria, from carbon exported as triosephosphate (DHAP) and re-imported into the chloroplast by shuttle systems. The glutamine synthetase system links not only the chloroplast carbon and nitrate metabolism but also that of the peroxisomal and mitochondrial photorespiratory nitrogen 'cycles'. Figure 8.4 (see Ch. 8) shows the nitrogen cycle associated with photorespiration. Energy for net nitrate reduction is probably less than 5 per cent of the energy used in CO_2 assimilation. The photorespiratory nitrogen cycle is, however, very active in C3 plants; the flux is some ten-fold greater than the net rate of nitrate reduction. Considerable reductant and ATP is consumed and the photorespiratory N and C cycles together consume over 30 per cent of total available energy. Photorespiratory nitrogen metabolism is important because it decreases the efficiency with which light is used without increasing the net assimilation of nitrate. However, it may be important in regulating the energy supply in metabolism and may protect metabolism from high NH_3 and reductant levels (p. 168). Nitrate assimilation consumes ATP and reductant and is, therefore, a photosynthetic process, similar in its requirements to carbon dioxide reduction and closely linked to it.

Synthesis of alanine, aspartate, glycine and serine in photosynthesis

These amino acids are rapidly labelled with ^{14}C when leaves assimilate $^{14}CO_2$. Alanine is formed from glutamate by pyruvate aminotransferase with pyruvate derived from 3PGA *via* phosphoenol pyruvate, possibly in the chloroplasts. The enzyme for transamination of oxaloacetic acid to aspartate is in the chloroplast. Glycine and serine are synthesized by three routes in photosynthesis. In the phosphorylated pathway 3PGA is dephosphorylated to glycerate and this is converted to hydroxypyruvate which is transaminated to give serine. Another route is *via* the glycolate pathway (p. 163), glycolate giving glycine; one glycine is decarboxylated and its β-carbon added to a second glycine giving serine; the reaction is catalysed by serine hydroxymethyl transferase in the mitochondria. A third route is *via* phosphoserine. Multiple routes for serine formation probably keep the supply of amino acid balanced when conditions, such as CO_2 supply, vary.

Amino acid synthesis requires relatively higher reductant-to-ATP ratio than triose formation. One molecule of aspartate requires 11 NAD(P)H and 10 ATP whereas triose requires 6 NADPH and 9 ATP. Nitrate reduction increases electron flow and accumulation of H^+ in the thylakoids, and may increase ATP synthesis which would benefit CO_2 fixation if the PCR cycle is limited by ATP (p. 124). Thus both CO_2 and NO_3^- reduction would increase together and stimulate assimilation.

Control of nitrate and ammonia assimilation

Nitrate reduction is controlled by nitrate reductase which is a rapidly synthesized and inactivated enzyme so nitrate metabolism is potentially sensitive to factors which affect protein synthesis. Should conditions become unsuitable, then enzyme synthesis slows and also ammonia production. A connection between NO_3^- reduction and the production of carbon skeletons would be provided by ATP or energy charge. Inadequate CO_2 supply would lead to a shortage of precursors from the PCR cycle for glutamate synthesis and other sources of carbon would be mobilized, perhaps starch. As ATP is required for glutamine and protein synthesis, shortage of ATP would inhibit both nitrate assimilation and protein synthesis as well as the PCR cycle. Change in the amount of enzyme may provide longer term control than regulation by nitrate concentration or reductant supply on the rate of the reaction. Nitrate reductase synthesis is stimulated by NO_3^-, enabling the plant to respond rapidly to NO_3^- availability. Another control point in NO_3^- reduction is at the transport of nitrate to the assimilatory sites. Concentration of Mg^{2+} and pH and energy charge may regulate glutamine synthetase; light may stimulate enzyme activity by mechanisms similar to those of the thioredoxin system on PCR cycle enzymes (see p. 138). Glutamate synthase does not appear to be inhibited by end products of the reaction. Light activation and conditions in the chloroplast would provide co-ordination between PCR cycle, NO_3^- reduction, light reactions and demand for products from secondary metabolism.

Dinitrogen fixation and photosynthesis

Higher plants cannot photosynthetically assimilate gaseous dinitrogen (N_2) because they lack a nitrogenase enzyme. However, photosynthetic bacteria and blue-green algae (e.g. *Nostoc*) reduce N_2 to ammonia. The nitrogenase is inhibited or destroyed by very small O_2 concentrations so that in the blue-green algae N_2 assimilation proceeds in heterocysts, cells with thick walls which are impermeable to oxygen and only have PSI activity so no O_2 is produced by water splitting. The electrons for reduction of N_2 (eqn 7.6) come from ferredoxin and are derived from outside the heterocysts, as they cannot be supplied from water. Sugars are transported into the heterocysts and metabolized to give electrons. Cyclic photophosphorylation probably provides ATP, the requirement for which may be as high as 15 molecules per N_2 because the N_2 molecule is particularly inert.

$$N_2 + 6\,e^- + 12\,ATP + 6\,H^+ \rightarrow 2\,NH_3 + 12\,ADP + 12\,P_i \qquad [7.15]$$

Hydrogen production during photosynthesis

Gaseous hydrogen (H_2) is produced by heterocysts in blue-green algae, some photosynthetic bacteria and primitive eukaryotic algae. Anaerobic environments are essential for the hydrogenase function. Nitrogenase also possesses hydrogenase activity and H_2 can be used as a source of electrons for N_2 reduction *via* ferredoxin. Electrons from PSI are supplied to H^+:

$$2\,H^+ + 2\,e^- \xrightarrow[\text{hydrogenase}]{\text{light}} H_2 \qquad [7.16]$$

No ATP is required and electrons come from reduced organic substrates, such as NADH or glucose, as water cannot be used.

Photosynthetic sulphur metabolism

Sulphur, absorbed as the sulphate ion and reduced in leaves by electrons from electron transport and ATP, is incorporated into the amino acids cysteine and methionine, and into sulphydryl groups of coenzymes and sulpholipids. Reduction of SO_4^{2-} to sulphite and sulphide and incorporation into cysteine in the chloroplasts (Fig. 7.13) is strongly stimulated by light. Sulphate is activated in two stages both requiring ATP; first SO_4^{2-} reacts with ATP giving adenosine phosphosulphate (APS). The enzyme, ATP: sulphate adenylytransferase (called ATP sulphurylase) which occurs in all organisms able to reduce sulphate, has several subunits, a broad alkaline pH optimum and requires Mg^{2+} (characteristic of chloroplast stromal enzymes). In the second stage, activation of APS by ATP gives phosphoadenosine phosphosulphate (PAPS) in an irreversible reaction catalysed by a specific transferase enzyme (APS kinase), possibly activated in the light by the ferredoxin–thioredoxin system (see p. 138). The enzyme may be activated by Mg^{2+} and alkaline pH and is bound to the thylakoid membrane. The ATP required for activation is generated by the chloroplast. Activated sulphate is transferred to a carrier protein thio group (X-S) forming a protein sulphite complex ($XS\text{-}SO_3H$) which is further reduced, by six electrons from ferredoxin, to bound sulphide. The enzyme, thiosulphonate reductase, also called sulphite reductase, is bound to

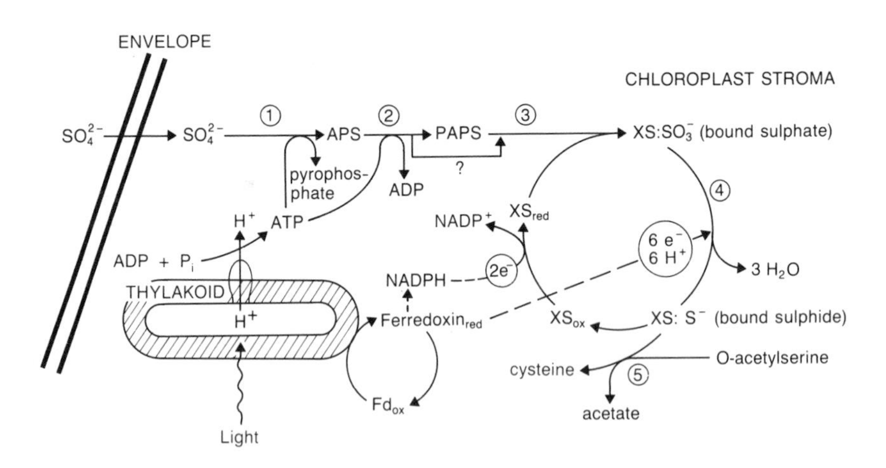

FIG. 7.13 Photosynthetic assimilation of sulphate in chloroplasts, shown schematically, with enzymes: (1) ATP sulphurylase; (2) APS kinase; (3) APS sulphotransferase; (4) thiosulphonate reductase; (5) cysteine synthase. ? denotes uncertainty of PAPS involvement in higher plant SO_4^{2-} reduction.

the thylakoid membrane. Reaction between bound sulphide and a serine derivative, O-acetylserine, may be the route of cysteine synthesis in chloroplasts. The protein carrier is oxidized and cysteine and acetate are formed. However, the protein carrier must be reduced to accept the SO_3^- from PAPS and this is achieved by two electrons from NADPH.

Thus SO_4^{2-} reduction is entirely chloroplastic; it consumes two ATP molecules and eight electrons, and is similar to reduction of NO_3^- with a six electron transfer reaction requiring the sirohaem prosthetic group of the respective reductases coupled to thylakoid membrane electron transport and ferredoxin.

Chloroplast shrinking and swelling

Chloroplasts are osmometers, allowing free and rapid movement of water across the envelope but excluding many solutes such as sugars (e.g. sucrose), polyhydroxyalcohols (e.g. mannitol), neutral amino acids and monovalent salts. However the 5-carbon sugar ribose, the lower polyhydroxyalcohol glycerol, and urea enter chloroplasts causing them to rupture due to the influx of water. Chloroplasts with intact envelopes are isolated in isotonic or hypertonic sugar solutions to prevent swelling and rupture; 0.4 M sucrose (buffered against pH changes) is a favoured medium, followed by centrifugation and resuspension of the chloroplasts in the same sucrose solution or sorbitol.

Chloroplast volume is greater in darkness than in light. Illumination decreases the thickness of the chloroplast from about 2.7 to 2.2×10^{-6} m and volume by 20–40 per cent; the half-time for the changes is about 3 minutes and they are fully reversible in darkness. As stromal volume is smaller in light, concentration of metabolites and enzyme active sites increases (provided there is no inhibition by increased concentrations of ions). Light causes extrusion of $K^+ + Cl^-$ from the stroma to the cytosol and water follows osmotically. Ion exchange and volume changes are linked to photophosphorylation; inhibitors of electron transport and phosphorylation stop chloroplast volume changes.

The volume of the thylakoid lumen decreases by 20 per cent in the light (because of water efflux associated with changes in ionic balance), H^+ moves into the lumen and Mg^{2+} and K^+ are extruded to maintain ionic balance. Some 300 nmoles Mg^{2+} per mg chlorophyll enter the stroma, an increase to 13 mM. Thylakoid membranes become thinner by 20–30 per cent possibly because H^+ binds to proteins causing the membrane to become more hydrophobic and 'shrink'. This may influence electron transport and the efficiency of light absorption because the spatial arrangements between molecules are modified. Uncouplers (see p. 121) of phosphorylation, like NH_3, cause massive influx of water into the lumen and thylakoid swelling.

Chloroplast volume changes can be demonstrated *in vitro* by a particle volume counter, such as a Coulter counter, and in the living cells by measuring the scatter of a light beam (of wavelength little absorbed by chlorophyll, e.g. 546 nm) passed through leaves. Chloroplast volume changes are therefore a result of passive water movement following fluxes of ions into or out of the thylakoid and stromal compartments due to electron transport.

Plants when desiccated lose water, and the osmotic concentration of photosynthetic cells and organelles increases. Chloroplasts do not function efficiently when desiccated. Thylakoid membranes are probably not damaged by mild desiccation for electron transport continues, but ATP synthesis is inhibited; it is currently thought that the increased osmotic and ionic concentration of the chloroplast stroma causes the disorganization of the subunits of CF_1. Although osmotic changes undoubtedly influence other processes in chloroplasts, there is little effect on enzyme activity, for example of RuBP carboxylase/oxygenase. Other environmental conditions may influence plant growth by changing chloroplast osmotic volume or ionic balance, for example plants not adapted to saline soils (glycophytes) grow poorly with salts because the osmotic and ionic concentration of their cells change, preventing the efficient functioning of thylakoids and chloroplasts.

References and Further Reading

Akazawa, T. (1979) Ribulose-1,5-bisphosphate carboxylase, pp. 208–99 in Gibbs, M. and Latzko, E. (eds), *Encyclopedia of Plant Physiology* (N.S.), vol. 6, *Photosynthesis II*, Springer-Verlag, Berlin.

Anderson, L. E. (1979) Interaction between photochemistry and activity of enzymes, pp. 271–81 in Gibbs, M. and Latzko, E. (eds), *Encyclopedia of Plant Physiology* (N.S.), vol. 6, *Photosynthesis II*, Springer-Verlag, Berlin.

Bassham, J. A. (1965) Photosynthesis: the path of carbon, pp. 875–902 in Bonner, J. and Varner, J. E. (eds), *Plant Biochemistry*, Academic Press, New York.

Bassham, J. A. (1979) Regulation of photosynthetic carbon metabolism and partitioning of photosynthate, pp. 139–59 in Atkinson, D. E. and Fox, C. F. (eds), *Modulation of Protein Function*, Academic Press, New York.

Bassham, J. A. (1979) The reductive pentose phosphate cycle and its regulation, pp. 9–30 in Gibbs, M. and Latzko, E. (eds), *Encyclopedia of Plant Physiology* (N.S.), vol. 6, *Photosynthesis II*, Springer-Verlag, Berlin.

Beck, E. (1979) Glycolate synthesis, pp. 327–37 in Gibbs, M. and Latzko, E. (eds), *Encyclopedia of Plant Physiology* (N.S.), vol. 6, *Photosynthesis II*, Springer-Verlag, Berlin.

Buchanan, B. B. (1979) Thioredoxin and enzyme regulation in photosynthesis, pp. 93–111 in Atkinson, D. E. and Fox, C. F. (eds), *Modulation of Protein Function*, Academic Press, New York.

Buchanan, B. B. (1980) Role of light in the regulation of chloroplast enzymes, *A. Rev. Plant Physiol.*, **31**, 341–74.

Calvin, M. and **Bassham, J. A.** (1962) *The Photosynthesis of Carbon Compounds*, Benjamin, New York.

Champigny, M.-L. (1978) Adenine nucleotides and the control of photosynthetic activities, pp. 479–88 in Hall, D. O., Coombs, J. and Goodman, T. W. (eds), *Proceedings of the 4th International Congress on Photosynthesis*, The Biochemical Society, London.

Cooper, T. G., Filmer, D., Wishnick, M. and **Lane, M. D.** (1969) The active species of 'CO_2' utilised by ribulose diphosphate carboxylase, *J. Biol. Chem.*, **244**, 1081–83.

Delieu, T. and **Walker, D. A.** (1981) Polarographic measurement of photosynthetic oxygen evolution by leaf discs, *New Phytol.*, **89**, 165–78.

Hatch, M. D. (1976) Photosynthesis: the path of carbon, pp. 797–845 in Bonner, J. and Varner, J. E. (eds), *Plant Biochemistry* (3rd edn), Academic Press, New York.

Heber, U. (1974) Metabolite exchange between chloroplasts and cytoplasm, *A. Rev. Plant Physiol.*, **25**, 393–421.

Heber, U. (1975) Energy transfer within leaf cells in Avron, M. (ed.), *Proceedings of the Third International Congress in Photosynthesis,* ***Rehovot,*** Elsevier Biomedical Press, Amsterdam.

Heldt, H. W. (1979) Light dependent changes of stromal H^+ and Mg^{2+} concentrations controlling CO_2 fixation, pp. 202–7 in Gibbs, M. and Latzko, E. (eds), *Encyclopedia of Plant Physiology* (N.S.), vol. 6, *Photosynthesis II*, Springer-Verlag, Berlin.

Jensen, R. G. (1980) Biochemistry of the chloroplast, pp. 274–313 in Tolbert, N. E. (ed.), *The Biochemistry of Plants,* vol. 1, *The Plant Cell*, Academic Press, New York.

Jensen, R. G. and **Bahr, J. T.** (1977) Ribulose 1,5-bisphosphate carboxylase oxygenase, *A. Rev. Plant Physiol.*, **28**, 379–400.

Kelly, G. J. (1978) Aspects of enzyme regulation illustrated by the properties of two-phosphofructokinases from spinach leaves, pp. 437–46 in Hall, D. O., Coombs, J. and Goodwin, T. W. (eds), *Proceedings of the 4th International Congress on Photosynthesis 1977*, The Biochemical Society, London.

Kelly, G. J., Latzko, E. and **Gibbs, M.** (1976) Regulatory aspects of photosynthetic carbon metabolism, *A. Rev. Plant Physiol.*, **27**, 181–205.

Lane, M. D. and **Miziorko, H. M.** (1978) Mechanism of action of ribulose bisphosphate carboxylate/oxygenase, pp. 19–40 in Siegelman, H. W. and Hind, G. (eds), *Photosynthetic Carbon Assimilation*, Plenum, New York.

Latzko, E. and **Kelly, G. J.** (1979) Enzymes of the reductive pentose phosphate cycle, pp. 239–50 in Gibbs, M. and Latzko, E. (eds), *Encyclopedia of Plant Physiology* (N.S.), vol. 6, *Photosynthesis II*, Springer-Verlag, Berlin.

Lorimer, G. H. (1981) The carboxylation and oxygenation of ribulose 1,5-bisphosphate: the primary events in photosynthesis and photorespiration, *A. Rev. Plant Physiol.*, **32**, 349–83.

Lorimer, G. H., Badger, M. R. and **Heldt, H. W.** (1978) The activation of ribulose-1,5-bisphosphate carboxylase/oxygenase, pp. 283–306 in Siegelman, H. W. and Hind, G. (eds), *Photosynthetic Carbon Assimilation*, Plenum, New York.

Miziorko, H. M. and **Lorimer, G. H.** (1983) Ribulose-1,5-bisphosphate carboxylase/oxygenase, *A. Rev. Biochem.*, **52**, 507–35.

Ogren, W. L. (1978) Increasing carbon fixation by crop plants, pp. 721–33 in Hall, D. O., Coombs, J. and Goodwin, T. W. (eds), *Proceedings of the 4th International Congress on Photosynthesis*, The Biochemical Society, London.

Paech, C., McCurry, S. D., Pierce, J. and **Tolbert, N. E.** (1978) Active site of ribulose-1,5-bisphosphate carboxylase/oxygenase, pp. 227–43 in Siegelman, H. W. and Hind, G. (eds), *Photosynthetic Carbon Assimilation*, Plenum, New York.

Rabinowitch, E. I. (1945) *Photosynthesis*, Interscience Publishers Inc., New York.

Robinson, S. P. and **Walker, D. A.** (1981) Photosynthetic carbon reduction cycle, pp. 193–236 in Hatch, M. D. and Boardman, N. K. (eds), *The Biochemistry of Photosynthesis*, vol. 8, *Photosynthesis*, Academic Press, New York.

Wilson, A. T. and **Calvin, M.** (1955) The photosynthetic cycle. CO_2 dependent transients, *J. Amer. Chem. Soc.*, **77**, 5948–57.

CHAPTER *8*

Metabolism of photosynthetic products

All the carbon for plant metabolism is provided, ultimately, by the PCR cycle, so it is difficult to decide where photosynthesis stops and 'secondary metabolism' starts. Purists may argue that the PCR cycle is the limit of photosynthesis and events that consume its products are 'secondary metabolism' and not photosynthesis. The argument is semantic although for convenience some limit must be drawn. Only some secondary processes are discussed because they directly consume products of the light reactions or of the PCR cycle.

In the steady state, carbon assimilated in excess of that needed to regenerate RuBP is used for synthesis of materials within the chloroplast, for example starch and PCR cycle products (particularly when the export of carbon is slow), lipids, amino acids and proteins coded in the chloroplast genome, particularly in young leaves. Mature leaves quickly export much of the assimilated carbon from the chloroplast as dihydroxyacetone phosphate (DHAP) and glycolate which are metabolized in the cytoplasm of C3 plants; there are considerable fluxes of organic acids in C4 plants. Dihydroxyacetone phosphate and glycolate are used, ultimately, in the synthesis of sucrose, the main product of photosynthesis, which is translocated throughout the plant (Fig. 8.1).

Starch synthesis

Starch accumulates in the chloroplast stroma of many species of plants during illumination, forming large granules. During darkness it is remobilized and consumed in respiration. Starch synthesis is restricted to illuminated chloroplasts, as the ability of leaves to form 'starch prints' shows. Starch prints are made by strongly illuminating a leaf (e.g. *Pelargonium*) through a photographic negative for several hours in air. The leaf is killed, chlorophyll removed and the starch stained blue-black with alcoholic potassium iodide. Only illuminated areas of the leaf (light areas of the negative) are deeply stained. The negative is reproduced in great detail with sharp distinction between chloroplasts from illuminated and darkened areas.

Starch is synthesized from fructose-6-phosphate (F6P) of the PCR cycle (Fig. 8.2) which is converted by hexose phosphate isomerase to glucose-6-phosphate (G6P) in a reaction of very small free energy change so that it is not a control step. Production of glucose-1-phosphate (G1P) from G6P by phosphoglucomutase is reversible and gives a ratio of G6P to G1P of about 20. Removal of G1P by subsequent reactions encourages synthesis of more G1P.

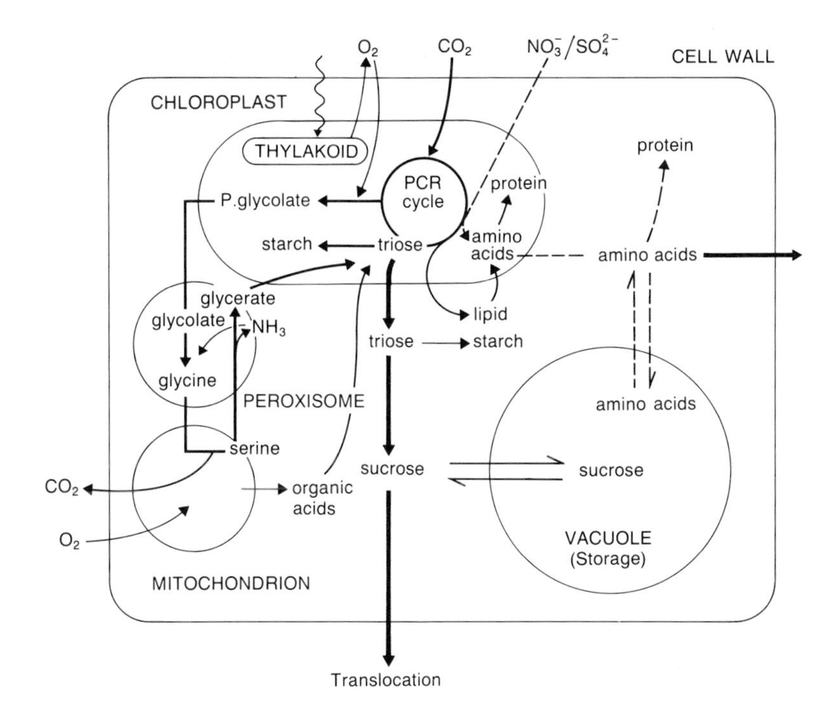

FIG. 8.1 Principal fluxes of carbon and their relation to nitrogen and sulphur assimilation in C3 plant leaf cells in the light.

As with many processes in carbohydrate metabolism, nucleotide sugars are involved in the formation of the starch polymer:

$$\text{ATP} + \text{glucose-1-phosphate} \rightarrow \text{ADP-glucose} + \text{PP}_i \quad \textbf{[8.1a]}$$

$$\text{ADP-glucose} + (\text{glucose})_n \rightarrow (\text{glucose})_{n+1} + \text{ADP} \quad \textbf{[8.1b]}$$

where $(\text{glucose})_n$ is a preformed polymer (an α-1,4-glucan primer) to which glucose residues are added.

Reaction [8.1a] is catalysed by the enzyme ADP-glucose pyrophosphorylase which regulates carbon flow; it has a molecular mass of 210 kD and the enzyme from bacteria has four similar subunits each of molecular weight about 50 kD. There is a lysine residue at the active site where two molecules of Mg-ATP bind to the four subunits and then four G1P molecules. The enzyme is controlled allosterically by products of the PCR cycle. The rate is increased 10–20 times by increased concentration of 3PGA and less so by F6P. These accumulate if in excess of that needed to regenerate RuBP, and 'signal' the enzyme to proceed with starch synthesis. If 3PGA is depleted, for example with low rates of photosynthesis or in darkness, the enzyme is inhibited. In addition ADP-glucose pyrophosphorylase is allosterically inhibited by P_i which interacts with 3PGA; a high ratio of 3PGA/P_i stimulates starch

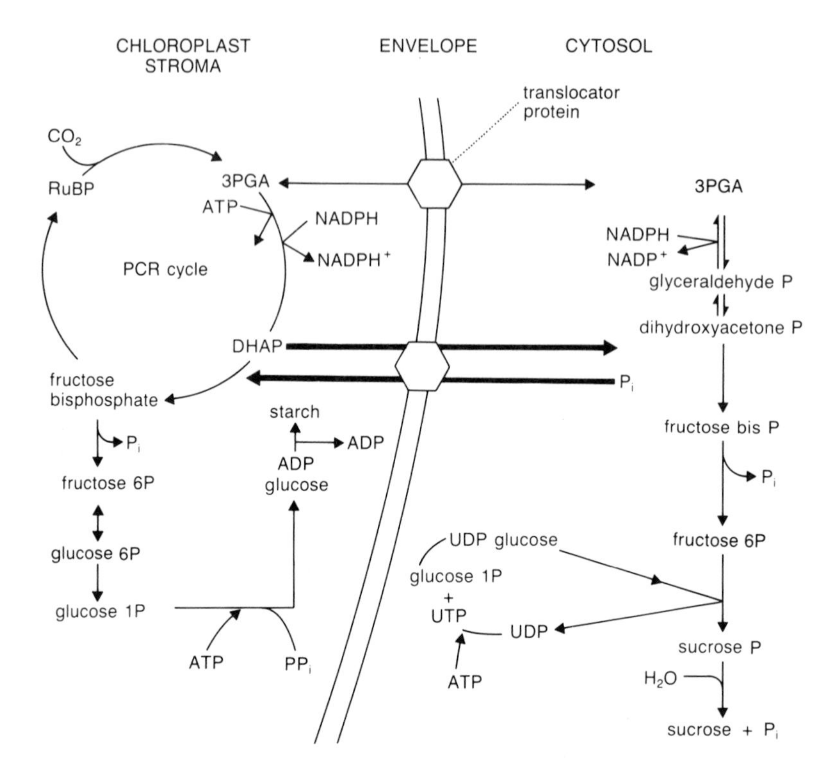

FIG. 8.2 Scheme of carbon metabolism in the chloroplast and cytosol in the light leading to starch and sucrose synthesis.

synthesis. When ATP synthesis is insufficient, 3PGA decreases and P_i accumulates, inhibiting starch synthesis and preventing PCR cycle depletion. If consumption of carbon outside the chloroplast is rapid then low 3PGA and increased P_i concentrations slow starch synthesis. Sequestering (binding) P_i into metabolites, for example by feeding leaves with mannose which is phosphorylated by hexokinase but not re-metabolized, decreases P_i concentration and stimulates synthesis of starch. A doubling of the P_i concentration from 1 mM in the medium of isolated intact chloroplasts increases the 3PGA/P_i ratio in the stroma ten-fold and the rate of starch synthesis forty-fold. Although starch synthesis liberates P_i, rate of the reaction is inadequate to maintain rapid photosynthesis but allows some CO_2 fixation to continue. Plants deprived of phosphate contain much starch which may be a symptom of nutritional inadequacy. Remobilization of starch may proceed *via* G1P (catalysed by phosphorylase) and G6P, F6P and fructose bisphosphate to DHAP. ATP is required but may be inadequate in the dark.

Sucrose synthesis

Sucrose is made by the dephosphorylation (by sucrose phosphate phosphatase, eqn 8.3) of sucrose phosphate, which is synthesized from uridine diphosphate glucose (UDPG) and fructose-6-phosphate by sucrose phosphate synthetase (systematic name; UDP-D-glucose:D-fructose-6-phosphate 2-glucosyltransferase):

$$\text{UDPG} + \text{fructose-6-phosphate} \xrightarrow[\text{synthetase}]{\text{sucrose phosphate}} \text{UDP} + \text{sucrose-6-phosphate} \quad [8.2]$$

$$\text{sucrose-6-phosphate} + H_2O \xrightarrow[\text{phosphatase}]{\text{sucrose phosphate}} \text{sucrose} + P_i \quad [8.3]$$

Sucrose is also made from UDPG and fructose by sucrose synthase but this enzyme is not believed to be important in synthesis of sucrose in leaves. It is now generally accepted that sucrose synthesis proceeds in the cytoplasm (possibly the enzymes are loosely bound to the chloroplast), for there is delay in ^{14}C labelling after exposure to $^{14}CO_2$ and separation of chloroplasts from the cytosol shows UDPG pyrophosphorylase (required for UDPG synthesis) to be cytoplasmic. Sucrose phosphate synthase is not pH-controlled, but is regulated by UDPG and F6P and the reaction rate increases rapidly as the concentration

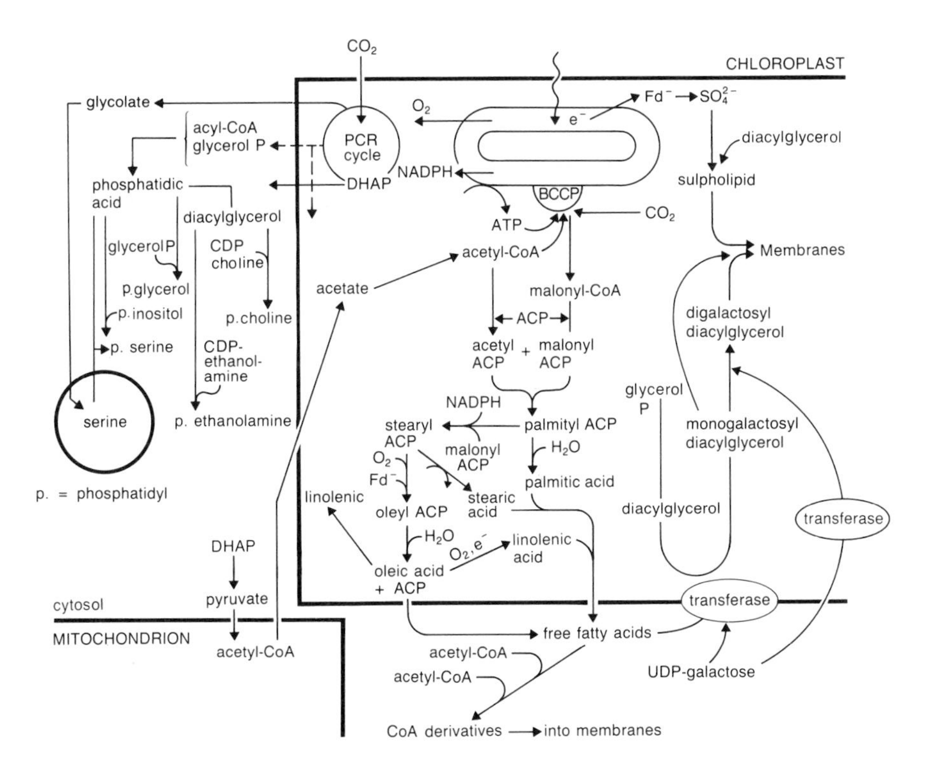

FIG. 8.3 Fatty acid synthesis in leaf cells requires products of the light reactions and the PCR cycle and co-operation between organelles.

increases. Free UDP, sucrose and other sugars stimulate sucrose phosphate phosphatase whilst sucrose phosphate synthetase and sucrose phosphate phosphatase may be regulated by Mg^{2+}. Translocation of sucrose out of the cell is beyond the discussion of photosynthesis although if export is slow, there may be feedback inhibition of photosynthesis, as the storage capacity of leaf cells is reached and intermediates accumulate.

Biosynthesis of chloroplast lipids

One third of the chloroplast dry mass is lipid, mainly synthesized during the development of the leaf, although new synthesis and replacement of lipids continues in mature leaves. Synthesis of fatty acids (Fig. 8.3) such as palmitic, oleic and linoleic acids, consumes PCR cycle products, reductant as ferredoxin and NADPH, and ATP. The light reactions, both PSII and PSI, are essential. Synthesis occurs in the stroma where all the required enzymes are found, but involves mitochondria and cytosol. Dihydroxyacetone phosphate moves from the stroma to the cytosol. Pyruvate is formed from the triosephosphate and enters the mitochondria where acetyl-CoA and acetate are produced. Acetate returns to the chloroplasts where acetyl-CoA is again made by a synthetase; it joins to an acyl carrier protein (ACP) that is found only in the stroma and is involved in most of the fatty acid synthesizing reactions.

Malonyl-CoA is also produced in the stroma by carboxylation of acetyl-CoA with CO_2 by the enzyme acetyl-CoA carboxylase (transcarboxylase) which requires Mg^{2+}. ATP activates the enzyme complex to form a CO_2-biotin carboxylase carrier protein (BCCP) which reacts with acetyl-CoA. BCCP is bound to the thylakoid lamellae, although the reason for such close proximity to the light reactions is not known. Malonyl-CoA is transferred to ACP giving malonyl-ACP. Condensation of 1 acetyl-ACP and 7 malonyl-ACP followed by reduction with NADPH and NADH gives palmityl-ACP, which is hydrolysed to palmitic acid or is condensed with more malonyl-ACP and reduced with NADPH to stearyl-ACP. This increases the chain length of the fatty acids from C 16:0 to C 18:0. Stearyl-ACP is reduced (by ferredoxin in the presence of O_2) to oleyl-ACP from which oleic acid is released by hydrolysis. Further modification of the fatty acids occurs in the chloroplast. Palmitoyl- and oleoyl-ACPs may be hydrolysed on the envelope and enter the cytosol for lipid synthesis. Fatty acids may not be formed directly from PCR cycle products and turnover of membrane lipids is slow, for only several hours after exposure to $^{14}CO_2$ is label found in them.

Synthesis of complex lipids occurs in several parts of the cell. Glycolipids, of which mono- (MGDG) and digalactosyl diacylglyceride (DGDG) are the major chloroplast lipids, are synthesized in the chloroplast. MGDG is synthesized by a transferase enzyme bound to the chloroplast envelope which attaches a galactose moiety from UDP-galactose to diacylglycerol, whilst DGDG synthesis is by a soluble, stromal enzyme catalysing transfer of two galactosyl residues from UDP-galactose to a monoacyl-glycerol. Sulpholipid synthesis (e.g. sulphoquinovosyl diglyceride) is linked to SO_4^{2-} reduction in the chloroplast.

Phosphatidyl glycerol is a major glycerophosphatide of leaves, constituting over 20 per cent of the total lipid of thylakoids and envelope membranes. Synthesis is from glycerol phosphate, produced by phosphorylation of glycerol, acylated by acyl-CoA to phosphatidic acid. Phosphatidic acid is dephosphorylated to form the diacylglycerols. Membrane glycerophosphatides include phosphatidylcholine and phosphatidylethanolamine, which require choline and ethanolamine derived from glycine and serine, possibly from the glycolate pathway. Phosphatidylinositol is formed from glycerol phosphate in the mitochondria, where phosphatidylglycerol is formed.

This brief summary of fatty acid and lipid synthesis serves to show the complex interaction between the light reactions, carbon assimilation and co-operation between cell organelles. Lipid synthesis is a photosynthetic process requiring much energy; 17 and 8 moles of ATP and NADPH respectively are needed for a C18 fatty acid. Regulation of lipid synthesis in relation to other chloroplast functions is not well understood.

Glycolate pathway and photorespiration

As a consequence of the RuBP oxygenase reaction (p. 133) phosphoglycolate is formed in the chloroplast stroma; it is dephosphorylated (Fig. 8.4) by phosphoglycolate phosphatase (optimum activity at alkaline pH) in the stroma or at the chloroplast envelope and P_i remains in the chloroplast. The RuBP oxygenase reaction is the main source of glycolate in chloroplasts. Glycolate may also come from the PCR cycle, glycoaldehyde–thyamine pyrophosphate complex reacting with photosynthetically-produced H_2O_2, but only slow rates of glycolate synthesis by this route have been found.

When algae form glycolate, they may excrete it into the medium, thus wasting the assimilated carbon and energy used in its production. In higher plants, glycolate is transported, possibly by a translocator on the chloroplast envelope, to the peroxisomes. There it is metabolized to glyoxylate which is transaminated to glycine by glutamate-glyoxylate aminotransferase with glutamate as amino-group donor. In the glycolate to glyoxylate reaction O_2 is taken up and H_2O_2 formed, which is destroyed by catalase before it can oxidize the photosynthetic apparatus. Glyoxylate can be converted to formic acid, and to oxalic acid which accumulates in many plants as insoluble crystals of calcium oxalate. Glycine is transported to the mitochondria where one molecule of glycine is decarboxylated and the 1-carbon fragment and one glycine are condensed to serine (by glycine decarboxylase and serine hydroxymethyltransferase respectively) and CO_2 is released and escapes from the mitochondrion; this is the CO_2 of photorespiration.

$$2\,\text{glycine} + H_2O \rightarrow \text{serine} + CO_2 + NH_3 + 2\,H^+ + 2\,e^- \quad [8.4]$$

Also, NH_3 is released; it reacts in the cytosol and possibly chloroplast with glutamate to give glutamine.

Electrons from the reaction can enter the mitochondrial electron transport chain and, *via* oxidative phosphorylation, synthesize three molecules of ATP

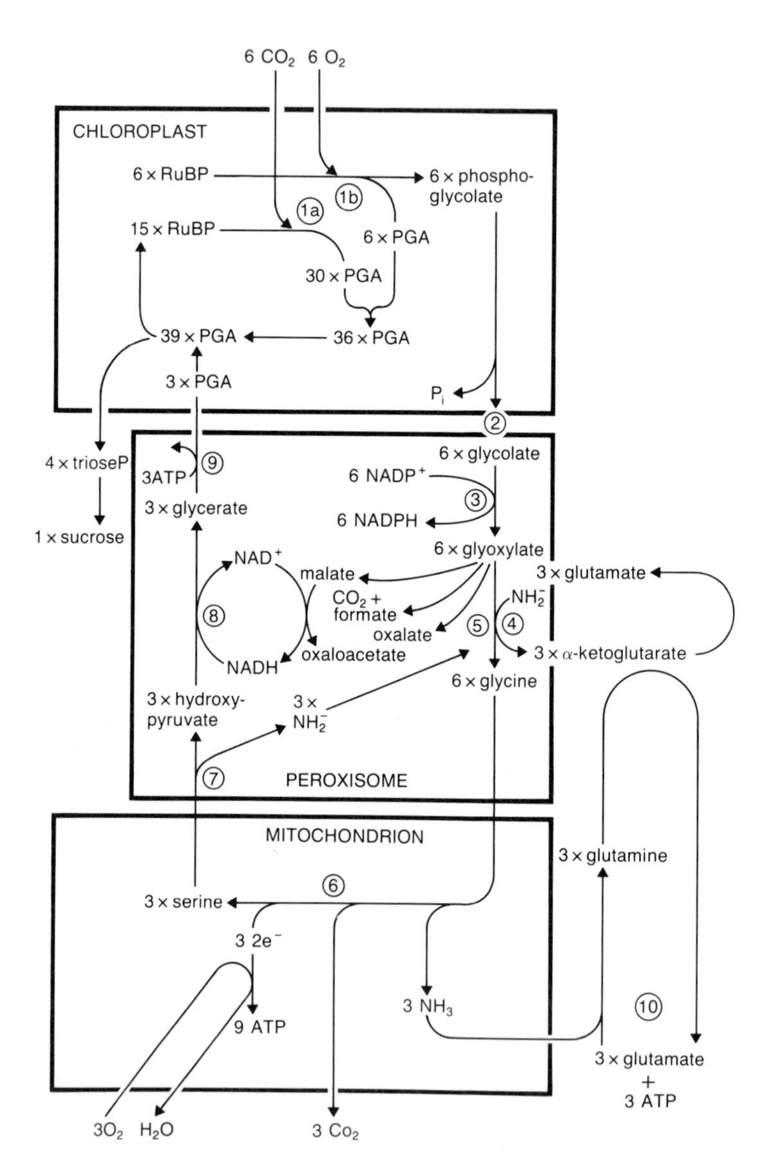

FIG. 8.4 Carbon, nitrogen and energy flows in the glycolate pathway and production of CO_2 when photorespiration is one-quarter of the rate of net photosynthesis. Numbers refer to the reactions listed below.

Enzymes of the glycolate pathway:
(1a) ribulose bisphosphate carboxylase; (1b) RuBP oxygenase; (2) phosphoglycolate phosphatase; (3) glycolate oxidase; (4) glutamate-glyoxylate aminotransferase; (5) serine-glyoxylate aminotransferase; (6) serine hydroxymethyl transferase (and glycine decarboxylase); (7) transaminase; (8) hydroxypyruvate reductase; (9) glycerate kinase; (10) glutamine synthetase, GOGAT

Important formulae:

Phosphoglycolate	CH_2OP COOH	Glycine	CH_2NH_2 COOH
Glycolate	CH_2OH COOH	Serine	CH_2OH $CHNH_2$ COOH
Glyoxylate	CHO COOH	Hydroxypyruvate	CH_2OH CO COOH
Glycerate	CH_2OH CHOH COOH	Phosphoglycerate	CH_2OP CHOH COOH

per two molecules of glycine decarboxylated. One ATP is needed to form glutamine in the cytosol and two ATP enter metabolism for sucrose synthesis, etc. Alternatively the electrons may be used in the reduction of oxaloacetate to malate, thereby establishing a shuttle exchange of reducing power with the cytosol.

Serine from the mitochondrial reaction is transferred into the peroxisomes, where it is deaminated by serine-glyoxylate aminotransferase. The resultant hydroxypyruvate is reduced by NADH to glycerate. In turn NAD^+ may be reduced by aspartate, malate and oxaloacetate. Glycerate is phosphorylated to 3PGA in the cytosol or, more probably, chloroplast. The flux of carbon through the pathway consumes NADPH (or NADH) derived from the light reactions.

Photorespiration

Photorespiration is important as an oxidative process consuming PCR cycle products and energy, and has been much studied in the last 20 years. Stoichiometry of the pathway and of glycine decarboxylation is such that for every two molecules (i.e. four carbons (4C)) of glycolate entering the peroxisomes one CO_2 (one carbon atom), or 25 per cent of the carbon is lost when two molecules of glycine (4C) are converted to serine (3C). The carbon flux to glycine depends on the ratio of RuBP oxygenase to RuBP carboxylase reactions (α) (Ch. 7, p. 136) and the O_2/CO_2 ratio. With CO_2 shortage, for example when stomata are closed because of water stress, the flux of carbon through the glycolate pathway increases relative to that through the PCR cycle (Fig. 8.6), and a greater proportion of newly assimilated CO_2 is lost by photorespiration. A simple model of the system (see Fig. 11.7) uses the characteristics of the RuBP carboxylase/oxygenase enzyme in relation to O_2 and CO_2 concentration. When gross photosynthesis and photorespiration are equal (in leaves at the CO_2 compensation concentration) α is 2 and the flux into the glycolate pathway is four times that in photosynthesis. Photorespiration and glycolate pathway metabolism are a drain on photosynthetic CO_2 fixation, and use extra ATP, despite ATP synthesis in the mitochondria, and NADPH. Table 8.1 shows

Table 8.1 ATP and NADPH requirements for carbon dioxide assimilation with different proportion of photorespiration (R_l) to gross photosynthesis (P_g) shown by α, the ratio of RuBP oxygenase to carboxylase (mol per mol CO_2). From Lawlor (1981).

P_g	R_l	α	C produced as triose	ATP	C/ATP	NADPH	C/NADPH
21	0	0.0	21	63	0.3	42	0.5
15	3	0.4	12	57	0.2	45	0.3
10	5.5	1.1	5	52	0.1	48	0.1
7	7	2.0	0	49	0.0	49	0.0
5.5	7.8	2.8	−3	47	0.0	50	0.0

how the requirements for ATP and NADPH change with O_2 and CO_2 concentration. Net carbon fixation decreases because photorespired CO_2 offsets gross assimilation. When O_2 is removed photorespiration stops and gross assimilation increases as the ATP and NADPH consumed by photorespiration become available to synthesize extra RuBP. The amount of NADPH and ATP required for synthesis of a molecule of sucrose depends on the pathway by which the precursors are formed; in an O_2-free atmosphere, 37 ATP and 24 NADPH are required in total for all reactions. However, in air, consumption is 58 and 45 molecules respectively. Sucrose synthesis with 50 per cent photorespiration demands 72 ATP and 59 NADPH. Competition for PCR cycle carbon between RuBP and sucrose synthesis increases as photorespiration increases and as a consequence carbon flux to sucrose is slowed. At very large α, a flux of carbon from storage into the PCR cycle would be necessary to keep the cycle running.

Biochemically, the glycolate pathway, involving three cellular organelles, may be a method by which C3 plants have partially overcome the effects of a large O_2 and small CO_2 concentration on the RuBP carboxylase. The pathway scavenges some of the carbon and produces some energy as ATP or reducing power in the cytosol, where it may be needed during assimilation. The RuBP oxygenase activity may be 'inevitable' due to the high O_2 concentration of the atmosphere, so that production of phosphoglycolate cannot be prevented. The oxygenation would be the best point for regulation of photorespiration for once phosphoglycolate is produced it must be used as productively as possible to avoid inhibition of photosynthesis and increase efficiency.

RuBPc/o probably evolved early in photosynthesis of CO_2, when the CO_2/O_2 ratio and CO_2 pressure were very large, and has remained relatively unchanged in mechanism despite the apparent disadvantages of its slow reaction rate (turnover of sites 2 s^{-1}) and the oxygenase activity. As the enzyme has a central role in assimilation, its structure has been strongly conserved; the multiple mutations in amino acid residues thought necessary to increase efficiency must all happen together and this would have a very low probability of occurring; single mutations might be lethal, as lack of photosynthesis would destroy the whole organism and so an improved active site structure is unlikely to have been selected. Other parts of the photosynthetic system seem to be more 'flexible' and have evolved around RuBPc/o, for example in C4 metabolism.

Attempts have been made to find chemicals which block the glycolate pathway and prevent waste of assimilated carbon by photorespiration. Inhibitors of glycolate metabolism have been extensively studied. Sodium bisulphate reacts with glycolate and forms the α-hydroxy sulphonate, disodium sulphoglycolate, which inhibits glycolate oxidase. Other α-hydroxysulphonates with the formula R-CHOH-SO_3 Na, such as hydroxymethanesulphonate or a pyridine aldehyde analogue, HPMS (α-hydroxy-2-pyridine methane sulphonic acid), are effective inhibitors although they may undergo reactions to produce glycolate bisulphate, which is the inhibitor. Isonicotinyl hydrazide (INH) causes accumulation of glycolate by blocking the glycine to serine conversion in the mitochondria. However, these inhibitors that block glycolate metabolism after RuBP oxygenase, stop carbon flow in the pathway and also inhibit photosynthesis because carbon accumulates as intermediates, preventing the autocatalytic resynthesis of RuBP. Thus increasing CO_2 assimilation cannot be achieved by blocking the glycolate pathway. It is necessary to stop the oxygenase activity either chemically or genetically by modifying the enzyme, if photosynthesis is to be increased. This goal is yet to be achieved despite much active work.

The glycolate pathway links the metabolism of nitrogen inside and outside the chloroplast. Ammonia released in the conversion of glycine to serine in mitochondria is reassimilated in the cytosol by glutamine synthetase. The glutamine formed is recycled into the chloroplast where it transaminates α-ketoglutarate, using reduced ferredoxin in the presence of glutamate synthase, giving two molecules of glutamate. For each CO_2 released in photorespiration one NH_3 is also produced. With photorespiration 25 per cent of gross photosynthesis, the NH_3 assimilation may be about 5 μmol NH_3 m^{-2} leaf s^{-1}. If reassimilation of ammonia is blocked by methionine sulphoxime (MSO), which inhibits glutamine synthetase, then ammonia accumulates in tissues and photosynthesis stops. Nitrate reduction is only a few per cent of assimilation so the nitrogen turnover during photorespiration is many times greater than the net nitrogen reduction in the chloroplast. Photorespiration thus involves combined carbon and nitrogen cycles which link the energy and metabolites in chloroplasts, mitochondria and cytosol.

The glycolate pathway may consume excess reduced pyridine nucleotide, 'burning off' reductant and allowing a faster turnover of NADH in the peroxisomes and NADPH in the chloroplast. The decarboxylation of glycine may provide a method of generating ATP in the mitochondria and cytosol in the light, or if the generated electrons can be used to form malate, a regulatory device to balance cellular functions. Consumption of excess reductant may be important when photosynthesis is severely restricted. For example, in water-stressed plants with closed stomata, the electron transport chain is fully reduced and chlorophyll absorbs excess energy forming excited states and dangerous products (e.g. superoxide, H_2O_2, p. 102) which damage the photochemical apparatus. Photorespiration decreases the energy burden, particularly on PSII which is sensitive to photoinhibition; it may function together with the carotenoids in quenching chlorophyll excited states and singlet O_2 (see Ch. 3, p. 43). C3 plants are predominantly of well watered, often dimly lit environments, where light harvesting may be a limiting factor in growth rather than CO_2 supply. Carbon lost in photorespiration may be relatively unimportant ecologically, compared to the 'safety valve' offered under temporarily adverse conditions. However, long exposure to bright illumination, hot conditions and low CO_2 concentrations cause excessive loss of CO_2 and photochemical damage to C3 plants.

Glycolate metabolism and C4 plants

C4 plants maintain a very large CO_2 and low O_2 concentration at the RuBP carboxylase active site so that RuBP oxygenase activity is small and little phosphoglycolate is formed. Leaves of C4 plants assimilating $^{14}CO_2$ in air produced little ^{14}C-labelled glycine and form serine by non-glycolate routes. This, together with absence of photorespiration, suggests no glycolate metabolism. However, C4 plants make and metabolize glycolate in small quantities. Conditions, such as water stress, which may decrease the internal CO_2 concentration stimulate the formation of glycine. Thus the glycolate pathway functions in C4 plants, but is only about 10 per cent of that in C3 species under comparable conditions. Any photorespiratory CO_2 is efficiently removed by the PEP carboxylase reaction.

Interaction of dark respiration and photosynthesis

Transition between light and dark occurs daily for plants and rapid fluctuations between bright and dim light are common in many habitats due to clouds or sunflecks in vegetation. As the light energy incident on the leaf changes so do the fluxes of energy and materials in photosynthetic cells and the ratio of assimilation to 'dark respiration'. How do photosynthesis and respiration interact and what are the controls operating during changes in photon flux and with darkness? Dark respiration, if continuing during photosynthesis, would consume assimilates and energy in a 'futile cycle' and slow accumulation of materials and growth. Respiration would deplete metabolites in chloroplasts and slow or inhibit the autocatalytic PCR cycle, which must retain intermediates to re-start rapidly after darkness (p. 139). Efficiency in dim light requires that respiration does not deplete metabolite pools at points within the cycle. Complexity of the PCR cycle, its multiple controls and size of metabolite pools all contribute to stability of photosynthesis under varying environmental and physiological conditions; however, understanding of the biochemistry in relation to growth of plants is limited, particularly the quantitative aspects under different conditions. It is not firmly established if respiration continues in the light during photosynthesis, or if its rate depends on the rate of assimilation. Control of the processes is, therefore, speculative although the general nature is established. This complex area of metabolic interaction has been reviewed by Graham (1980) in detail. Here only some aspects of the problem are considered but before that a brief outline of 'dark respiration' is given to remind the reader of important aspects of the process and its control.

Dark respiration is the consumption of carbohydrates and other compounds to produce energy and carbon substrates (e.g. organic acids) for cellular synthetic processes. The process consumes O_2 and produces CO_2 and H_2O; starch and sugars are degraded to pyruvate by glycolysis, generating ATP. The pyruvate is converted to acetyl-CoA by pyruvate dehydrogenase, situated in the mitochondria. There, acetyl-CoA enters the tricarboxylic acid cycle (TCA cycle or Krebs or citric acid cycle) where it is metabolized to organic acids (e.g. citrate, fumarate, α-ketoglutarate), releasing CO_2 and forming NADH. NADH is oxidized by the mitochondrial electron transport chain with O_2 as terminal acceptor, giving water. Electron transport is coupled to phosphorylation, and this is the source of ATP in respiration. In addition to glycolysis, the oxidative pentose phosphate pathway consumes glucose *via* glucose-6-phosphate, which is oxidized by $NADH^+$ to 6-phosphogluconate by glucose-6-phosphate dehydrogenase. The 6-phosphogluconate is oxidized and decarboxylated to ribulose-5-phosphate.

TCA cycle enzymes are in the mitochondrial matrix. Rates of respiration (O_2 uptake or CO_2 evolution) depend on the rate of glycolysis and acetyl-CoA synthesis, enzyme activity, and the supply of ADP and P_i for phosphorylation, which control the rate of electron flow and hence oxidation of NADH from inside or outside the mitochondrion. High substrate concentration, O_2 and ADP + P_i give rapid respiration but if any of these is deficient respiration falls. Control may be provided by several factors, for example high ATP/ADP ratio or NAD^+/NADH ratio or low substrate concentration inhibits respiration; the respiratory states are discussed by Wiskich (1980). Also, accumulation of organic acids inhibits cycle activity, oxaloacetate slows succinate dehydrogenase for example. High ATP/ADP and NADH levels inhibit malate and isocitrate dehydrogenases and these are major control points.

Evidence for respiration during photosynthesis is subject to many uncertainties; both produce fluxes of O_2 and CO_2 but in opposite direction so that, for example, the evolution of CO_2 by respiration cannot be distinguished from consumption by assimilation without tracers and even then, involves many assumptions about refixation and recycling.

It is well established that a form of respiration proceeds in the light, for C3 plants photorespire, but it is not clear if photosynthesis or photorespiration or light itself inhibits dark respiration or if it occurs concomitantly. Photorespiration is three to eight times greater than dark respiration (measured in darkness) and involves the mitochondrial electron transport chain, so that dark respiration could proceed during photosynthesis as the chain *per se* is not inhibited by light. NADH produced in the glycolate pathway could also be oxidized by the mitochondria, which do metabolize exogenously applied NADH in darkness. There appears to be no evidence that dark respiration could not take place during photosynthesis, however there are many points in glycolysis, TCA cycle and electron transport at which products of light reactions and the PCR cycle may inhibit mitochondrial activity.

Respiration of leaves, measured in darkness, is 5–10 per cent of net photosynthesis in plants of brightly lit environments, but from those of shade it may be a much larger proportion. Dark respiration is saturated at low O_2 concentration (1–2 kPa) but photorespiration increases with increasing O_2. Measurements of respiration in the light at different O_2 concentrations extrapolated to zero O_2 suggest that 'dark' respiration is inhibited, but measurements of CO_2 evolution with increasing light at low CO_2 concentration which prevents net photosynthesis, suggests that dark respiration continues. However, at very small photosynthetic carbon flux in the cell, respiration may be stimulated, paricularly if the ATP/ADP ratio is low. Respiration may depend on the rate of photosynthesis, for the estimated specific radioactivity of CO_2 evolved when rapidly photosynthesizing leaves are fed $^{14}CO_2$ at higher CO_2 concentration, approaches that of the feeding gas, suggesting that little CO_2 is evolved by respiration. When leaves assimilate more slowly, due to CO_2 shortage or water stress for example, the specific activity of CO_2 evolved decreases, possibly because dark respiration contributes a greater proportion. There is evidence that with deficient assimilation, carbon from storage is consumed by dark respiration in the light. However, all these techniques suffer from the uncertainty caused by photosynthetic refixation of CO_2 from respiration which may be 20–100 per cent of that evolved.

Measurements of incorporation of metabolites into TCA cycle intermediates in the light have been made to clarify the role of dark respiration but, like the physiological measurements, they are equivocal. $^{14}CO_2$ from photosynthesis enters some organic acids and alanine, which is formed *via* pyruvate, if assimilation is rapid. However, it does not quickly (30 min) enter other organic acids or all amino acids. Thus glutamate becomes radioactive only in some experiments. Long term pulse chase studies suggest that ^{14}C enters TCA cycle intermediates in the light. However, transfer of $^{14}CO_2$ labelled leaves from light to darkness causes an influx of ^{14}C into the TCA cycle, as expected if the flux is inhibited in the light. Leaves fed with ^{14}C-acetyl-CoA or pyruvate (from glycolysis) form labelled TCA cycle acids, glutamate and aspartate, consistent with an active TCA cycle. Other TCA acids are metabolized in the light to TCA intermediates, and amino acids to organic acids as expected from active dark respiration. Possibly the metabolic controls of the TCA cycle are bypassed when large concentrations of substrate are added to photosynthesizing tissues, allowing respiration to proceed even if normally blocked or much reduced.

Inhibitors of the TCA cycle (e.g. fluoroacetate) have been used to block respiration. The distribution of $^{14}CO_2$ into assimilates shows continuation of the cycle. Inhibition of photosynthesis by DCMU showed that repiration occurs in the light; however preventing assimilation could stimulate respiration because photosynthetic carbon and energy fluxes are inhibited. Graham (1980) concluded that the TCA cycle continues in the light; with rapid photosynthesis respiration is small and slower photosynthesis encourages TCA cycle activity. Respiration seems to decrease in the first few minutes of illumination but increases later.

The TCA cycle provides carbon skeletons for amino acid synthesis, particularly α-ketoglutarate for glutamate, and this function may be required in the light, for although chloroplasts can synthesize most amino acids, organic acids cannot be made. This may be more important in young tissues than in old. Succinyl-CoA from mitochondria is also required for porphyrin synthesis. Mitochondria supply cells in darkness with ATP and NADH but in the light photophosphorylation and photosynthetic electron transport provide ATP and NADPH. Photorespiration gives NADH which may donate electrons to O_2 *via* the mitochondrial electron transport chain coupled to phosphorylation.

Adenylates in cell compartments

Control of adenylate composition in cellular compartments during dark–light transitions has been examined by rapidly killing and separating chloroplasts, mitochondria and cytosol from leaves or, with greater ease and precision, from protoplasts. Protoplasts are prepared by incubating leaf slices in an enzyme preparation which digests the cellulose walls. The released protoplasts are separated by centrifugation in density gradients of sucrose and sorbitol under controlled conditions. The protoplasts are exposed to experimental conditions (e.g. light or dark) for required times and then rapidly disrupted by centrifugation through a nylon net and filters which break the protoplasts. Cell organelles are separated on density gradients and, by measuring marker enzyme complements, the purity of cell fractions is checked. Interconversion of adenylates is inhibited by these rapid separation methods allowing adenylates to be measured (see p. 110) in organelles; corrections for cross contamination of fractions are made. Although there is not complete agreement about the state of adenylates within the cell compartments, some general observations are possible.

In darkness (Fig. 8.5) the ATP/ADP ratio is larger in the cytosol than in the

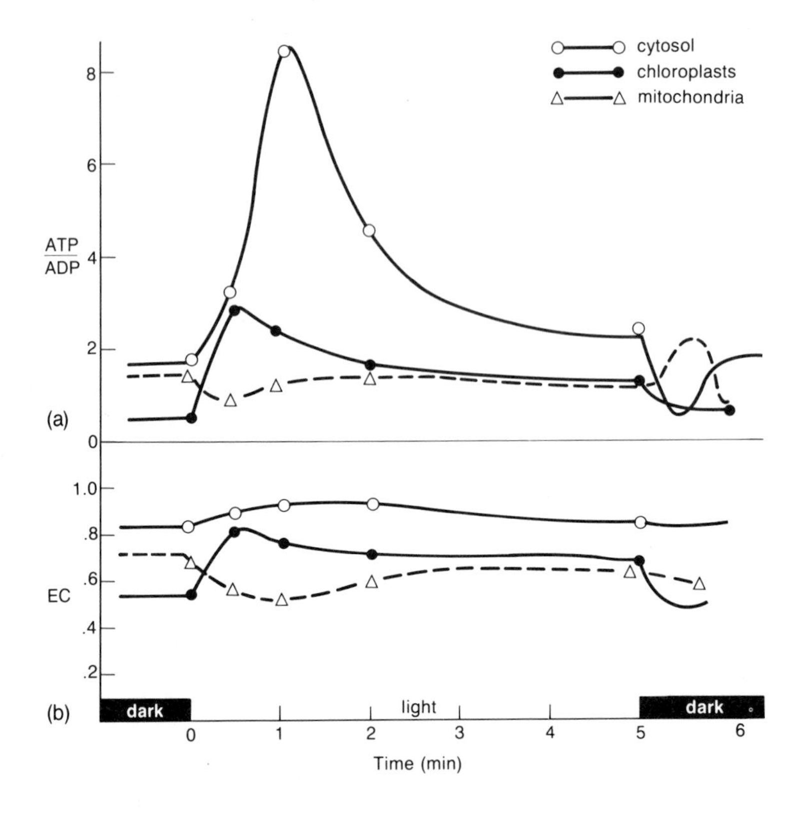

FIG. 8.5 Adenylate (ATP/ADP) ratio (a) and energy charge (EC) (b) in cell (chloroplasts, cytosol and mitochondria) during the light to dark transitions in photosynthetic tissues; semi-schematic.

mitochondria and chloroplast stroma; oxidative phosphorylation in the mitochondria maintains a very high ATP/ADP ratio and energy charge (EC) in the cytosol, although the mechanism by which this is achieved is not known. Possibly shuttles of adenylate from the mitochondria to cytosol maintain the high concentration of ATP there. Shuttles of triosephosphate and 3-phosphoglycerate between chloroplast stroma and cytosol regulate the chloroplast's adenylate levels without direct transfer of ATP. ATP in chloroplasts in darkness is required to maintain ion balance, levels of photosynthetic intermediates and some synthetic processes such as protein synthesis, although the rates may be small in mature chloroplasts. However, protein synthesis continues in darkness in the cytosol, so high EC is necessary. On illumination photophosphorylation in the chloroplast stroma increases rapidly (within 30 s) so that the ATP/ADP ratio increases, but AMP decreases with the result that EC increases. However, the sum of ATP, ADP and AMP remains constant showing that little direct adenylate exchange occurs between chloroplast and cytosol. The cytosolic ATP/ADP ratio and EC increase greatly and those in the mitochondria decrease, suggesting that the cytosol exercises control over respiration. However, after a few minutes illumination, the cytosolic ATP/ADP ratio and EC fall to values greater than in darkness and the mitochondrial values rise but to less than those observed in darkness. On darkening cells there is a decrease in the ATP/ADP ratio and EC values in the chloroplast stroma, and concomitantly a transient decrease in the cytosol and a transient increase in mitochondrial ATP/ADP ratio, before they return to values seen after long periods in darkness. Mitochondrial adenylate would seem to be controlled by cytoplasmic EC which remains high at all times, possibly because there is no AMP in the cytosol and no adenylate kinase. The mitochondrial and chloroplast ATP synthesis is coupled to the cytosol to keep a high EC although the mechanism is not clear. Even in the light the chloroplast EC is lower than the cytosolic, so the photosynthetic carbon cycle and other stromal activities (e.g. protein synthesis) presumably operate at lower EC than those in the cytosol. The reason for and consequences of this are not yet understood.

In the light a high cytosolic ATP/ADP ratio and high EC are generally thought, with little evidence, to depress mitochondrial oxidative phosphorylation and glycolysis. However the changes required between light and dark to provide control may be larger than those measured, thus raising doubt about the control mechanisms. Possibly the adenylate control mechanism does not operate *in vivo* and an alternative, as yet unknown, system regulates cellular adenylate charges. Although photorespiration has been thought to use the mitochondrial electron transport chain in the light, other evidence strongly suggests that the electrons from glycine oxidation may be transferred into the cytosol by malate and aspartate shuttles and not consumed by the mitochondrial electron transport chain. Thus it may not be essential for mitochondria to function in the light.

A working model of the interaction between dark respiration and photosynthesis and the adenylate systems is that the TCA cycle operates in darkness and transfer of adenylate energy keeps the cytosol EC high and also maintains a minimal EC in the stroma. In light chloroplasts produce ATP, increasing the chloroplast ATP/ADP ratio and EC and transferring energy to the cytosol which experiences greatly increased ATP/ADP ratio and a rise in EC which decreases or stops the activity of the mitochondria, either by direct adenylate control or another unknown system. This inhibition is only transient, allowing the ATP/ADP ratio and EC to re-equilibrate. In dim light a small adenylate supply from the chloroplast might not completely inhibit mitochondria which would provide some energy to the cytosol. ATP is required for many metabolic processes and if demand is large, respiration may continue. Perhaps under conditions where ATP synthesis may be inhibited (e.g. with water stress) respiration increases and maintains cellular adenylate levels. Under such conditions CO_2 from stored compounds contributes a much larger proportion of the total CO_2 released in photorespiration. If the demand for energy and substrates between parts of the cell becomes unbalanced then the dynamic equilibria between cell compartments provides for that most important of cellular characteristics, stability. Control of such a complex metabolic network is poorly understood and the above account is speculative. Probably respiration and photosynthesis interact dynamically according to the supply and demand for energy and metabolites in the cell and are not simply switched on and off.

Amino acid synthesis in chloroplasts

Chloroplasts contain enzymes for synthesis of most protein amino acids, with nitrogen derived from nitrate reduction. The primary products of the glutamine synthetase and glutamate synthase reactions in the stroma are glutamine and glutamate (p. 151), and these provide amino groups for formation of other amino acids. Chloroplasts are the main site (and for many amino acids the only site) of biosynthesis in leaves. Glutamate is a precursor of arginine and most steps in the conversion are chloroplastic but the last step in the pathway (arginosuccinate lyase) is cytoplasmic. It is not established if the enzymes of proline biosynthesis, also derived from glutamate, are chloroplastic. Proline accumulates in large amounts in water-stressed leaves when photosynthesis is inhibited but the mechanisms and controls are unknown. Aspartate is also metabolized in chloroplasts to lysine, threonine and homocysteine, but the last methylation of this amino acid to methionine occurs outside the chloroplast. Aromatic amino acids (e.g. phenylalanine and tyrosine) are made in chloroplasts from photosynthetic assimilate. Serine is made in chloroplasts, as well as in the mitochondria during photorespiration. Although mature chloroplasts are virtually self-sufficient in amino acids, they are not independent of the supply of organic acids from the mitochondria. Occurrence of some control steps in amino acid synthesis in the cytosol may ensure close integration of chloroplast, cytosol and mitochondrial activity, so that cellular production is closely linked to energy supply during respiration and photosynthesis in relation to the demands of protein synthesis. Mature chloroplasts supply the plant with all amino acids, although not directly but *via* shuttle mechanisms and interconversions at many points in metabolism.

Chloroplast genome and protein synthesis

Chloroplasts contain DNA, which codes for ribosomal RNA (rRNA), transfer RNA (tRNA), and for messenger RNAs (mRNAs) for synthesis of some but not all thylakoid and stromal proteins. The chloroplast genome is a circular, supercoiled molecule of 82–96 million daltons (MD) (varying with species) with many identical copies in each chloroplast; it may be attached to chloroplast membranes. The contour length of the molecule is 40–50 μm. Chloroplast DNA differs from nuclear DNA in the absence of associated histones. Chloroplast DNA codes for RNAs (16, 23, 5 and 4.5 S). The 16 S rRNA is in the 30 S subunit of chloroplast ribosomes and the 23, 5 and 4.5 S rRNAs are in the 50 S subunit. The large and small subunits of the chloroplast ribosomes contain about 38 and 24 proteins respectively. Polysomes, assemblies of ribosomes on mRNA, are both free and membrane bound in chloroplasts.

Chloroplast ribosomes are smaller than cytoplasmic ribosomes, sedimenting at 70 S and 80 S respectively and their rRNAs are smaller than the equivalent units in the cytoplasmic ribosomes. The mechanism of protein synthesis in chloroplasts, but not in the cytosol, is sensitive to chloramphenicol and resembles the bacterial process. Chloroplasts and bacteria may have had a common ancestor and it has been suggested that photosynthetic bacteria invaded non-photosynthetic eukaryotic cells during evolution.

Chloroplasts also code for more than 30 tRNAs which activate amino acids and orientate them on mRNA for protein formation. Proteins known to be made by the chloroplast are the large subunit of RuBP carboxylase, a 32 kD (herbicide binding?) protein in the thylakoid membrane and the α, β and ε subunits of CF_1 and subunits I and III of CF_0 of coupling factor. Some other PSI and PSII components, cytochrome *f* and some polypeptides of unknown function, are also specified by and made in the chloroplast. However, there is no evidence for the enzymes of chlorophyll or carotenoid synthesis, ferredoxin, or any PCR cycle enzymes being coded in the chloroplast; those for which a site of synthesis is known are made in the cytosol on 80 S ribosomes after transcription of nuclear DNA. In addition enzymes of fatty acid synthesis, chloroplast RNA polymerase (involved in formation of RNA in the chloroplast) and proteins of the chloroplast 70 S ribosomes and most of the major thylakoid proteins, together with the small subunit of RuBP carboxylase, are made in the cytosol and transported into the chloroplast. The small subunit of RuBP carboxylase is made as a larger (20 kD) precursor, which is carried into the chloroplast across the membrane before the 5 kD transfer protein is removed. In experimentally altered growth conditions, for example at high temperature, the small subunit accumulates in chloroplasts but the production of the large subunit is inhibited.

Light-harvesting chlorophyll protein is encoded in the nucleus and is under control of the phytochrome system. Changes in the production of chlorophyll *a* but not chlorophyll *b* are linked to inhibition of the polypeptides of the complex. Thus long term adjustment of the composition and structure of the chloroplast and balance between thylakoid and stroma is controlled by conditions acting on a very carefully regulated genetic mechanism. Perhaps this complex control system arose as a consequence of evolution from the symbiotic origins of chloroplasts, and provides for integration of cellular processes to ensure that the structure of the system remains stable over a wide range of conditions. The photosynthetic system maintains its integrity, and often its capacity, even under conditions which greatly affect the size of the plant.

Chloroplast development and protein synthesis

Although mature chloroplasts are virtually self-sufficient in amino acids and probably ATP for protein synthesis, the main period of synthesis occurs during development and so, necessarily, precedes assimilation. Chloroplasts originate in meristem cells as proplastids, colourless, undifferentiated plastids with a double membrane. As leaves grow and cells expand and multiply, the proplastids differentiate into chloroplasts, which divide by binary fission. The daughter plastids are transmitted to the cells as they divide and the chloroplast DNA carries genetic information to daughter cells. Conifers develop chlorophyll in darkness but most angiosperms require light for plastids to develop into functional, green chloroplasts; only dim light of red wavelength is needed and its effect is reversed by far red light, suggesting that chloroplast development is controlled *via* the phytochrome system. Light regulates the differentiation of thylakoids, changing the number, area and degree of stacking; this is achieved by alterations in the light-harvesting complex (LHC) and chlorophyll *b* synthesis. Transcription of the nuclear gene for the LHC protein, but not chlorophyll *a/b* apoprotein or chlorophyll synthesis, is under phytochrome control and in intermittent light no LHC protein or chlorophyll *b* is made, but chlorophyll *a* is synthesized. Only if intermittent-light grown plants are placed in continuous light is LHC protein made rapidly together with chlorophyll *b*. The 32 kD protein is also under photocontrol. However, light may not be directly required for chloroplast division. The mechanism of light modification of chloroplast number, size and structure is not understood; it is important as it is a major control on the productivity of plants under different conditions.

In darkness proplastids develop into etioplasts, forming some prothylakoid membranes and a regular lattice of linked tubules called the prolamellar body, of crystalline appearance, which develops into thylakoids if illuminated. Etioplasts accumulate starch and grow for up to 14 days in darkness after which they degenerate. In light, etioplasts start to develop rapidly into chloroplasts after a lag phase (45 minutes). Sheets of thylakoids develop from the prolamellar body and there is massive (400 per cent) increase in proteins of the thylakoids over 50 hours; the large subunit of RuBP carboxylase is made at the rate of 10 molecules per plastid per hour. Stromal enzymes increase in parallel to the thylakoid proteins during this period. Protein synthesis during this rapid growth phase is at the cost of stored or translocated substrates and respiration provides the energy. Thus even with severe deficiency of substrates, direction of materials into new leaves ensures development of photosynthetically competent tissues. One result of nutritional deficiencies is that less leaf is made but often of relatively high photosynthetic capacity. During later stages of chloroplast development, photosynthetic processes become functional at different rates, for example NADPH reduction may precede ATP synthesis, but neither the significance of such imbalance for the development of the chloroplast, nor the control processes are known.

After rapid growth of chloroplasts in young leaves mature tissues have slower turnover of proteins, probably at the expense of amino acids and energy from photosynthesis. The chloroplast genome is repressed as the leaf approaches maturity and there is a period of many days over which photosynthesis declines from a peak at full leaf expansion. This loss of activity is related to cessation of chloroplast protein synthesis (e.g. of RuBP carboxylase). In older leaves proteins are not replaced and are eventually degraded by proteolytic enzymes, which are formed in the cytosol, probably from mRNA produced earlier but latent until the leaf is aged, so the chloroplast genome has little function during senescence. Environmental conditions such as poor nitrogen nutrition, water stress and high temperatures may induce premature senescence. The induction may be by effects on specific enzymes or parts of the photosynthetic apparatus; thus N shortage would slow the formation of rapidly turned over enzymes and trigger senescence. There is rapid remobilization of amino acids from proteins which may be a mechanism to provide developing leaves with essential nutrients to form a viable photosynthetic system. The control of the rate at which plants lose photosynthetic competence is poorly understood but is very important in relation to plant productivity.

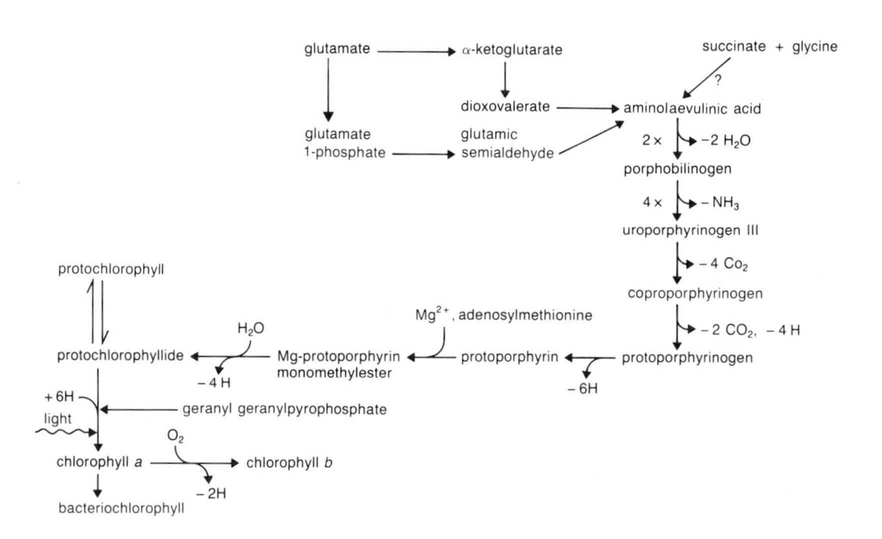

FIG. 8.6 Outline of the biosynthesis of chlorophylls and related molecules.

Synthesis of chlorophylls and carotenoids

Synthesis of chlorophylls is shown in Fig. 8.6. The precursors of chlorophyll are relatively simple organic compounds of intermediary metabolism. In higher plants succinyl-CoA and glycine were believed to be the precursors, as they are of tetrapyrroles in animals and bacteria. However, more recently the precursor has been demonstrated to be glutamate which is deaminated giving α-ketoglutarate which is reduced to γ, δ-dioxovalerate and transaminated to yield δ-aminolaevulinic acid. Synthesis requires ATP and NADPH.

Removal of water from two molecules of aminolaevulinic acid produces a porphobilinogen with the pyrrole ring structure, basic form of the tetrapyrroles. When four molecules of porphobilinogen are condensed, with loss of four NH_3 groups, the uroporphyrinogen III molecule is formed. Removal of four CO_2 molecules from the acetic acid residues on the molecule, gives coproporphyrinogen; decarboxylation and dehydrogenation of the proprionic acid groups on rings I and II forms vinyl groups. Further dehydrogenation of the protoporphyrinogen IX (the common precursor of haems, bilins and chlorophylls) and addition of magnesium and adenosylmethionine to protoporphyrin forms Mg-protoporphyrin monomethyl ester. Ring III on this undergoes dehydration and reduction, and ring II a hydrogenation, producing protochlorophyllide and further addition of H gives chlorophyllide *a*. Light is absorbed by protochlorophyllide which is reduced to chlorophyll *a*; chlorophyll *b* is probably formed from a special pool of chlorophyll *a* in a reaction requiring O_2 and NADPH, suggesting that it arose during later, aerobic evolution. Addition of geranyl-geranyl pyrophosphate to the chlorophyllide forms the phytyl 'tail'. The complex process is essentially a conversion of a water-soluble, linear molecule to a more aromatic structure with special characteristics for light capture in the lipid environment of the thylakoid. Magnesium 'tunes' the absorption spectrum. Early organisms possibly captured light with porphobilinogens. Subsequently the pigments evolved towards more efficient light capture, by changes in chemistry and therefore in the enzymes of the synthetic pathways. Protoporphyrin is the precursor for many biological redox compounds in respiration. Addition of iron to protoporphyrin gives cytochrome; catalase, peroxidase and phycobilin are related. Bacteriochlorophyll appears to be derived from chlorophyll suggesting that it was a late evolutionary development.

Carotenoids are synthesized from acetyl-CoA *via* mevalonic acid and then geranyl pyrophosphate, farnesyl pyrophosphate and geranyl-geranyl pyrophosphate to phytoene. The process increases the length of the chain giving the alternative double bond structure, the isoprene unit. From phytoene, successive reductions produce lycopene, and then cyclization, with formation of the α and β-ionone rings at the ends of the molecule, gives carotenoids. The basic structure of the carotenoid molecule is an isoprene unit and extension of the chain produces monoterpenes, sesquiterpenes (which include farnesol with hormone properties), and redox carriers in the photosynthetic electron transport chain (e.g. plastoquinone), together with abscisic acid, another plant hormone. Diterpenes include the gibberellins, another important group of plant hormones. Steroids are triterpenes whilst carotenoids are tetraterpenes. This important family of compounds was formed early in the evolution of plants. Carotenoids probably functioned as light-harvesting pigments in early photosynthesis and later, after development of the chlorophylls, became more important in regulating the energy state of the chemical reaction centres. When oxygen accumulated in the atmosphere carotenoids assumed a protective role.

References and Further Reading

Bonner, J. and **Varner, J. E.** (1965) The path of carbon in respiratory metabolism, pp. 213–30 in Bonner, J. and Varner, J. E. (eds), *Plant Biochemistry*, Academic Press, New York.

Bradbeer, J. W. (1981) Chloroplast development: a perspective, pp. 745–54 in Akoyunoglou, G. (ed.), *Photosynthesis V. Chloroplast Development*, Balaban International Science Services, Philadelphia.

Bray, C. M. (1983) *Nitrogen Metabolism in Plants*, Longman, London.

Buetow, D. E. (1982) Molecular biology of chloroplasts, pp. 43–88 in Govindjee (ed.), *Photosynthesis*, vol. II, *Development, Carbon Metabolism, and Plant Productivity*, Academic Press, New York.

Douce, R. and **Joyard, J.** (1980) Plant galactolipids, pp. 321–63 in Stumpf, P. K. (ed.), *The Biochemistry of Plants*, vol. 3, *The Plant Cell*, Academic Press, New York.

Ellis, R.J. (1984) *Chloroplast Biogenesis*, Cambridge University Press, Cambridge.

Giersch, C. H., Heber, U. and **Krause, G. H.** (1980) ATP transfer from chloroplast to the cytosol of leaf cells during photosynthesis and its effects on leaf metabolism, pp. 65–82 in Spanswick, R. M., Lucas, W. J. and Dainty, J. (eds), *Plant Membrane Transport: Current Conceptional Issues*, Elsevier/North-Holland Biomedical Press, Amsterdam.

Goller, M., Hampp, R. and **Ziegler, H.** (1982) Regulation of the cytosolic adenylate ratio as determined by rapid fractionation of mesophyll protoplasts of oat. Effect of electron transfer inhibitors and uncouplers, *Planta*, **156**, 255–63.

Graham, D. (1980) Effects of light on 'dark' respiration, pp. 525–79 in Davies, D.D. (ed.), *The Biochemistry of Plants*, vol. 2, *Metabolism and Respiration*, Academic Press, New York.

Hampp, R., Goller, M. and **Ziegler, H.** (1982) Adenylate levels, energy charge, and phosphorylation potential during dark-light and light-dark transition in chloroplasts, mitochondria, and cytosol of mesophyll chloroplasts from *Avena sativa* L., *Plant Physiol.*, **69**, 448–55.

Heber, U. and **Heldt, H. W.** (1981) The chloroplast envelope: structure, function and role in leaf metabolism, *A. Rev. Plant Physiol.*, **32**, 139–68.

Heber, U., Enser, U., Weis, E., Ziem, U. and **Giersch, C.** (1979) Regulation of the photosynthetic carbon cycle, phosphorylation and electron transport in illuminated intact chloroplasts, pp. 113–38 in Atkinson, D. E. and Fox, C. F. (eds), *Modulation of Protein Function*, Academic Press, New York.

Hewitt, E. J., Hucklesby, D. P. and **Notton, B. A.** (1976) Nitrate metabolism, pp. 633–82 in Bonner, J. and Varner, J. E. (eds), *Plant Biochemistry* (3rd edn), Academic Press, New York.

Hewitt, E. J. (1975) Assimilatory nitrate–nitrite reduction, *A. Rev. Plant Physiol.*, **26**, 73–100.

Keys, A. J., Bird, I. F., Cornelius, M. J., Lea, P. J., Wallsgrove, R. M. and **Miflin, B. J.** (1978) Photorespiratory nitrogen cycle, *Nature*, **275**, 741–43.

Lawlor, D. W. (1981) Photorespiration and its control; is there a role for plant growth regulators? pp. 111–21 in Jeffcoat, B. (ed.), *Aspects and Prospects of Plant Growth Regulators*, Monograph 6, British Plant Growth Regulator Group, Wantage.

Lea, P. J. and **Miflin, B. J.** (1979) Photosynthetic ammonia assimilation, pp. 445–56 in Gibbs, M. and Latzko, E. (eds), *Encyclopedia of Plant Physiology* (N.S.), vol. 6, *Photosynthesis II*, Springer-Verlag, Berlin.

Miflin, B. J. and **Lea, P. J.** (1979) Amino acid metabolism, *A. Rev. Plant Physiol.*, **28**, 299–329.

Mudd, J. B. (1980) Phospholipid biosynthesis, pp. 250–82 in Stumpf, P. K. (ed.), *The Biochemistry of Plants*, vol. 3, *The Plant Cell*, Academic Press, New York.

Ogren, W. L. and **Chollet, R.** (1982) Photorespiration, pp. 191–230 in Govindjee (ed.), *Photosynthesis*, vol. II, *Development, Carbon Metabolism and Plant Productivity*, Academic Press, New York.

Preiss, J. and **Kosuge, T.** (1976) Regulation of enzyme activity in metabolic pathways, pp. 277–336 in Bonner, J. and Varner, J. E. (eds), *Plant Biochemistry* (3rd edn), Academic Press, New York.

Robinson, S. P. and **Walker, D. A.** (1981) Photosynthetic carbon reduction cycle, pp. 193–236 in Hatch, H. D. and Boardman, N. K. (eds), *The Biochemistry of Plants*, vol. 8, *Photosynthesis*, Academic Press, New York.

Roughan, P. G. and **Slack, C. R.** (1982) Cellular organisation of glycerolipid metabolism, *A. Rev. Plant Physiol.*, **33**, 97–132.

Schmidt, A. (1977) Photosynthetic assimilation of sulfur compounds, pp. 481–96 in Gibbs, M. and Latzko, E. (eds), *Encyclopedia of Plant Physiology* (N.S.), vol. 6, *Photosynthesis II*, Springer-Verlag, Berlin.

Schnarrenberger, C. and **Fock, H.** (1976) Interactions among organelles involved in photorespiration, pp. 185–234 in Stocking, C. R. and Heber, U. (eds), *Encyclopedia of Plant Physiology*, vol. 3, *Transport in Plants*, Springer-Verlag, Berlin.

Schiff, J. A. and **Hodson, R. C.** (1973) The metabolism of sulfate, *A. Rev. Plant Physiol.*, **24**, 381–414.

Schwenn, J. D. and **Trebst, A.** (1976) Photosynthetic sulfate reduction by chloroplasts, pp. 315–34 in Barber, J. (ed.), *The Intact Chloroplast*, Elsevier/North-Holland Biomedical Press, Amsterdam.

Stitt, M., McC. Lilley, R. and **Heldt, H. W.** (1982) Adenine nucleotide levels in the cytosol, chloroplasts and mitochondria of wheat leaf protoplasts, *Plant Physiol.*, **70**, 971–77.

Stumpf, P. K. (1980) Biosynthesis of saturated and unsaturated fatty acids, pp. 177–204 in Stumpf, P. K. (ed.), *The Biochemistry of Plants*, vol. 3, *The Plant Cell*, Academic Press, New York.

Thomson, W. W., Mudd, J. B. and **Gibbs, M.** (1983) *Biosynthesis and Function of Plant Lipids*, Proceedings of the 6th Annual Symposium in Botany, American Society of Plant Physiologists, Baltimore.

Tolbert, N. E. (1971) Microbodies – peroxisomes and glyoxysomes, *A. Rev. Plant Physiol.*, **22**, 45–74.

Tolbert, N. E. (1979) Glycolate metabolism by higher plants and algae, pp. 338–52 in Gibbs, M. and Latzko, E. (eds), *Encyclopedia of Plant Physiology* (N.S.), vol. 6, *Photosynthesis II*, Springer-Verlag, Berlin.

Tolbert, N. E. (1980) Photorespiration, pp. 488–525 in Davies, D. D. (ed.), *The Biochemistry of Plants*, vol. 2, *Metabolism and Respiration*, Academic Press, New York.

Turner, J. F. and **Turner, D. H.** (1975) The regulation of carbohydrate metabolism, *A. Rev. Plant Physiol.*, **26**, 159–86.

Vennesland, B. and **Guerro, M. G.** (1979) Reduction of nitrate and nitrite, pp. 425–44 in Gibbs, M. and Latzko, E. (eds), *Encyclopedia of Plant Physiology* (N.S.), vol. 6, *Photosynthesis II*, Springer-Verlag, Berlin.

Walker, D. A. (1976) Plastids and intracellular transport, pp. 85–136 in Stocking, C. R. and Heber, U. (eds), *Encyclopedia of Plant Physiology* (N.S.), vol. 3, *Transport in Plants*, Springer-Verlag, Berlin.

Whittingham, C. P., Keys, A. J. and **Bird, I. F.** (1979) The enzymology of sucrose synthesis in leaves, pp. 313–26 in Gibbs, M. and Latzko, E. (eds), *Encyclopedia of Plant Physiology* (N.S.), vol. 6, *Photosynthesis II*, Springer-Verlag, Berlin.

Wilson, A. T. and **Calvin, M.** (1955) The photosynthetic cycle. CO_2 dependent transients. *J. Amer. Chem. Soc.*, **77**, 5948–57.

Wilson, L. G. and **Reuveny, Z.** (1976) Sulfate reduction, pp. 559–632 in Bonner, J. and Varner, J. E. (eds), *Plant Biochemistry* (3rd edn), Academic Press, New York.

Wiskich, T. (1980) Control of the Krebs cycle, pp. 243–78 in Davies, D. D. (ed.), *The Biochemistry of Plants*, vol. 2, *Metabolism and Respiration*, Academic Press, New York.

CHAPTER 9

C4 photosynthesis and crassulacean acid metabolism

The PCR cycle is the only system giving a net increase in chemical energy yet found in living organisms and the mechanism and its control is very similar in all. As 3PGA, a 3-carbon compound, is the first product of the PCR cycle, this mode of carbon assimilation is called C3 photosynthesis (Fig. 9.1a). However, in many families of higher plants, additional metabolic systems have evolved for accumulating CO_2, and passing it to the PCR cycle, increasing the efficiency of photosynthesis, particularly under adverse environmental conditions. One group (see Fig. 9.1b), which includes many tropical grasses, assimilates CO_2 into the 3-carbon precursor phosphoenol pyruvic acid (PEP) to produce 4-carbon carboxylic acids as primary product, and hence are called C4 plants. The acids are formed in mesophyll cells and transferred to bundle sheath cells where they are decarboxylated and the CO_2 assimilated by the PCR cycle. The other major modification (Fig. 9.1c), found in many succulent plants including, but not confined to, the family Crassulaceae is called crassulacean acid metabolism (CAM). Four-carbon organic acids are formed from CO_2 and PEP as in C4 plants, but this synthesis proceeds in darkness using energy accumulated during previous illumination. The acids are decarboxylated in the subsequent light period and the CO_2 is assimilated by the PCR cycle with light energy. CAM plant leaves are not structurally differentiated into tissues with different biochemistry, but CO_2 accumulation and PCR cycle assimilation are separated in time. The changes associated with C4 and CAM photosynthesis are ecologically important as CAM is related to water conservation and C4 to use of intense light under low atmospheric CO_2 concentration. About 3000 species of plant possess C4 metabolism and some 250 exhibit CAM compared with 300 000 with C3 metabolism (putative!). Particularly within the C4 species, there are substantial differences in the detailed metabolism (enzymes, light harvesting, metabolite exchange) and structure, suggesting that they have different evolutionary origins, that is C4 photosynthesis is polyphyletic.

C4 photosynthesis

C4 assimilation occurs in monocotyledons, particularly the tropical panicoid grasses such as sugar cane (*Saccharum officinarum*), maize (*Zea mays*) and sorghum (*Sorghum vulgare*), and in dicotyledons, for example in the

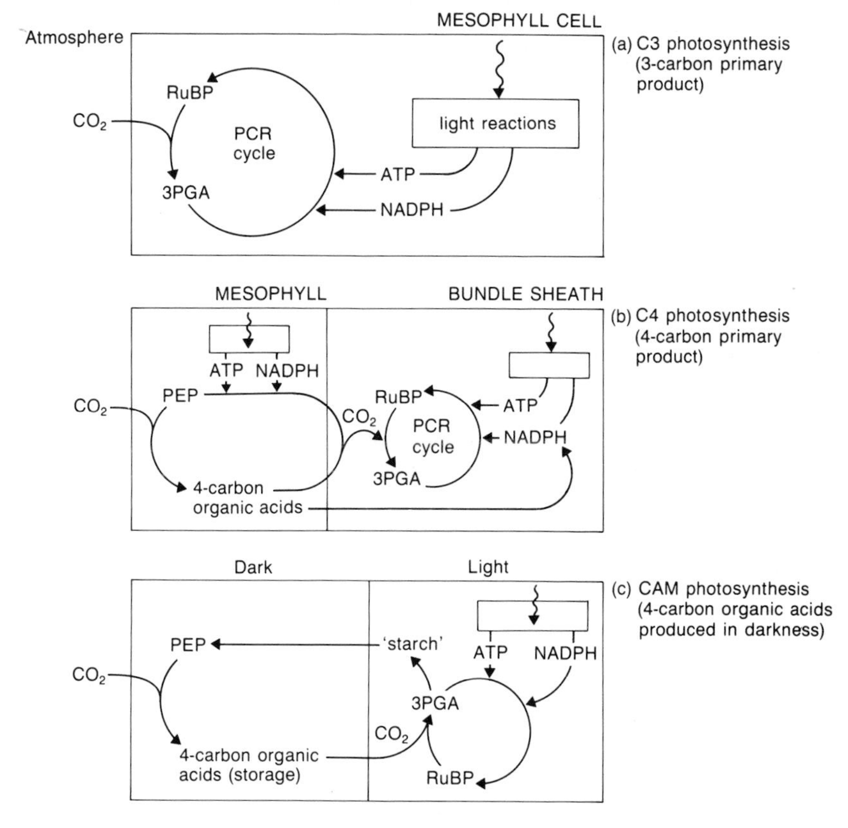

FIG. 9.1 Variation in photosynthetic mechanisms in higher plants; (a) C3 photosynthesis by the photosynthetic carbon reduction cycle (PCR), and synthesis of dicarboxylic acids in (b) C4 photosynthesis and (c) crassulacean acid metabolism related to the PCR cycle.

Chenopodiaceae and Compositae. Even within each genus some species may be C4 whilst others are C3 plants. In *Atriplex* (Chenopodiaceae) some species (e.g. *A. sabulosa*) have C4 and others (e.g. *A. hastata*) C3 photosynthesis. C4 plants generally have higher rates of CO_2 assimilation than C3 plants; photosynthesis does not saturate in bright light and continues at very low concentrations of CO_2. Associated with this are rapid growth rates and higher dry matter yields than C3 plants and C4 crops are markedly more productive than C3. Within the general group of plants with C4 photosynthesis there is considerable variation in anatomy (which is discussed in detail for maize in Ch. 4) but the bundle sheath, a ring of large, closely packed cells around the vascular tissue, is distinctive. Arrangement of the chloroplasts in bundle sheath cells, is related to the type of C4 metabolism. Large chloroplasts with many thylakoids are

Table 9.1 Characteristics of different types of C4 plants, and examples of species with this form of metabolism, including distribution of chloroplasts and enzymes in bundle sheath and mesophyll cells, compared with C3 species

	C4			
Type of photosynthesis	NADP-ME	PCK	NAD-ME	C3
Examples	*Zea mays* *Sorghum sudanense* *Saccharum officinarum*	*Panicum maximum* *Chloris gayana* *Sporobolis fimbriatus*	*Atriplex spongiosa* *Portulaca oleracea* *Amaranthus edulis*	*Triticum aestivum* *Glycine max* *Pisum sativum*
Bundle sheath	Yes	Yes	Yes	No
Cell size (μm)	113 × 18		40 × 24	
No. chloroplasts/cell	42		39	
bs chloroplast position	Centrifugal	Centrifugal	Centripetal	—
Grana	Very reduced	Yes	Yes	—
Mitochondria	Few	Few?	Many	—
Mesophyll				
Cell size (μm)	56 × 16		38 × 8	60 × 20
No. chloroplasts/cell	31		10	90
Grana	Yes	Yes	Yes	Yes
Major organic acid	Malate	Aspartate	Aspartate	—
Enzyme activity				
NADP malate dehydrogenase	High meso chloropl	Low meso & bs	Low meso & bs chloropl	Very low
NADP malic enzyme	High bs chloropl	Low meso & bs	Low meso & bs	Very low
Aspartate amino-transferase	Low meso chloropl	Very high cytosol	High meso & bs cytosol & mitochondria	Very low
PEP carboxykinase	Low meso & bs	High bs cytosol	Low meso & bs cytosol	Very low
NAD malic enzyme	Low	Low	High bs mitochondria	Very low
NAD malate dehydrogenase	—	Low	bs mitochondria	Very low
Alanine amino-transferase	Low	High bs & meso cytosol	High bs & meso cytosol	Very low
PEP carboxylase	High meso cytosol	High meso cytosol	High meso cytosol	Very low
Pyruvate, P_i dikinase	High meso chloropl	High meso chloropl	High meso chloropl	None
3PGA kinase	Meso & bs chloropl	High meso chloropl	meso & bs chloropl	Chloropl
RuBP carboxylase	bs chloropl	bs chloropl	bs chloropl	Chloropl
PCR cycle enzymes	bs chloropl	bs chloropl	bs chloropl	Chloropl
Light reactions	PSI, II meso PSI bs	PSI, II meso PSI, II bs	PSI, II meso PSI, II bs	PSI, II —

bs, bundle sheath; meso, mesophyll; chloropl, chloroplast

arranged (Table 9.1) around the outer wall of the cells (centrifugal) in some aspartate formers (e.g. *Panicum maximum*) and malate formers (e.g. *Zea mays*). Chloroplasts are on the walls nearest to the vascular tissue (centripetal) in NAD-ME types (e.g. *Amaranthus edulis*) which produce aspartate and contain many more large mitochondria close to the vascular tissue than other C4 plants. Bundle sheath cells are larger than mesophyll cells and have a much smaller surface-to-volume ratio which may restrict diffusion of O_2 into them. The peripheral reticulum (p. 70) increases the internal membrane surface area for transport of assimilates. In some malate producers (e.g. maize) only stromal thylakoids are present. Some aspartate-producing C4 plants (e.g. *Sporobolis aeroides*) have pronounced grana. In both malate- and aspartate-forming types the mesophyll chloroplasts are granal. Many large plasmodesmata link the bundle sheath and mesophyll cells (see Ch. 4) and are important for transport of material, for bundle sheath cells have in their walls a dense, probably impermeable layer, which may prevent diffusion of gases.

C4 metabolism

There are four main processes in C4 photosynthesis (see Fig. 9.2a). First, carboxylation of PEP produces organic acids in the mesophyll cells without a net gain in energy. Second, the organic acids are transported to the bundle sheath cells. Third, they are decarboxylated there, producing CO_2 which is assimilated by RuBP carboxylase and the PCR cycle with a net gain in energy. Fourth, compounds return to the mesophyll to regenerate more PEP by pyruvate, P_i dikinase in the mesophyll chloroplasts. Variations in C4 photosynthesis are now established but the steps are common to all types. However, the decarboxylation enzymes and the compounds transported between mesophyll and bundle sheath differ between C4 species. These conclusions are based on the proportions of $^{14}CO_2$ in assimilate products and time course of labelling (Fig. 9.3) and on the distribution of enzymes between cell organelles and tissue (Table 9.1). When maize photosynthesizes in $^{14}CO_2$, the primary radioactive products are oxaloacetic, malic and aspartic acids (Hatch 1976). Later, ^{14}C accumulates in PCR cycle compounds (e.g. 3PGA), suggesting flow of carbon from organic acids to the PCR cycle. Extrapolation of the ^{14}C content (measured at steady state, when the fluxes of carbon between pools are constant) back to zero time, shows that more than 95 per cent of ^{14}C enters organic acids and less than 5 per cent enters 3PGA directly. Thus in the C4 leaf, RuBP carboxylase is, effectively, separated from the atmosphere.

Metabolic processes in mesophyll and bundle sheath have been shown by measuring the distribution of enzymes between them using several techniques. Mesophyll and bundle sheath cells are separated by grinding leaves to disrupt the 'softer' mesophyll tissues but not the more 'resistant' bundle sheath; the cells are separated by centrifugation. Also, enzymes are used to disrupt the cell walls of the tissues at different rates. The cells of mesophyll and bundle sheath are distinguished by different enzymes, chlorophyll *a/b* ratios, etc. Chloroplasts

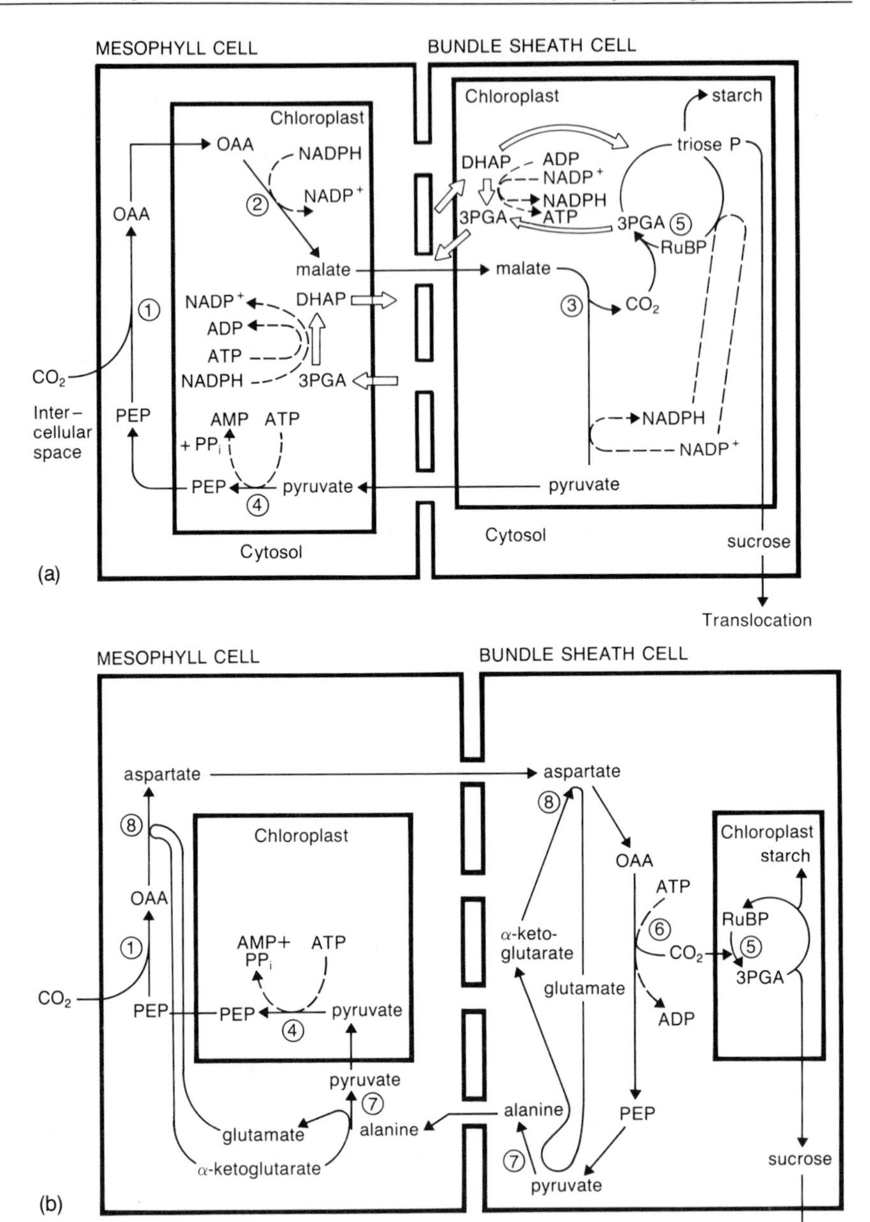

FIG. 9.2 Photosynthetic metabolism of C4 plants, compounds transferred between mesophyll and bundle sheath cells and method of decarboxylation with: (a) NADP requiring malic enzyme or 'NADP-ME' type; (b) aspartate forming and PEP carboxykinase or 'PCK' type of C4 metabolism; (c) aspartate forming and NAD requiring malic enzyme 'NAD-ME' type of C4 metabolism. Numbers refer to enzymes listed below: (1) PEP carboxylase; (2) NADP malate dehydrogenase; (3) NADP malic enzyme; (4) pyruvate, P_i dikinase; (5) RuBP carboxylase/oxygenase; (6) PEP carboxykinase; (7) alanine amino transferase; (8) aspartate amino transferase; (9) NAD malate dehydrogenase; (10) NAD malic enzyme.

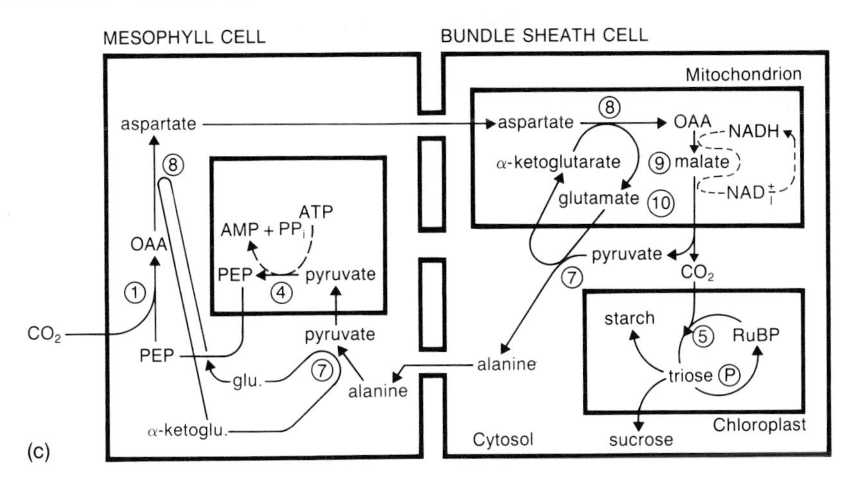

(c)

Fig. 9.2 continued

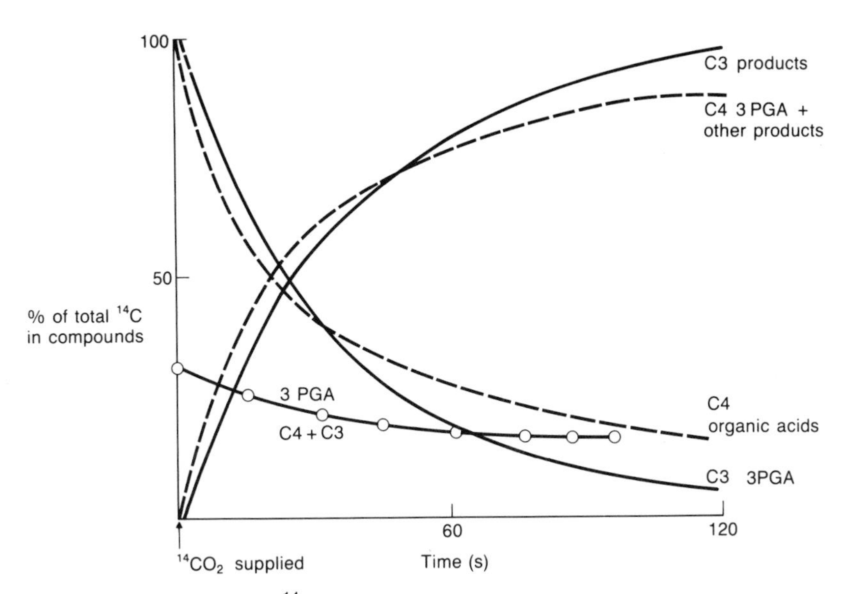

FIG. 9.3 Distribution of ^{14}C in products of photosynthesis by C4 plants; C3 plants; hypothetical distribution expected if C4 plants have substantial (30%) CO_2 fixation by C3 photosynthesis (3 PGA, C4 + C3). (Modified from Hatch 1976b, Figures 3 and 4).

and other organelles from the cells can be obtained by normal methods of cell disruption and separation. Distributions of enzymes in tissues and cell organelles clearly show the spatial separation of the carboxylating and decarboxylating processes. PCR cycle and decarboxylating enzymes are in the bundle sheath but not in mesophyll cells, where the only PCR cycle enzymes are for reduction of 3PGA to DHAP.

In all C4 plants carboxylation occurs in the mesophyll cell cytosol, catalysed by PEP carboxylase (PEPc), which consumes bicarbonate ions in reaction with PEP, forming oxaloacetic acid (OAA):

$$PEP + HCO_3^- + H^+ \xrightarrow[\text{carboxylase}]{\text{PEP}} OAA + P_i \qquad [9.1]$$

The reaction is irreversible with a large negative free energy change (−30 kJ mol^{-1}) and is therefore a control step in C4 assimilation. It has a pH optimum of about 8 (that of the mesophyll cell cytosol) and an affinity (K_m) of about 300 μM for bicarbonate. Such concentrations are in equilibrium with the internal CO_2 concentration of C4 leaves in air. As PEPc removes HCO_3^-, CO_2 from the atmosphere is rapidly converted to HCO_3^- by carbonic anhydrase even in low CO_2 in which C4 plants can photosynthesise but C3 plants cannot. The apparent K_m for CO_2 is 2 μM and PEPc can remove bicarbonate ions from water in equilibrium with an atmosphere containing 0.5–1 Pa CO_2. PEPc has no oxygenase activity, in contrast to RuBPc/o, to offset CO_2 fixation. Thus the enzyme is very efficient in assimilating CO_2, and gives very small compensation concentration, that is, when enclosed in an air-tight chamber in the light C4 plants remove almost all CO_2, in contrast to C3 plants. The enzyme has a large K_m for PEP, the other substrate, which is in considerable concentration in actively photosynthesizing mesophyll cells. The PEP reaction acts as an effective 'CO_2 pump', supplying CO_2 to the PCR cycle, and is efficient at high temperatures, to which many C4 plants are adapted. PEP is allosterically controlled by many photosynthetic assimilates (i.e. they change the rate of reaction without entering into it). Glucose- and fructose-6-phosphates and ribulose-5-phosphate increase the enzyme reaction, but OAA and malate slow it. If metabolism of the primary products is inhibited, e.g. by slow PCR cycle turnover, malate accumulates and PEP consumption slows thus decreasing the drain of carbon from the PCR cycle. The complex response of PEPc to metabolites probably regulates carbon fluxes in the leaf. Large concentration of glucose-6-phosphate, an important metabolite in starch synthesis (which can be made from PEP), activates PEPc, thus increasing photosynthesis and decreasing the drain to storage.

Oxaloacetate is not translocated to the bundle sheath but is first reduced to malate by NADP-dependent malate dehydrogenase in the mesophyll. Malate dehydrogenase, which maintains the reaction towards malate synthesis, has a high affinity for NADP and OAA, requires light or reduced substances to become active and is specific for NADP. Formation of malate is characteristic of maize and sugar cane, which have NADP malate dehydrogenase; malate is transported to the bundle sheath (Fig. 9.2c) and decarboxylated in the chloroplast by NADP-specific malic enzyme (controlled by Mg^{2+} and pH); hence this is 'NADP-ME type' of C4 assimilation (see Hatch 1976):

$$\text{malate} + NADP^+ \xrightarrow[\text{enzyme}]{\text{NADP malic}} \text{pyruvate} + CO_2 + NADPH \qquad [9.2]$$

Pyruvate from decarboxylation of malate is recycled to form PEP, by a reaction with ATP and P_i; the enzyme, pyruvate, P_i dikinase, and the reaction are unique to C4 plants. The enzyme, which is phosphorylated before reacting with pyruvate, is light activated and requires Mg^{2+} and alkaline pH. It is inhibited by products of the reaction, thus regulating the activity to the supply of CO_2 or ATP and keeping the PEP cycle and PCR cycle reactions balanced:

$$\text{pyruvate} + \text{ATP} + P_i \xrightarrow[\text{dikinase}]{\text{pyruvate, } P_i} \text{PEP} + \text{AMP} + PP_i \qquad [9.3]$$

Although the PEPc- and pyruvate, P_i dikinase-mediated reactions are common to all C4 species, there are differences in the metabolic routes by which pyruvate is formed, both in the enzymes and their location. A major difference is in the transport of aspartate rather than malate from mesophyll to bundle sheath. Aspartate is rapidly formed in the mesophyll cells by transamination of oxaloacetate by aspartate aminotransferase, with glutamate as the amino group donor.

$$\text{OAA} + \text{glutamate} \xrightarrow[\text{aminotransferase}]{\text{aspartate}} \text{aspartate} + \alpha\text{-ketoglutarate} \qquad [9.4]$$

The enzyme has a pH optimum of 8, a high affinity for substrates, and is inhibited by malate. In bundle sheath cells it deaminates aspartate. Aspartate is transported to the bundle sheath by a dicarboxylate shuttle (see p. 145).

There is distinction between C4 plants in the enzyme decarboxylating organic acids in the bundle sheath (Table 9.1). In malate formers, as already mentioned, decarboxylation is by NADP-requiring malic enzyme in the chloroplast of bundle sheath cells (NADP-ME type). In one group of aspartate formers, called the PCK type because of the role of PEP carboxykinase, aspartate is probably deaminated in the bundle sheath cell cytosol by aspartate aminotransferase, giving OAA. This is phosphorylated by PEP carboxykinase to PEP, releasing CO_2 which is fixed in the PCR cycle (Fig. 9.2b) and PEP is converted to pyruvate, which is transaminated to alanine (by a specific enzyme with glutamate as amino group donor). Alanine returns to the mesophyll cell where it is deaminated to pyruvate and phosphorylated to PEP.

The 'NAD malic enzyme (or NAD-ME) type' converts aspartate to oxaloacetate which is reduced to malate by the bundle sheath cell mitochondria and decarboxylated by an NAD-specific malic enzyme.

$$\text{malate} + \text{NAD}^+ \xrightarrow[\text{enzyme}]{\text{NAD malic}} \text{pyruvate} + CO_2 + \text{NADH} \qquad [9.5]$$

This contrasts with the NADP-ME type located in the chloroplast. Pyruvate is converted to alanine by the specific aminotransferase and returned to the mesophyll cell cytosol for conversion to the acceptor.

Energetics of C4 photosynthesis

Compared with C3 plants, C4 plants require an extra 2 or 3 molecules of ATP for fixation of 1 CO_2, *viz.* 5 or 6 ATP and 2 NADPH because of the conversion of pyruvate to PEP. The NADP-ME type (Table 9.2) requires 2 ATP and 1 NADPH in the mesophyll cells whereas the NAD-ME and PCK type require only 2 ATP and the PCK type uses 1 ATP in the bundle sheath. Photochemical competence differs between forms of C4 plants (Table 9.1). Bundle sheath chloroplasts of NADP-ME types have little PSII and may synthesize ATP by cyclic photophosphorylation. This greatly decreases oxygen accumulation in bundle sheaths. NADPH is supplied in NADP-ME types from the mesophyll by a malate shuttle; for each malate decarboxylated 1 NADPH is released. If NADPH supply is insufficient for the PCR cycle, reductant might be provided by a shuttle of DHAP from the mesophyll chloroplast into the bundle sheath where it is oxidized to 3PGA giving NADPH and ATP.

$$\text{DHAP} + \text{ADP} + \text{NADP}^+ \rightarrow \text{3PGA} + \text{ATP} + \text{NADPH} \qquad [\mathbf{9.6}]$$

The 3PGA from the bundle sheath is recycled back to the mesophyll to provide the substrate.

Bundle sheath cells of NAD-ME types have some non-cyclic electron flow to $NADP^+$, those of PCK types have less PSII activity and more cyclic photophosphorylation. Mesophyll chloroplasts of NAD-ME type have larger capacity for cyclic ATP synthesis than the NADP-ME type, which like the PCK type have both PSI and II. Aspartate formers probably have a normal

Table 9.2 Energy requirements of mesophyll (meso) and bundle sheath (bs) cells for ATP and NADPH per CO_2 fixed in C4 species (modified from Edwards *et al.* (1976) with permission). C4, is requirement in that part of pathway; PCR cycle, requirement in photosynthetic reduction cycle

	NADP-ME		NAD-ME		PCK	
	meso	bs	meso	bs	meso	bs
ATP						
C4	2	0	2	0	2	1
PCR cycle	1	2	0.5	2.5	1	2
Total	3	2	2.5	2.5	3	3
NADPH						
C4	1	0	0	0	0	0
PCR cycle	1	0	0.5	1.5	1	1
Total	2	0	0.5	1.5	1	1

balance of PSI and II activity in the bundle sheath chloroplast and appear not to require transport of reductant.

The advantages of the modifications to the photochemical apparatus and of the complex shuttles of reductant and ATP are not clear. C4 plants are able to absorb incident radiation at high intensity and use it for CO_2 assimilation but they are less efficient at low intensity. Light harvesting and distribution to the different photosystems may be more efficient if the functions are separated. ATP production by cyclic photophosphorylation (minimizing the production of O_2 and reductant) may overcome a limitation to the rate of PCR cycle activity, necessary for the efficient CO_2 accumulating system of PEP carboxylase to be fully exploited. Decreased PSII in bundle sheath chloroplasts decreases O_2 production, so that RuBPc/o, which has somewhat higher K_m for CO_2 in C4 compared to C3 plants, functions at low O_2 and high CO_2. PEPc is very efficient at assimilating CO_2 at very dilute concentrations allowing large CO_2 accumulation in the sheath. When organic acids from ten mesophyll cells supply one bundle sheath cell, decarboxylation gives a ten-fold increase in CO_2 concentration. In air this may give a partial pressure of 150–300 Pa of CO_2. As bundle sheath cells are impermeable and O_2 production inside them (in the NADP-ME type at least) is small, the CO_2/O_2 ratio is large. Thus C4 photosynthesis is insensitive to O_2 and photorespiration is very small with smaller light and CO_2 compensation concentration than C3 plants. If C4 and C3 plants (e.g. maize and bean) are illuminated together in a chamber the C4 plants remove CO_2 from the atmosphere more efficiently than the C3 species, which deplete their carbohydrate reserves and eventually die. As Clayton (1980) says 'CO_2 becomes transferred from the bean plant to the corn plant: the latter "eats" the former'. Thus efficient photosynthesis and C4 metabolism, which eliminates or minimizes photorespiration, are linked. However, some phosphoglycolate is formed and there is carbon flux through the glycolate pathway. Under conditions such as water stress, where CO_2 supply is limited by stomatal closure, an increase in glycine and serine formation indicates that more carbon enters the glycolate pathway but photorespiration, measured as CO_2 production and from the difference in assimilation between low (1 kPa) and high (21 kPa) O_2, does not increase. As PEPc is efficient, any CO_2 produced is re-assimilated and only under extreme conditions does CO_2 escape from maize leaves in the light. C4 plants have a greater photosynthetic rate than C3 plants at smaller stomatal conductance, so that they have a larger water use efficiency (WUE = photosynthesis/transpiration) and smaller total water loss in drought-prone environments. This slows the onset of water-stress and may decrease its severity under conditions of intermittent rainfall.

What is the biological rationale of C4 metabolism? It requires considerable movement of assimilates across cell membranes, imposing an additional energy burden, but probably transport between cells is by simple diffusion. C4 photosynthesis is superior to C3 because it allows efficient CO_2 assimilation, even in dilute CO_2. Stomatal conductance is smaller in C4 than C3 plants, helping to conserve water, but the resulting low internal CO_2 concentration (see Chs 10 and 11) has little effect on photosynthesis as PEPc is so efficient at gathering CO_2. Differences in photosystems, chloroplast structure etc., in C4 types probably relate to the requirements for balanced energy supply and reductant and the importance of the CO_2/O_2 ratio in their different environments. Most C4 plants are productive under bright sunlight in hot, often dry but not desert, conditions. However, C4 species are also found in temperate and in shaded conditions. The C4 temperate grass *Spartina townsendii* is found in salt marshes, where C4 metabolism may be an advantage under salt and osmotic stress.

The efficiency of C4 species has been achieved by extensive modification of the 'normal' C3 plant. Anatomy is differentiated at subcellular and cellular level, the enzyme complement is modified quantitatively (e.g. more PEP carboxylase), and qualitatively, with a unique enzyme, PEP carboxykinase. Chloroplast structure and light reactions are modified to optimize ATP and NADPH supply. The genome of C3 predecessors has been, presumably, modified as a result of a long evolutionary process by selection under environmental pressure. If C4 photosynthesis has arisen more than once (as the different metabolic types suggest) then present day C3 species may include individuals with more C4-like characteristics, which could be selected. Such plants might be more productive agricultural crops in hot, dry or saline conditions. Selection has been attempted by growing varieties of C3 species such as wheat together with C4 plants in an enclosed atmosphere so that they were subjected to CO_2 concentrations below the compensation concentration: those with C4-like assimilation would have maintained activity, or lived longer than the C3 types. Thousands of individual plants of many different wheat varieties have been tested but no C4 or high efficiency plants have been obtained.

Intermediates between C3 and C4 exist in the genera *Mollugo* and *Panicum*, with intermediate gas exchange rates, compensation values and anatomy; CO_2 fixation seems to be C4, with PEP carboxylase. Intermediate forms are of interest for the light they may shed on the evolution of C4 photosynthesis. Crossing C3 and C4 species of *Atriplex* (e.g. *A. triangularis* (C3) and *A. rosea* (C4)) gives fertile F_1 hybrids, which are intermediate morphologically and anatomically, with a weakly developed bundle sheath. The enzymes are C4, with PEP carboxylase and RuBP carboxylase, but the enzymes are not compartmented as in the C4 parent, for example RuBP carboxylase is in all cells and not restricted to the bundle sheath. Also C4 acids are synthesized but not effectively metabolized. This leads to inefficient photosynthesis and growth.

Crassulacean acid metabolism

CAM plants assimilate CO_2 and synthesize organic acids, mainly malate, during darkness but assimilate little CO_2 in the light (Fig. 9.4). In darkness stomata open, offering a low resistance to CO_2 diffusion into leaves. Carboxylation by PEPc, in the β position of PEP, forms oxaloacetate, probably in the

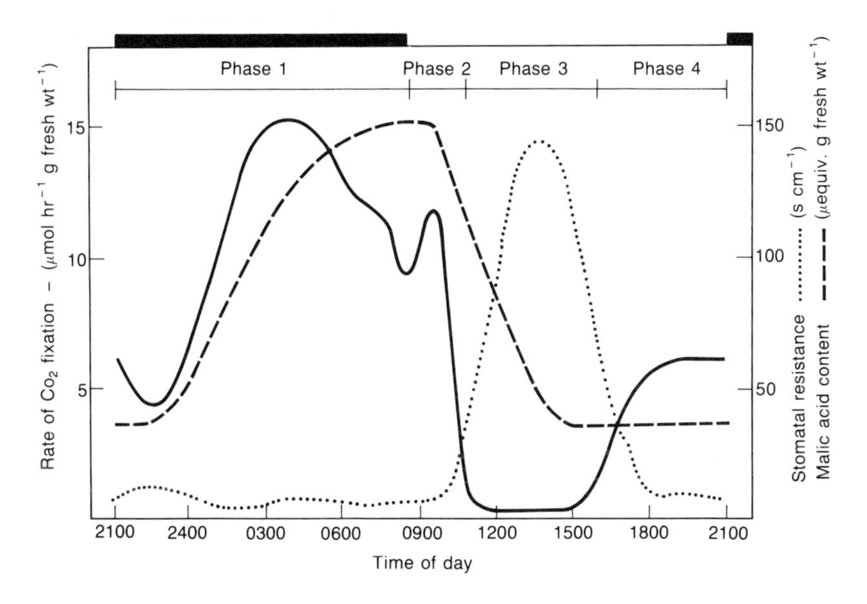

FIG. 9.4 Diurnal course of assimilation, stomatal opening and malate accumulation in a CAM plant. (Modified from Osmond 1978, Figure 1).

cytosol (Fig. 9.5). Oxaloacetate is reduced by NADH to malate by cytoplasmic NAD^+ malate dehydrogenase. Malic acid is transported as hydrogen malate across the tonoplast and accumulates (150 μequivalents per gram fresh weight) in the vacuole; there may be other pools in leaf cells. The pH of the vacuole becomes very acidic although metabolism is protected by compartmentation. PEP is derived from storage carbohydrate, probably glucans (starch and a dextran, a glucose polymer), mobilized to produce 3PGA and PEP; ATP is consumed and synthesized in these reactions so that the energy consumption is balanced. NADH from oxidation of glyceraldehyde phosphate reduces oxaloacetate; as with ATP, storage carbohydrate supplies the energy for CO_2 fixation in the dark. Early in the dark period malate accumulation and CO_2 fixation are slow, but increase in the middle of the dark period before decreasing. This pattern results from stimulation of CO_2 fixation by a higher PEP concentration which overcomes the inhibition of PEPc by malate. Phosphofructokinase which converts fructose-6-phosphate to fructose bisphosphate is a hundredfold less sensitive to PEP than that of C3 plants. Control of processes in CAM has been reviewed by Osmond (1978) and Kluge (1979).

With illumination of CAM leaves, CO_2 fixation from the atmosphere decreases rapidly to a very low rate and stomata close. Malate is transported out of the storage compartment and is decarboxylated by NADP malic enzyme, and perhaps NAD malic enzyme, which is found in many families with CAM, or by PEP carboxykinase which is more important in others (e.g. Bromeliaceae,

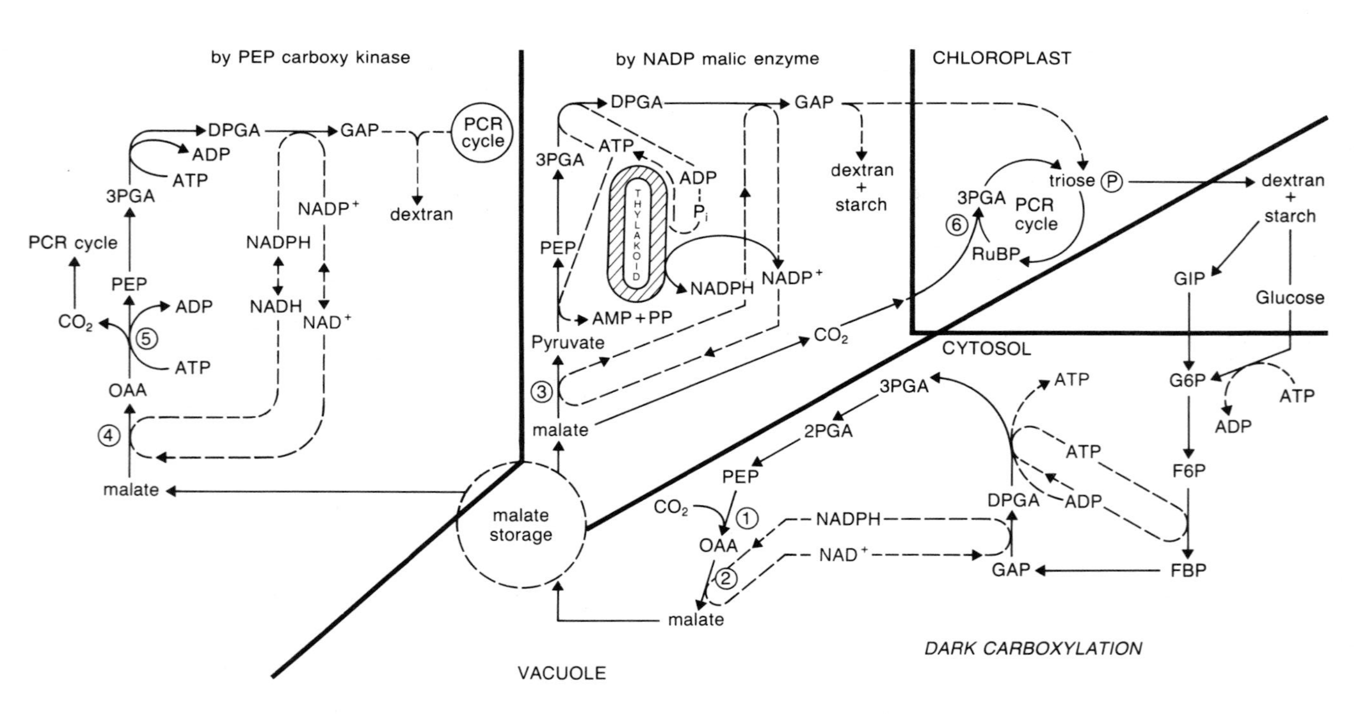

FIG. 9.5 Scheme of crassulacean acid metabolism (CAM) showing the reactions responsible for dark CO_2 assimilation with formation of organic acid, storage and subsequent decarboxylation in the light. Numbers refer to the enzymes listed below: (1) PEP carboxylase; (2) malate dehydrogenase; (3) NADP malic enzyme; (4) NAD malic enzyme; (5) PEP carboxykinase; (6) RuBP carboxylase/oxygenase.

Liliaceae). They are designated NADP malic enzyme and PEP carboxykinase types respectively. The CO_2 released, probably in the cytosol, enters the chloroplast and is assimilated by the PCR cycle with ATP and NADPH from electron transport. Pressure of CO_2 in the tissue may reach 1000 Pa and oxygen rises from 21 to 26 kPa but the RuBP carboxylase reaction is favoured and photorespiration minimized. Malate decarboxylation is controlled by a low $NADP^+$/NADPH ratio, as $NADP^+$ is the acceptor for reductant. Decarboxylation could occur with $NADP^+$ in the dark but ATP is required for decarboxylation and is available (in the required amount) only in light. Pyruvate is phosphorylated to PEP and recycled to triosephosphate and thence to carbohydrates (mainly glucan) which are stored. PEP carboxykinase CAM plants generate oxaloacetate from malate and NADH by malate dehydrogenase in the cytosol (or possibly chloroplast).

Early in the light period there is little CO_2 fixation because PEP carboxylase is inhibited by malate, but as this is decarboxylated the PEPc fixes more CO_2 in the light. After long periods in the light and extensive decarboxylation, CO_2 from the atmosphere is assimilated directly by RuBPc/o and the PCR cycle, giving rise to storage glucans. Photorespiration occurs in CAM plants as a result of a low CO_2/O_2 ratio, when decarboxylation has depleted the storage materials.

CAM permits CO_2 accumulation in darkness when stomata are open, without substantial loss of water, for the plants are predominantly of arid habitats with hot, dry and very bright days, which would cause much water loss if stomata were open. A very high ratio of CO_2 fixation to water loss is achieved under conditions lethal to most C3 plants. Energy accumulated in the light is used to generate the substrate for the PEPc reaction at night when low temperatures (characteristic of desert and alpine environments) favour malate accumulation. Large vacuoles enable CAM plants to store both malate and water, obviously important as a buffer against desiccation.

Photosynthetic discrimination against carbon isotopes

Atmospheric CO_2 is composed of the stable isotopes of carbon, 98.9 per cent ^{12}C and 1.1 per cent ^{13}C. The thermodynamic and kinetic properties of compounds containing the different isotopes differ due to mass, and they participate to different extents in physical and chemical reactions, that is there is discrimination, and the proportion of isotopes in products of the reactions changes. The $^{13}C/^{12}C$ ratio gives important information about the reaction and can be used as a tracer for the flow of carbon.

The ratio is determined by mass spectrometry of CO_2 (organic matter is first oxidized) and compared with a standard of known ratio, usually a belemnite (fossil shell) earth from the Peedee formation in South Carolina, USA and called the PDB standard. The ratio of unknown to standard isotope distribution is $\delta^{13}C$:

$$\delta^{13}C\ (^{o}/_{oo}) = \frac{(^{13}C/^{12}C)\text{ unknown substance} - (^{13}C/^{12}C)\text{ standard}}{(^{13}C/^{12}C)\text{ standard}} \times 1000$$

Negative $\delta^{13}C$ shows ^{13}C enrichment; atmospheric CO_2 is $-7^{o}/_{oo}$. Plant material $\delta^{13}C$ ranges from -10 to $-40^{o}/_{oo}$; C3 plants have -22 to $-34^{o}/_{oo}$ (mean $-27^{o}/_{oo}$), C4 plants -10 to $-18^{o}/_{oo}$ (mean $-13^{o}/_{oo}$) and CAM plants are in the range of C3 and C4 plants. These differences arise as RuBPc/o discriminates more strongly ($-28^{o}/_{oo}$) against $^{13}CO_2$ than PEP carboxylase ($-9^{o}/_{oo}$) due to the chemical reaction; RuBPc/o uses CO_2 and PEPc HCO_3^- ion. In C4 plants, CO_2 is concentrated in the bundle sheath and $^{13}CO_2$ is fixed more effectively than from the atmosphere. CAM plants use PEPc and also RuBP carboxylase directly, hence the wide range of $\delta^{13}C$ values. Measurement of $\delta^{13}C$ provides confirmation of the metabolic origin of organic carbon; it suggests, for example, that ferns and bryophytes from the Carboniferous period had C3 photosynthesis, and it distinguishes between sucrose from sugar cane (C4) and sugar beet (C3), and between artificial and natural vanilla.

Sun and shade plants

Metabolism and characteristics of C3 plants, their light-harvesting systems, chemistry of CO_2 fixation and rate of photosynthesis, have been presented as if without variation. Even within the distinct types of C4 and CAM plants, uniformity of response has perhaps been implied. However, this is not the case. C3 plants show great diversity of response to the environment, particularly to the major determining factor in plant growth – light. C4 plants, although predominantly adapted to bright light, also vary. Distinction may be made between plants with high rates of photosynthesis and growth in very intense light, so-called 'sun plants', which are inefficient, with poor photosynthesis and survival in dim light, and 'shade plants' which photosynthesize and survive in dim light but are unable to function efficiently in bright light (low maximum rates of photosynthesis and photochemical damage). Divisions are rather blurred when a wide range of species is examined. Those plants clearly in either category are obligate sun or shade plants, genetically adapted and incapable of adjustment to the other extreme condition. However, many species show flexibility in response to light intensity. They are facultative sun or shade species and may grow in high and low intensity illumination but lack the ability to adapt to the extremes.

Sun plants, which include many crops and plants of tropical regions, achieve maximum rates of photosynthesis greater than 30 μmol CO_2 m^{-2} s^{-1} and respiration rates in darkness of 2 μmol CO_2 m^{-2} s^{-1}. Shade plants may have photosynthesis rates less than 10 μmol CO_2 m^{-2} s^{-1} at light intensity perhaps one-tenth of sun species and may be damaged by light intensities above half that of full sunlight.

Extreme adaptation to low light is shown by plants from the floor of tropical forests, where the photosynthetically-active radiation (PAR) may be less than

3 per cent of the radiation at the top of the forest canopy and greatly enriched in wavelengths of green, far red and infra-red, which are not absorbed by the foliage above. Sunflecks, small spots of intense light, which pass between the leaves in the canopy, contain over half the PAR in rainforests. Many taxonomically different plants are able to grow in poorly illuminated habitats, species which have been studied in detail such as *Alocasia* and *Cordyline* are from such environments. Even in temperate conditions, dim light within the canopy of a single tree is also related to the development of shade leaves.

Anatomically, shade plants have characteristics which are, without undue teleology, adapted to increase light absorption. Leaves are thin, with small weight per unit leaf area, thin epidermis and large intercellular spaces; they have few cell layers in the mesophyll and small, thin-walled cells. Sun plants have thick palisade and spongy mesophyll tissue so that there is a two- to three-fold increase in the number of cell layers. Mesophyll cells are large and thick-walled, and there is less intercellular air space and a greater (up to five-fold) surface area of cells to total leaf than shade plants. Leaf thickness appears to be an important variable in many plants and allows flexibility in the use of light and CO_2 etc., without changing the other physiological properties of the leaf. However, some shade plants have thick leaves; it is not clear what ecological advantage this may have.

Shade and sun plants and leaves of the same plant from different illuminations differ in chloroplast membranes and light-harvesting and electron transport mechanisms. Chloroplasts are usually more numerous in mesophyll cells of shade plants, and arranged near the upper leaf surface whilst cells of the lower mesophyll have few chloroplasts. However as the number of cell layers is smaller, shade plants often have less chlorophyll per unit leaf area. Thylakoid membranes are stacked into many grana with many lamellae and less stromal lamella. *Alocasia* grown in dim light has four times as many granal partitions per stack as spinach grown in bright light, thus increasing the area of light-capturing membrane. Grana are often irregularly orientated which may increase capture of diffuse or variably orientated light.

Shade plants may contain four to five times more chlorophyll *a* and *b* per unit volume of chloroplast and have a higher *b/a* ratio than sun plants because the light-harvesting complex (LHC) increases. There are more LHC particles on the protoplasmic freeze fracture faces of stacked areas (EF_s) without an equivalent increase in the large particles, so that there is relatively more LHC in shade plants compared with sun plants. Both PSI and PSII decrease with shade but the ratio of antenna chlorophyll to reaction centres is slightly larger than in high-light grown plants of sun species. Thus shade enhances the capacity for light capture and energy transfer to the reaction centres. However the capacity of the electron transport chain in shade plants is not increased, as there is relatively much less (one-fifth) cytochrome *f*, plastoquinone, ferredoxin and carotenoid per unit of chlorophyll than in sun plants. Shade plants have, therefore, more light-collecting apparatus, but a smaller complement of electron carriers than sun plants. In dim light the rate of electron transport is limited by the number of photons falling on the leaf; it would be no advantage for shade plants to produce a large capacity electron transport chain. With a pool of plastoquinone receiving electrons from several PSII reaction centres, this rate limiting step is minimized in the shade plants, except when the plant is exposed to light of a brightness outside its normal range.

Sun plants have less developed thylakoid systems, fewer granal stacks and partitions and less LHC so they are less efficient at absorbing light energy at low photon flux than shade plants and so have lower quantum yield. However, photon capture does not limit photosynthesis in bright light and electron transport, or more probably photophosphorylation, may be limiting at very large photon flux for many species. Hence the larger capacity of the electron transport acceptors, particularly plastoquinone. Electron transport in sun plants may be 15–30 times faster than in shade species (uncoupled rate). However the quantum yields of electron transport to DCIP, are very similar in very dim light, so there is no difference in photochemical efficiency, that is the reaction centres are able to transfer electrons at the same rate. Yet the uncoupled electron transport rates are saturated at 600–800 μmol quanta $m^{-2} s^{-1}$ for sun plants and 50–100 μmol quanta $m^{-2} s^{-1}$ for shade species. Clearly the main differences between the two groups of plants is the capacity of the light-harvesting system and of electron transport. The light absorbing system in shade plants makes them very effective at gathering the light available and passing it to the reaction centres, especially in dim light, but they are limited in bright light by the rate of electron transfer; sun plants in contrast are very efficient at transporting electrons but not at gathering weak light.

Of course it is not electron transport *per se* which determines CO_2 assimilation but NADPH and ATP synthesis. Light capture is the dominant problem in shade plants and their photosynthesis may be limited by NADPH synthesis; possibly extensive granal stacking is essential to obtain sufficient rates of electron flow to reduce $NADP^+$. Cyclic electron flow could drive ATP synthesis to match $NADPH^+$ production and the extensive grana may restrict coupling factor to a small area of thylakoid to increase the ΔpH gradient near CF_1. Sun plants probably have adequate $NADP^+$ reduction, but may be limited by ATP supply; this applies also to all C4 plants. Cyclic electron transport then may produce the required ATP without $NADP^+$ reduction. With PSI concentrated in the unstacked thylakoids of sun plants, cyclic flow could be enhanced. Until the rates of ATP and NADPH synthesis are determined for plants of different light environments the controlling factor will remain unknown. The photosynthetic systems of strongly illuminated shade plants may be irreversibly damaged by very intense light, whereas sun plants are apparently insensitive. Slow movement of electrons through plastoquinone at high rates of electron flow in bright illumination causes ‘backing up’ of electrons and reaction centres cannot use excitation energy so that the high energy states of chlorophyll accumulate and damage increases. Photosystem II appears more sensitive to photoinhibition than PSI. The structure of shade plant thylakoids makes them more easily damaged. Possibly the structure of the photosystems, or carotenoid complement, which reduces the energy load and provides a 'safety valve' is inadequate in shade plants. Sun plants have relatively smaller light-harvesting systems so are inefficient in weak light, but the greater number of cell layers and the greater capacity of the electron transport chain for electron flow and excess energy dissipating systems contribute to their greater efficiency and capacity to assimilate CO_2 in intense illumination.

Assimilation is expected to be limited by CO_2 supply when photosynthesis is rapid; sun plants have more stomata per unit area of leaf and larger stomatal conductances than shade species. Internally the cell surface area is greater, increasing the conductance of the CO_2 supply pathway. Sun plants also have more enzyme capacity in a greater stromal volume. There is more RuBP carboxylase per unit of chlorophyll; as this enzyme is a major protein of leaves, the larger protein content of sun leaves and the lower content in shade plants is expected. In very brightly illuminated sun and shade plants, with saturated rates of electron transport, the capacity of the enzyme system might be insufficient to exploit all the NADPH and ATP synthesis. However this remains to be tested; there is a parallel between rate of photosynthesis and the amount of $RuBP_{c-o}$ under such circumstances.

Dark respiration in shade plants is much less than in sun plants, *c.* 0.15 compared with 2 μmol $m^{-2} s^{-1}$. Also, photorespiration is probably less because although the RuBP oxygenase to carboxylase ratio of the protein is similar, the CO_2/O_2 ratio in the tissue is more favourable and dimly lit environments are cooler. Light compensation, the photon flux at which respiratory CO_2 production and photosynthetic CO_2 assimilation are equal, and net photosynthesis is zero, is very much smaller in shade than sun plants. Of course the ratio of respiration to CO_2 assimilation at saturating light intensities is smaller in sun plants compared with shade plants and so the growth rate of sun plants is much greater than shade species.

Effects of growth conditions on facultative sun and shade plants are similar in type if not degree to those described for obligate species. Species such as *Solidago virgaurea* and *Solanum dulcamera* adapt to light conditions during growth, developing more grana and partitions in dim light, rather more small particles compared with large particles on the thylakoids and more LHCP. Bright light encourages plants to form more RuBP carboxylase, a larger proportion of stroma per chloroplast and more chloroplasts. Also stomatal resistance per unit area of leaf is smaller in bright light because the number of stomata increases. Strong illumination during development leads to a greater amount of electron transport components, particularly plastoquinone and cytochrome *f* and a higher intrinsic rate of transport, without greater quantum efficiency. Thus the changes in plants of one species parallel those in obligate sun and shade plants. Possibly there is a parallel development between the light-harvesting and electron transport systems and the 'dark reaction' mechanism to ensure balanced activity. This may be genetically determined for maximum efficiency in use of resources, such as nitrogen for protein production. However the degree of change is usually restricted, and most species cannot adapt fully to the extremes, presumably due to genetic regulation.

References and Further Reading

Black, C. C. Jr. (1973) Photosynthetic carbon fixation in relation to net CO_2 uptake, *A. Rev. Plant Physiol.*, **24**, 253–86.

Boardman, N. K. (1971) The photochemical systems in C3 and C4 plants, pp. 309–22 in Hatch, M. D., Osmond, C. B. and Slatyer, R. O. (eds), *Photosynthesis and Photorespiration*, Wiley Interscience, New York.

Brown, R. H. (1976) Characteristics related to photosynthesis and photorespiration of *Panicum milioides*, pp. 311–25 in Burris, R. H. and Black, C. L. (eds), *CO_2 Metabolism and Plant Productivity*, University Park Press, Baltimore.

Canvin, D.T. (1979) Photorespiration: Comparison between C3 and C4 plants, pp. 368–98 in Gibbs, M. and Latzko, E. (eds), *Encyclopedia of Plant Physiology* (N.S.), vol. 6, *Photosynthesis II*, Springer-Verlag, Berlin.

Chollet, R. and **Ogren, W. L.** (1975) Regulation of photorespiration in C3 and C4 species, *Bot. Rev.*, **41**, 137–79.

Clayton, R. K. (1980) *Photosynthesis: Physical Mechanisms and Chemical Patterns*, I.U.P.A.B. Biophysics Series, Cambridge University Press, London.

Coombs, J. (1979) Enzymes of C4 metabolism, pp. 251–62 in Gibbs, M. and Latzko, E. (eds), *Encyclopedia of Plant Physiology* (N.S.), vol. 6, *Photosynthesis II*, Springer-Verlag, Berlin.

Coombs, J. (1976) Interactions between chloroplasts and cytoplasm in C4 plants, pp. 279–313 in Barber, J. (ed.), *The Intact Chloroplast*, Elsevier/North-Holland Biomedical Press, Amsterdam.

Dittrich, P. (1979) Enzymes of crassulacean acid metabolism, pp. 263–70 in Gibbs, M. and Latzko, E. (eds), *Encyclopedia of Plant Physiology* (N.S.), vol. 6, *Photosynthesis II*, Springer-Verlag, Berlin.

Edwards, G. E. and **Walker, D.** (1983) *C3, C4: Mechanisms, and Cellular and Environmental Regulation of Photosynthesis*, Blackwell Scientific/Oxford, University of California Press, Berkeley.

Edwards, G. E., Huber, S. C., Ku, S. B., Natham, C. K. M., Gutierrez, M. and **Mayne, B. C.** (1976) Variations in photochemical activities of C4 plants in relation to CO_2 fixation, pp. 83–112 in Burris, R. H. and Black, C. L. (eds), *CO_2 Metabolism and Plant Productivity*, University Park Press, Baltimore.

Hatch, M. D. (1976) The C4 pathway of photosynthesis: mechanism and function, pp. 59–181 in Burris, R. H. and Black, C. L. (eds), *CO_2 Metabolism and Plant Productivity*, University Park Press, Baltimore.

Hatch, M. D. (1976b) Photosynthesis: the path of carbon, pp. 797–845 in Bonner, J. and Varner, J. E. (eds), *Plant Biochemistry*, Academic Press, New York.

Hatch, M. D. and **Osmond, C. B.** (1976) Compartmentation and transport in C4 photosynthesis, pp. 144–84 in Stocking, C. R. and Heber, U. (eds), *Encyclopedia of Plant Physiology*, vol. 3, *Transport in Plants*, Springer-Verlag, Berlin.

Kluge, M. (1979) The flow of carbon in crassulacean acid metabolism (CAM), pp. 113–25 in Gibbs, M. and Latzko, E. (eds), *Encyclopedia of Plant Physiology* (N.S.), vol. 6, *Photosynthesis II*, Springer-Verlag, Berlin.

Laetsch, W. M. (1974) The C4 syndrome: a structural analysis, *A. Rev. Plant Physiol.*, **25**, 27–52.

Mayne, B. C., Edwards, G. E. and **Black, C. C.** (1971) Light reactions in C4 photosynthesis, pp. 361–71 in Hatch, M. D., Osmond, C. B. and Slatyer, R. O. (eds), *Photosynthesis and Photorespiration*, Wiley Interscience, New York.

Osmond, C. B. (1978) Crassulacean acid metabolism: a curiosity in context, *A. Rev. Plant Physiol.*, **29**, 379–414.

Osmond, C. B. and **Holturn, J. A. M.** (1981) Crassulacean acid metabolism, pp. 283–328 in Hatch, M. D. and Boardman, N. K. (eds), *The Biochemistry of Plants*, vol. 8, *Photosynthesis*, Academic Press, New York.

Ray, T. B. and **Black, C. C.** (1979) The C4 pathways and its regulation, pp. 77–101 in Gibbs, M. and Latzko, E. (eds), *Encyclopedia of Plant Physiology* (N.S.), vol. 6, *Photosynthesis II*, Springer-Verlag, Berlin.

Ting, I. P. and **Gibbs, M.** (1982, eds) *Crassulacean Acid Metabolism*, American Society of Plant Physiology, Waverly Press, Baltimore, Maryland.

Troughton, J. H. (1979) $\delta^{13}C$ as an indicator of carboxylation reactions, pp. 140–9 in Gibbs, M. and Latzko, E. (eds), *Encyclopedia of Plant Physiology* (N.S.), vol. 6, *Photosynthesis II*, Springer-Verlag, Berlin.

CHAPTER *10*

Carbon dioxide supply for photosynthesis

Carbon dioxide, the major substrate for photosynthesis, is supplied from the medium in which an organism lives, water for aquatic bacteria, algae and some higher plants and the atmosphere for terrestrial plants. Only transport of carbon dioxide from the atmosphere to higher plant leaves is considered here although the basic principles are applicable to other situations. To understand the factors controlling the fluxes of CO_2 the relationships between mass and volume of gases, the effects of temperature and pressure and units of expression of gas concentration are briefly considered, for it is often essential to interconvert volume, mass and pressure of a gas in studies of photosynthesis.

For an ideal gas, and employing S.I. units, the relation between the number of moles of gas (n) volume (V, m^3) and pressure (P) in Pascal (1 Pa = 1 Nm^{-2}) at a temperature (T, in Kelvin) is given by the ideal gas equation:

$$PV = nRT \qquad [\mathbf{10.1}]$$

The molar gas constant (R) is 8.31 J mol^{-1} K^{-1}. Standard atmospheric pressure is 101 325 Pa = 1.013 bar = 760 mm mercury. A mole of gas at 273 K (0 °C) and 101 325 Pa occupies 0.0224 m^3 (22.414 dm^3 or litres in the frequently employed but non-S.I. unit) and contains Avogadro's number of particles (6.023×10^{23}).

Gas concentrations are expressed in several ways as volume or mass or number of moles of gas per unit volume, or as partial pressure or mol fraction. Expression of gas concentration as volume per unit volume of a mixture of gases is frequent and confusing, e.g. m^3 m^{-3}, litre $litre^{-1}$ or μl $litre^{-1}$; 1 cm^3 m^{-3} and 1 μl $litre^{-1}$ are equivalent to one volume per 10^6 volumes and are called (volume) parts per million, abbreviated to vpm or ppm. Volume units are also expressed as (volume) %, e.g. 1 vpm = 0.0001 volume %. S.I. units of volume concentration are m^3 (or cm^3) m^{-3}. Air contains about 340 cm^3 CO_2 m^{-3}, equivalent to 340 vpm CO_2 or 0.034% CO_2 and 0.21 m^3 O_2 m^{-3} or 210 000 vpm or 21 (volume) % O_2. Volume of gas may be converted to mass from eqn 10.1; air at 0 °C and 101 325 Pa contains 668 mg CO_2 m^{-3} or 15 mmol CO_2 m^{-3} and 300 g or 9.4 mol O_2 m^{-3}.

The proportion of a gas in a mixture is also described by its pressure. Dalton's law of partial pressure states that in a mixture of gases (which do not interact) each gas has the same pressure as if occupying its volume alone and the total pressure is the sum of the partial pressures; for air (ignoring rare gases):

$$P_{total} = P_{O_2} + P_{N_2} + P_{CO_2} + P_{H_2O} \quad [10.2]$$

Partial pressure of CO_2 in air is 34 Pa or 344 μbar and that of O_2 is 21.3 kPa or 210 mbar. Partial pressure and mole fraction of each gas are numerically equal (from eqn 10.1) and remain constant with changing temperature and pressure. To conform with S.I. units, partial pressures of gases are used in the text. Table 10.1 gives conversion factors for units of concentration and pressure.

Table 10.1 Conversion factors for CO_2 and O_2 concentrations in air at 20 °C and 101 315 Pa

1 μl CO_2 litre^{-1} = 41.6 × 10^{-6} mol CO_2 m^{-3} = 1.83 × 10^{-3} g CO_2 m^{-3}

1 mg CO_2 m^{-3} = 0.0227 mmol m^{-3} = 0.554 μl CO_2 litre^{-1}

1 mol CO_2 m^{-3} = 44 g m^{-3} = 24.4 cm^{-3} litre^{-1}

21% O_2 = 210 cm^3 O_2 litre^{-1} = 0.21 m^3 O_2 m^{-3}

1 mol O_2 m^{-3} = 32 g m^{-3} = 24.4 cm^3 litre^{-1}

Pressure

1 μl litre^{-1} = 1.01 μbar = 0.101 Pa

340 μl CO_2 litre^{-1} = 340 μbar = 34.4 Pa

21% O_2 = 210 mbar = 21.3 kPa

It is important to appreciate the great difference in concentration between CO_2 and O_2 in the atmosphere; O_2 is 600 times more concentrated than CO_2 which favours even inefficient oxygenation reactions compared with carboxylation reactions, for example RuBP oxygenase compared with carboxylase. Plants must accumulate CO_2 from very dilute concentration and, at the same time, function at large O_2 concentration with large gradients of water vapour pressure between leaf and atmosphere. A leaf at 23 °C with the internal air saturated with water vapour (100% relative humidity) has a vapour pressure of about 2.8 kPa compared with 1.5 kPa of air at 25 °C and 50% relative humidity; this corresponds to a gradient of water content of about 12 g m^{-3}. Thus leaves must absorb CO_2 whilst limiting loss of water vapour. Stomata function as 'control valves' which regulate these conflicting interests.

CO_2 in the atmosphere

Carbon dioxide concentration in the earth's atmosphere changes from year to year and with season, whereas O_2 concentration varies little. Carbon dioxide is removed from the atmosphere by photosynthesis with carbon accumulating in wood, peat and humus, and CO_2 is absorbed by the world's oceans where chemical reactions lead to deposition of carbonates. Global patterns of CO_2 fixation are therefore determined by seasonal processes such as photosynthesis and sea temperature. Carbon dioxide is produced by respiration of photosynthetic products, by burning wood and, of increasing importance since the 18th century, the fossil fuels, oil, coal and gas.

The world's atmosphere contains some 7×10^{14} kg of carbon as CO_2 and there is 1×10^{15}–2×10^{15} kg of carbon in organic material both living and dead in the biosphere; the oceans hold 4×10^{16} kg of carbon as CO_2 and bicarbonate ions which are the major store of relatively rapidly remobilized carbon and an enormous buffer for the world's atmospheric CO_2. However, the rates of CO_2 exchange between atmosphere, oceans and biosphere are too slow to maintain a constant atmospheric CO_2 concentration. Over the last century atmospheric CO_2 has increased from less than 28 Pa to 34 Pa, some 0.7 Pa per decade. The present annual rate of increase is 0.12 Pa, due to fossil fuel consumption and also to destruction of the great tropical forests which is proceeding at an historically unparalleled rate, liberating CO_2 and decreasing photosynthesis. Of the total plant mass on the land, 82 per cent is as wood in forests, mainly in the tropics (50 per cent of all forests) and these account for 60 per cent of terrestrial photosynthesis. The rate of global photosynthesis is approximately 8×10^{13} kg C year^{-1}, compared with the production of 5×10^{12} kg of C (as CO_2) year^{-1} from burning fossil fuels and forests and 5.6×10^{13} kg C from respiration of heterotrophs and fires.

Effects of CO_2 on world climate

Carbon dioxide absorbs infra-red radiation; short wave solar radiation warms the earth and plants, which emit long wave infra-red radiation. This is absorbed by CO_2 and the atmosphere heats up, producing a 'greenhouse or glasshouse effect', so-called because it is analogous to heating in glasshouses caused by poor transmission by glass of infra-red compared to short wave radiation. Increased atmospheric CO_2 concentration will therefore increase atmospheric temperature. However, the extent of global warming is hotly debated, as counterbalancing factors may minimize the effect, for example a warmer atmosphere increases evaporation and clouds which reflect solar radiation back into space and decrease the input of energy.

Increased CO_2 would benefit plants, giving more substrate for assimilation and offsetting the effects of O_2 concentration on C3 plants (Ch. 11). Photosynthesis would increase with warmth and reduce the CO_2 effect by accumulation of organic carbon, such as wood. The present CO_2 concentration is low compared to earlier periods in the earth's history, the result of accumulation of inorganic matter which also allowed O_2 to increase. A large CO_2 concentration in Carboniferous times, associated with high temperatures may have allowed rapid photosynthesis and massive accumulation of organic carbon which decreased CO_2 in the atmosphere.

Rapid increase in CO_2 may affect many aspects of the human environment, by melting polar ice and raising sea level for example. Agriculture might benefit by warmer, longer, growing seasons (particularly in temperate zones) and increased CO_2. However, if world climate zones are affected, rainfall could be seriously disrupted and deserts might enlarge; these changes may already be occurring although difficult to detect against the fluctuations in world climate. The long term effects of changing the balance between CO_2 production, photosynthesis and geochemical processes and the implications for mankind, are discussed in the books listed at the end of the chapter.

Solubility of CO_2 and O_2 in water

Photosynthesis consumes forms of CO_2 dissolved in water, as enzymatic reactions proceed only in the aqueous state. Ribulose bisphosphate carboxylase uses dissolved CO_2, and PEP carboxylase bicarbonate ions. Solubility of CO_2 and O_2 depends on solvent temperature, partial pressure of the gas, and chemical nature of gas and solvent. For dilute solutions, Henry's law states that at constant temperature the volume or mass of gas dissolved in a given volume of liquid is directly proportional to the pressure. Table 10.2 gives the solubility of CO_2 and O_2 in water. For air 6.51 cm^3 O_2 dissolve in 1000 cm^3 or 0.291 mol m^{-3} (291 μM). For CO_2 at a partial pressure of 34 Pa and with solubility of 0.888 m^3 CO_2 m^{-3} water, 0.29 cm^3 dissolve in 1000 cm^3 or 12 mmol CO_2 m^{-3} (12 μM) (see Table 10.4).

Gases are less soluble, generally, at warm temperatures than cold (Table 10.2). Use of the mol fraction or partial pressure (including solvent vapour) accounts for the temperature and pressure effects in the gas phase and solution temperature effects are included in the solubility. Solubility of gases is inversely proportional to the concentration of solutes; in complex biological fluids proteins and salts decrease solubility similarly for all gases.

Table 10.2 Solubility of CO_2 and O_2 in pure water at 101 325 Pa pressure of the gas as a function of temperature

T(°C)	mol CO_2 m^{-3}	m^3 CO_2 m^{-3}	mol O_2 m^{-3}	m^3 O_2 m^{-3}
0	76.4	1.71	2.17	0.049
10	51.4	1.19	1.68	0.038
15	43.1	1.02	1.50	0.034
20	36.5	0.87	1.36	0.031
25	31.1	0.76	1.23	0.028
30	26.7	0.67	1.12	0.026
40	21.2	0.53	1.01	0.023

Carbon dioxide in solution

In pure water over 99 per cent of CO_2 is in solution, but some is hydrated to carbonic acid, H_2CO_3, which dissociates to give the bicarbonate ion, HCO_3^-, and H^+, decreasing pH:

$$\underset{\text{(gas)}}{CO_2} \rightleftharpoons \underset{\text{(dissolved)}}{CO_2} + H_2O \rightleftharpoons \underset{\text{(solution)}}{H_2CO_3} \rightleftharpoons \underset{\text{(solution)}}{H^+ + HCO_3^-} \qquad [10.3]$$

Formation of carbonate ions, CO_3^{2-}, from $HCO_3^- \rightleftharpoons CO_3^{2-} + H^+$ is not considered further. If H_2CO_3 or HCO_3^- are removed by chemical reactions then the apparent solubility of CO_2 changes. As H_2CO_3 in solution is only 1/400 of the other components it may be neglected. The pH of solutions of CO_2 in water depends on the molar concentration of CO_2, the gas phase CO_2 partial pressure, the dissociation constant (pK) and bicarbonate ion concentration. This is expressed in the Henderson–Hasselbach equation:

$$pH = pK + \log [HCO_3^-] - \log [CO_2] \qquad [10.4]$$

The pH determines the balance between CO_2 dissolved and the bicarbonate ion concentration (Table 10.3); H^+ ions drive the reaction in favour of CO_2 (by the law of mass action) so that in acid solutions there is little bicarbonate (acids are used to remove bicarbonate from solution and produce CO_2).

Changes in CO_2 and bicarbonate concentration are important to photosynthetic organisms, algae, for example, grow in acid or alkaline waters differing greatly in temperature. With increasing acidity and temperature the concentrations of bicarbonate ions and dissolved CO_2 decrease, and with them the supply of substrate for photosynthesis. Also, within photosynthesizing tissues the cellular fluids such as the chloroplast stroma, have a variable pH which, together with temperature, influences the CO_2 concentration. The solubility of CO_2 and O_2 and their ratio in equilibrium are shown in Table 10.4. Solubility of oxygen is not affected by pH so the CO_2/O_2 ratio rises greatly with increased alkalinity. As temperature increases, solubility of O_2 decreases less than that of CO_2 so the ratio of O_2/CO_2 increases substantially.

Table 10.3 Amounts of bicarbonate ion (HCO_3^-) in solution at different pH, in equilibrium with air at 30 Pa CO_2 (or 12 μM of CO_2 in solution)

pH	CO_2 concentration (μM)	HCO_3^- (μM)	Ratio HCO_3^-/CO_2
5		0.58	0.05
6		5.78	0.5
6.3		11.6	1.0
7.0	12	57.0	5.0
7.3		116	10.0
8.0		578	50.0
8.3		1156	100.0

Table 10.4 Solubility of CO_2 and O_2 in water when in equilibrium with air containing 32 Pa (320 μbar) of CO_2 and 21 kPa (21%) O_2 at atmospheric pressure as a function of temperature

T(°C)	μM CO_2	cm^3 CO_2 (m^3 water)$^{-1}$	μM O_2	m^3 O_2 (m^3 water)$^{-1}$	Molar ratio O_2/CO_2
0	23	515	458	0.0102	19.9
10	16	371	356	0.0079	22.2
20	12	290	291	0.0066	24.3
25	10	244	263	0.0059	26.3
30	9	223	245	0.0055	27.2
40	7	179	216	0.0048	30.9

After Šesták, Čatský and Jarvis (1971).

CO_2 and bicarbonate equilibria and photosynthesis

The chloroplast stroma is alkaline during illumination and bicarbonate ions predominate, but RuBP carboxylase uses CO_2 as substrate, not HCO_3^-. This is shown by supplying the enzyme with $^{14}CO_2$ or $H^{14}CO_3^-$ at high pH and low temperature (10 °C), where HCO_3^- formation from CO_2 is very slow. With $^{14}CO_2$ the reaction can proceed and ^{14}C is incorporated into 3PGA but $H^{14}CO_3^-$ is not used. By adding the enzyme carbonic anhydrase which increases the equilibration rate greatly, $H^{14}CO_3^-$ is used as well, as it is converted to $^{14}CO_2$. During photosynthesis, in order to avoid starving RuBP carboxylase of substrate CO_2, the rate of supply of CO_2 must match the rate of reaction. The rate of conversion of bicarbonate to CO_2 in alkaline conditions is slow, so that both the CO_2 concentration in solution and the rate of supply to RuBP carboxylase could limit assimilation at high pH. However, the rate of conversion is increased a hundred-fold in tissues by carbonic anhydrase; as CO_2 is depleted it is rapidly produced from bicarbonate. Conversely if CO_2 is available in solution but HCO_3^- is needed (by PEP carboxylase) then the rate of HCO_3^- formation is also increased by carbonic anhydrase.

Carbonic anhydrase, a protein of 180 kD molecular mass and containing zinc, is found in all photosynthetic tissues, often in very large amounts, particularly in the chloroplast and at cell membranes. It also occurs in non-photosynthetic tissues and in animals. The enzyme, which has an extremely large turnover number (10^6 s^{-1}) facilitates the diffusion of CO_2 by speeding up the formation of HCO_3^- in the cytosol and maintaining a large gradient of CO_2 concentration. Also, carbonic anhydrase allows HCO_3^- to act as a buffer (albeit small) to provide CO_2 if the supply temporarily fails. In the chloroplast, carbonic anhydrase prevents depletion of CO_2 and probably maintains a high partial pressure of CO_2 at the active site of RuBP carboxylase, only slightly below that of the intercellular spaces, and minimizes the effects of low atmospheric CO_2. Plants or algae grown in high concentrations of CO_2 contain less carbonic anhydrase than those grown in low concentrations. Also, chloroplasts of C3 plants contain more carbonic anhydrase than C4 plants, which have a CO_2 concentrating system, although the enzyme is active in the C4 mesophyll. Algae, particularly in alkaline natural waters, have carbonic anhydrase at the cell surface which increases the rate of supply of CO_2 to the cells.

Carbon dioxide movement to photosynthetic cells

Carbon dioxide in the chloroplast stroma is removed by the RuBP carboxylase reaction and a gradient of CO_2 develops across the chloroplast envelope, cytosol, cell membranes and walls to the intercellular spaces and *via* the stomata, to the ambient air. The gradient is the driving force for CO_2 diffusion, but the rate at which this occurs to the reaction site depends on the conductances (reciprocal of resistance) to CO_2 diffusion in the gas and liquid phases in the leaf and atmosphere and on external CO_2 concentration. In the atmosphere CO_2 diffuses toward and O_2 away from the leaf during illumination and the fluxes are reversed in darkness; water vapour diffuses away from the wet internal surfaces in the leaf to the dry atmosphere under most conditions. In the turbulent air gas concentrations are uniform, due to rapid mixing by mass flow. In the intercellular gas spaces and also in the liquid spaces in cell wall and cytosol, the movement of gases is by diffusion, not by mass mixing. In the 'boundary layer' surrounding the cell or leaf there is incomplete mixing by mass flow and gases move partially by diffusion. Diffusion is the movement of molecules of a substance from higher to lower concentration and diffusion coefficients depend on molecular species, the medium, and temperature and pressure. Graham's law states that the diffusion coefficient for a gas in air is approximately equal to the reciprocal of the square root of the molecular mass. Thus H_2O (mass 18) and CO_2 (mass 44) have, in air, diffusion coefficients of 0.257 and 0.16 $cm^2\ s^{-1}$ respectively so that the ratio of D_{H_2O}/D_{CO_2} is 1.6. Diffusion coefficients increase for all gases by approximately 1.3 between 0 and 50 °C so the ratios are almost constant over the physiological temperature range. The magnitude of the diffusion coefficients and their ratio are important in calculating fluxes and for determining the diffusion of one molecular species from measurements of another. Diffusion depends upon the temperature and pressure of the medium and is very rapid over short distances; it is faster in gases than liquids. Diffusion of CO_2 in air (0.16 $cm^2\ s^{-1}$) is four orders of magnitude greater than in water ($0.16 \times 10^{-4}\ cm^2\ s^{-1}$ at 15 °C). Diffusion of gases and ions in the liquid phases of cell walls and cytosol is particularly important for its slows the flux of materials and may limit the rate of photosynthesis.

The flux per unit area and time (also called the flux density) of a gas is determined by the gradient of concentration, the length of the diffusion path and the diffusion coefficient by Fick's law. For CO_2:

$$F_{CO_2} = -D_{CO_2} \frac{[CO_{2ambient}] - [CO_{2internal}]}{x} \qquad [10.5a]$$

and for water vapour:

$$F_{H_2O} = -D_{H_2O} \frac{[\chi_{internal}] - [\chi_{ambient}]}{x} \qquad [10.5b]$$

where χ is the absolute humidity. The length of the diffusion pathway, x, is an 'equivalent length' of still air between the ambient air and the internal site. Thus in leaves, stomatal closure decreases the area through which gases diffuse, increasing the effective but not the actual length of the pathway. Flux may be expressed as mass or mol $m^{-2} s^{-1}$.

Resistance to gaseous movement

From eqns 10.5a and 10.5b flux depends on the length of the path and the ratio x/D is the resistance (R) offered by the pathway. Units of R are $m/m^2 s^{-1}$ or $s m^{-1}$ which may be thought of as the time taken by a gas molecule to traverse unit path length. The conductance of the pathway is the reciprocal of resistance ($g = 1/R$, $m s^{-1}$). Resistance increases with the effective length of the pathway and decreases with greater diffusion coefficient. It is inversely related to the flux of material whereas conductance is proportional to it. The concept of resistance has been important in the analysis of the exchange of matter and heat between plants and their environment. Fluxes may be analysed in a manner analogous to the flow of electricity in electrical circuits; Fick's law of diffusion is analogous to Ohm's law, although movement of electricity and transport of gases are physically different. Resistances may be added and manipulated as in electrical circuits and compared for different parts of the flux pathway. Resistances calculated for one substance may be converted to resistances for another from the ratio of their diffusion coefficients. Thus resistance to CO_2 movement may be calculated from water vapour flux ($R_{CO_2} = R_{H_2O} \times 1.6$) determined experimentally; by measuring stomatal resistance to water vapour flux the resistance for CO_2 flux may be obtained.

An alternative concept of the relationship between transpiration and the gradient of water vapour has been developed by Cowan (1977). With the driving force expressed as the partial pressure (e) or mol fraction (m), which avoids errors due to temperature and pressure (P) differences, and a molar water flux, F(mol $m^{-2} s^{-1}$), the relationship is:

$$F = \frac{D}{PV} \cdot \frac{(e_{internal} - e_{ambient})}{L_a + L_{st}} = \frac{D}{V} \cdot \frac{(m_{internal} - m_{ambient})}{L_a + L_{st}} \qquad [10.6]$$

where L is an 'equivalent length' of the pathway in the boundary layer (L_a) and stomatal pore (L_{st}) and D is the diffusion coefficient of water vapour in air. Setting $g = D/LV$ gives $F = g\,(e_{internal} - e_{ambient})/P = g\,(m_{internal} - m_{ambient})$ and conductance (g) has units of mol $m^{-2} s^{-1}$ and is independent of temperature and pressure; g of 1 mol m^{-2} s^{-1} is equal to 2.5 cm s^{-1} at 25 °C and atmospheric pressure. Conductances of parts of the pathway for CO_2 can be calculated from conductance to water vapour, g_a and g_{st} are boundary layer and stomatal conductances respectively. In the following sections conductances of parts of the pathway to CO_2 are designated $g' = g/1.6$. However in the boundary layer, due to the transition from turbulent to diffusional transfer, $g' = g/1.37$. Transpiration causes convection in the stomatal pores and a correction (*c.* 5%) to the stomatal conductance for CO_2 is required for accurate measurements. Expression of conductances as moles per area and time, which is a permeability, relates the stoichiometry of CO_2 and water vapour fluxes to biochemical processes and shows explicitly the effects of pressure and temperature. Use of conductance terminology gives a direct proportionality to fluxes of material, not an inverse one as with resistance.

Pathway of water vapour and CO_2 movement in leaves

Water vapour evaporates from the wet cell surfaces lining the internal air spaces and diffuses out to the atmosphere *via* the substomatal cavities, stomatal pores and cuticle, and across the boundary layer. Carbon dioxide enters by the same route and dissolves in the water films in the cell wall and diffuses, in the liquid phase, across the membranes and cytosol to the chloroplast stroma. Figure 10.1 illustrates the flux pathway and gives an electrical analogue in a two-dimensional resistance model. Boundary layer thickness and therefore conductance (g_a) at the leaf surface varies with wind speed, surface dimensions and characteristics such as hairiness. At wind speeds of 0.5–10 m s^{-1}, g_a is 0.2–4 mol m^{-2} s^{-1}. Effects of g_a on CO_2 assimilation depend on its magnitude relative to other conductances in the system. In general g_a is large compared to g_{st} (2 compared to 0.4 mol m^{-2} s^{-1} with open stomata) and so has little control over water or CO_2 flux. To avoid errors in experimental determinations of CO_2 and H_2O fluxes and calculation of stomatal conductances, measurements are made in standard conditions, for example in vigorously stirred leaf chambers (Fig. 10.2) to maximize g_a and keep it constant. The waxy cuticle is a very effective barrier to water loss with small conductance (g_c, 0.005 mol m^{-2} s^{-1} or smaller) but it also prevents the entry of CO_2. Plants appear to be in an evolutionary impasse, with large surface area required for light and CO_2 capture but losing water as a consequence. Desiccation is a major limitation to growth of terrestrial plants which have not developed a material allowing diffusion of CO_2 but not of water vapour.

Stomatal pores in the cuticle offer a high conductance pathway (g_{st}) for CO_2 flux into the leaf, but allow H_2O to escape. Cuticular and stomatal pathways operate in parallel and both are in series with the boundary layer (Fig. 10.1). Stomata occur on leaf surfaces in variable numbers (0–3 × 10^8 m^{-2}) on both surfaces (amphistomatous) or only on upper (ab-) or lower (adaxial) surfaces. The ratio of total stomatal pore area to leaf surface is about 1 per cent so that water vapour diffusion out of, and CO_2 diffusion into, the leaf occur only

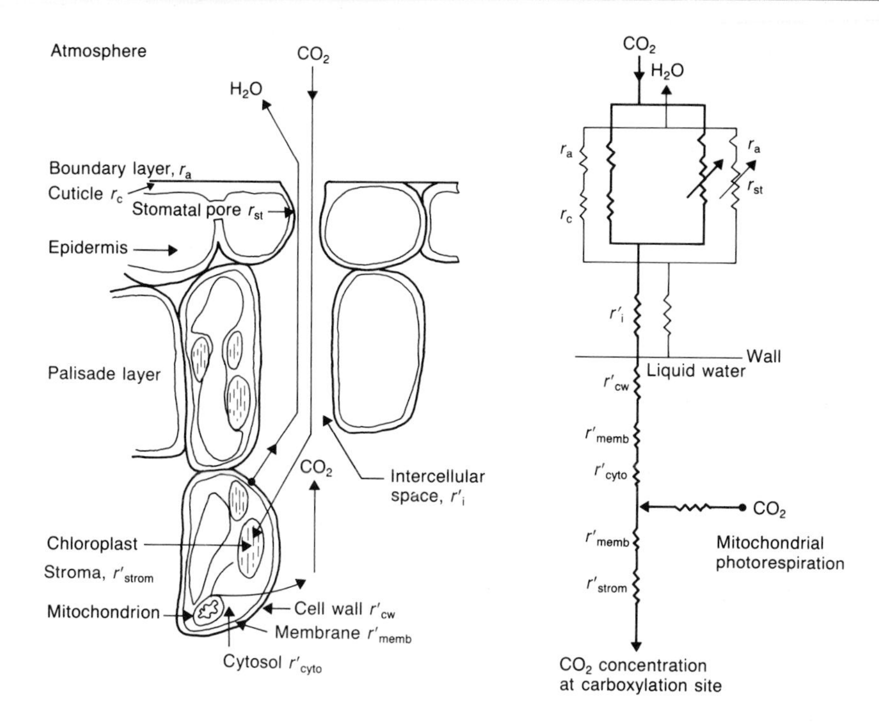

FIG. 10.1 Pathway for water vapour loss from, and carbon dioxide entry into a leaf in the light, shown as an electrical analogue with resistances denoting sites of restricted diffusion related to leaf structure.

through a very small part of the leaf area; the maximum g_{st} of mesophytic C3 crop plants is 1.2–0.4 mol m^{-2} s^{-1} for H_2O vapour, but many trees and C4 plants have smaller maximum stomatal conductance than C3 plants. The stomatal pores are bounded by guard cells which regulate the width and area of the aperture *via* changes in turgor pressure, controlling CO_2 entry and, possibly of more importance for land plants, water vapour loss. Conductance of closed stomata is of the order of 0.01 mol m^{-2} s^{-1} and probably approaches cuticular conductance. As the maximum g'_{st} is much greater than g'_c the flux of CO_2 into leaves occurs *via* the stomatal pore. Conditions which encourage stomatal opening – adequate water supply, bright light and high humidity – stimulate rapid photosynthesis.

Internal or substomatal conductance (g_{ss}) depends on the geometry of the mesophyll air spaces and is large (4–0.8 mol m^{-2} s^{-1}) and probably constant except in severely wilted leaves. Leaves with larger photosynthetic rates have relatively large internal spaces allowing rapid diffusion of gases. A large cell surface/leaf surface area ratio (20/40 in some mesophytes) maximizes g_{ss}. Evaporation from the water-saturated cell walls is the source of transpiration and the vapour pressure at their surface is taken as that of pure water at surface

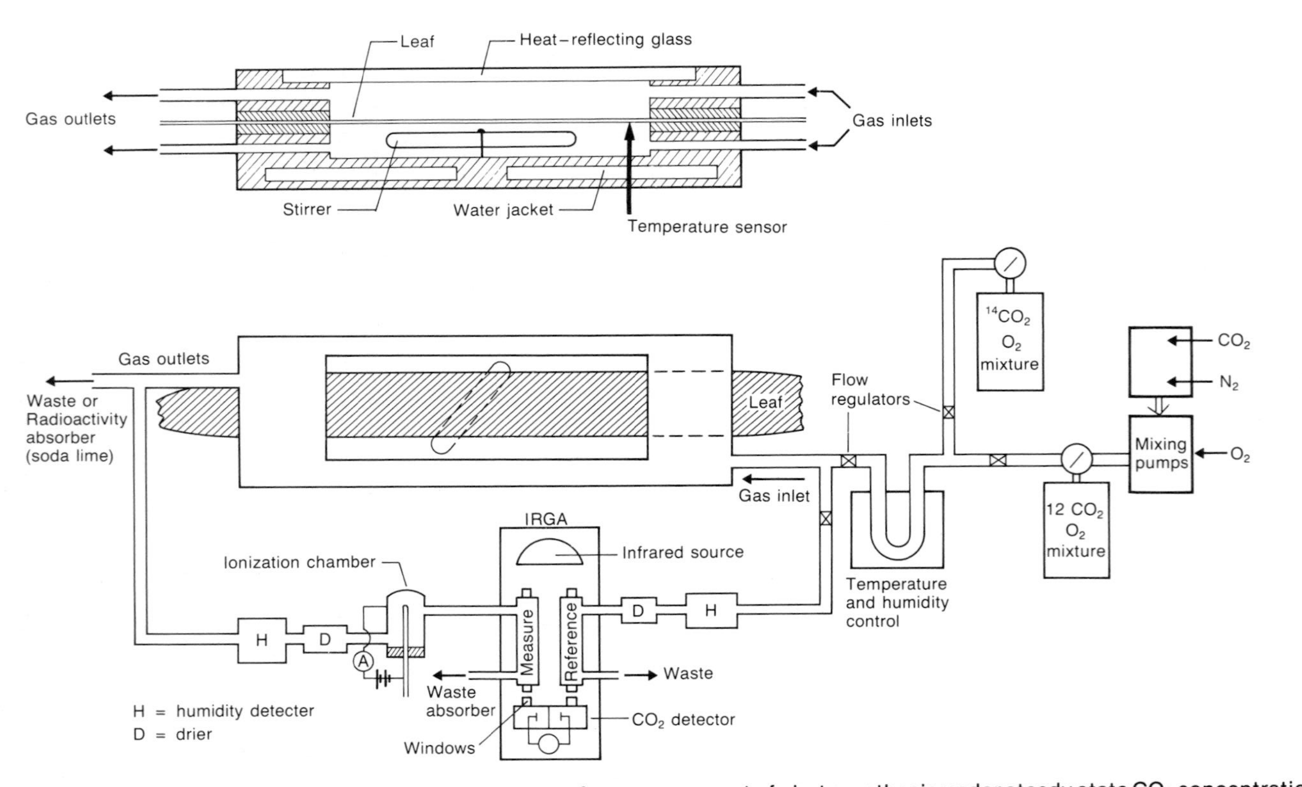

FIG. 10.2 Construction of a leaf chamber and gas exchange system for measurement of photosynthesis under steady state CO_2 concentration with regulated O_2 and humidity. Addition of $^{14}CO_2$, pre-mixed in a pressurized cylinder, and ionization detector enables ^{12}C and ^{14}C uptake to be measured at the same time and the leaf may also be sampled for analysis of radioactive assimilation products.

temperature. As this cannot be directly measured the saturation vapour pressure at bulk leaf temperature is taken; most mesophyte leaves are thin and isothermal conditions are assumed throughout. Thus the gradient of water vapour pressure ($e_{internal} - e_{ambient}$) and the conductances of the pathway can be calculated from measurements of water flux, temperature and atmospheric humidity. Stomata are the main control point regulating CO_2 in the tissue and water balance within the plant, maintaining a complex interaction between plant water status and environmental conditions.

Conductance and CO_2 assimilation

Conductance of the pathway for CO_2 is calculated, as described, from water vapour flux; net photosynthesis, P_n, is related to C_a and C_i, the partial pressures of CO_2 in the atmosphere and internal spaces respectively, and to c_a and c_i the mol or volume fractions in the same position by:

$$P_n = g'(C_a - C_i)/P = g'(c_a - c_i) \qquad [10.7]$$

From measurement of P_n, g for water vapour, C_a and P (pressure), the internal CO_2 partial pressure may be calculated:

$$c_i = c_a - P_n/g' \qquad [10.8]$$

Transpiration to the air 'sweeps' CO_2 from the intercellular spaces and a correction is applied:

$$P_n = g'(c_a - c_i) - (c_i + c_a)/2 \cdot F_{H_2O} \qquad [10.9]$$

with F_{H_2O} the flux of water vapour. By measuring P_n with changing external CO_2, CO_2 response curves relating P_n to c_i may be constructed, independent of stomatal conductance, allowing the relation between metabolism and CO_2 supply to be analysed (p. 218).

Assimilation is also proportional to the conductance of the liquid pathway (g'_{mes}) which is the sum of the conductances of the cell wall, membranes, cytosol, and chloroplast. With a gradient of CO_2 partial pressure from internal spaces to chloroplast ($c_i - c_c$):

$$P_n = g'_{mes}(c_i - c_c) \qquad [10.10]$$

and the CO_2 pressure at the carboxylation site is given by:

$$c_c = c_i - P_n/g'_{mes} \qquad [10.11]$$

These liquid phase conductances are not purely physical but depend on metabolism in the cell. Conductance of 0.04–0.2 mol m^{-2} s^{-1} has been calculated for C3 plants, similar to conductances of other parts of the pathway. Calculated values for C4 plants are larger than for C3, as their assimilation rate is greater and the gradient of CO_2 is not proportionally smaller. Although c_c was once thought to be zero or very small at the carboxylation site, it is probably close to c_i, as the characteristics of RuBP carboxylase *in vitro* may be used to calculate assimilation rates *in vivo* without major correction for g_{mes}, and there is also discrimination against $^{13}CO_2$ during assimilation, which cannot occur if g_{mes} is large.

Carboxylation conductance may be calculated for the enzyme reactions, although they are physically quite different mechanisms; the greater the efficiency of carboxylation the higher the conductance. It may be calculated from the slope of the relationship between assimilation and internal CO_2, when the process is CO_2 dependent. Carboxylation efficiency (μmol CO_2 assimilated m^{-2} s^{-1} Pa^{-1}) of C3 plants is lower than for C4 and is dependent on growth conditions; nitrogen deficiency, for example, decreases the amount of RuBP carboxylase in leaves and the efficiency. Analysis of the CO_2 response curves provides much information about the combined metabolic processes in leaves. RuBP carboxylation rate determines the demand for CO_2; the relationship between internal CO_2 and assimilation (discussed further in Ch. 11) is called the 'demand function'. Assimilation is linearly related to CO_2 at low partial pressure but rapidly saturates at high partial pressures. Internal CO_2 supply must equal the demand function to maintain a given rate of assimilation; this is called the 'supply function'. Thus, for complete saturation of photosynthesis, c_i must equal or exceed the saturation point of RuBP carboxylase. If gas and liquid phase conductances were infinite then internal and external CO_2 would be equal. However, with decreasing stomatal conductance the partial pressure of CO_2 in the air must rise to maintain saturating CO_2 in the leaf; that is, the 'draw down' of CO_2 inside the leaf depends on assimilation rate and conductance. At high stomatal conductance and ambient CO_2, carboxylation is saturated and stomata no longer control assimilation. However, water vapour loss is regulated by stomata. In low CO_2, assimilation is dependent on internal CO_2 and therefore regulated by stomatal conductance. Stomata may operate to maintain high assimilation and minimize water loss; as the pathway for CO_2 flux is longer and of lower conductance than for water vapour, stomatal closure has greater control of transpiration than of assimilation. This is expressed in the water use efficiency (= dry matter gained/water used). Plants with high water use efficiency maintain relatively larger assimilation and low water consumption. However, plants may have relatively high water use efficiency but low production, depending on their absolute rate of assimilation. High rates of assimilation in air cannot be achieved with low g_{st} without decreasing internal CO_2 substantially, which considerably decreases potential productivity. Increasing stomatal conductance would permit faster photosynthesis but allow excessive water loss. Plants with inherently small assimilation rates (because of limited metabolic capacity) may operate with smaller stomatal conductance and internal CO_2 than those with large capacity. If the carboxylation capacity is small the system may be efficiently used and water saved. Plants from habitats deficient in water and nutrients may have low carboxylation capacity and save energy and materials in construction of leaves and decrease water loss at the cost of low assimilation and growth rate. Under normal conditions stomata probably limit assimilation rather little and regulate water balance in relation to environment. Stomatal closure for prolonged periods is unusual in mesophytes, possibly because of photochemical or other metabolic damage, or increased photorespiration. Thus stomata may regulate carbon, water and energy balance in cells to protect the biochemical processes.

Control of stomata in relation to photosynthesis

Control of stomatal behaviour is not understood in detail. Low CO_2 partial pressure inside the leaf and humid air increase g_{st} in most species and increase photosynthesis. Light stimulates opening, although probably not because increased photosynthesis decreases internal CO_2. Spectral quality of light is important; blue light promotes opening more effectively than green or red but interactions with photosynthesis are not understood. Stomata close with high CO_2 concentrations, at which assimilation is saturated. This saves water but has little effect on assimilation. Water deficits and loss of turgor override the CO_2 response of stomata. Abscisic acid, which accumulates with water stress, may be produced in the chloroplasts of water stressed leaves when assimilation is much reduced. It may maintain stomatal closure during stress and although decreasing photosynthesis it serves to slow water loss and to maintain cellular turgor. Thus photosynthesis and water loss are regulated in response to environmental and plant processes. Their integration ensures the productivity and ecological success of the plant; full analysis of this complex system requires application of mathematical control theory.

Measurement of water vapour and CO_2 exchange

A leaf of known projected surface area (A, m^2) is enclosed (Fig. 10.2) in a transparent chamber through which air (or other gas) flows at volume (V, m^3 or mol) per unit time (s^{-1}). The air is vigorously stirred to minimize boundary layer thickness. Water vapour pressure is measured by a suitable sensor in air entering and leaving the chamber and the water vapour flux per unit area of leaf (mol m^{-2} s^{-1}) is calculated from flow rate and leaf area. Water vapour pressure in the chamber air is calculated from inlet and outlet humidity, and water vapour in the intercellular spaces is the saturated vapour pressure at leaf temperature. Conductivity of the gas phase (g) is calculated from:

$$F_{H_2O} = g(e_{internal} - e_{ambient})/P \cdot V/A \quad [10.12]$$

where $e_{ambient}$ and $e_{internal}$ are the vapour pressures in the atmosphere and at the cell wall respectively. The boundary layer is estimated either by using a wet filter paper replica which only has boundary layer resistance, or from leaf energy balance. In the same manner the stomatal conductance of leaves under 'natural' conditions is also measured rapidly, from water vapour flux, with porometers which clip onto the leaf surface. The rate at which water evaporates from the leaf into dry air enclosed in the porometer cup is measured, and used to compute conductance. The air is vigorously stirred.

Carbon dioxide exchange can also be measured at the same time as water vapour flux. Infra-red gas analysers (IRGA) are used to determine the molar concentration of CO_2. In this technique the gas entering the leaf or porometer chamber and the gas which has passed over the leaf and is partially depleted of CO_2, flow in parallel through two tubes of the IRGA. Each tube has end windows made of fused silica or calcium fluoride, which are transparent to infra-red radiation. Radiation from a source is passed through the gas in each tube. Absorption of infra-red radiation (4 μm) depends on the concentration of CO_2 molecules in the tube. Radiation that is not absorbed passes from the end windows into detectors filled with CO_2 where it is absorbed and heats the detectors. Any pressure difference between the two detectors caused by the differential heating is measured electronically, providing a continuous signal proportional to the difference in CO_2 concentration between the gas streams. Water vapour also absorbs infra-red radiation and, as it is at much greater concentration than CO_2, must be removed, or special optical filters used to block the water absorption bands. IRGAs are calibrated with CO_2 of known concentration, produced by mixing CO_2 with CO_2-free air or with O_2 and N_2, using gas mixing pumps or mass flow regulators.

References and Further Reading

Budyko, M. I. (1974) *Climate and Life* (English Edition, ed. by D. H. Miller), Academic Press, New York.

Flohn, H. (1980) *Possible Climatic Consequences of a Man-made Global Warming*, International Institute for Applied Systems Analysis, Laxenberg, Austria.

Kellogg, W. W. and **Schware, R.** (1981) *Climate Change and Society*, Westview Press, Boulder, Colorado.

Lemon, E. R. (1982) CO_2 and plants. The response of plants to rising levels of atmospheric carbon dioxide, *American Association for the Advancement of Science, Symposium 84*, Westview Press, Baltimore.

Liss, P. S. and **Crane, A. J.** (1983) *Man-made Carbon Dioxide and Climatic Change. A Review of Scientific Problems*, Geo Books, Norwich.

Poincelot, R. P. (1979) Carbonic anhydrase, pp. 230–8 in Gibbs, M. and Latzko, E. (eds), *Encyclopedia of Plant Physiology* (N.S.), vol. 6, *Photosynthesis II*, Springer-Verlag, Berlin.

Šesták, Z., Čatský, J. and **Jarvis, P.G.** (1971 (eds) *Plant Photosynthetic Production, Manual of Methods*, Dr W. Junk N.V., Publishers, The Hague.

Umbreit, W. W. (1957) Carbon dioxide and bicarbonate, pp. 18–27 in Umbreit, W. W., Burris, R. H. and Stauffer, J. F., *Manometric Techniques* (3rd edn), Burgess Publishing Co., Minneapolis.

Woodwell, G. M. (1978) The carbon dioxide question, *Scientific American*, **238**, 34–43.

CHAPTER *11*

Photosynthesis by leaves

The rate of photosynthesis by the whole leaf and its response to environmental conditions differs greatly between species of plant and is correlated with habitat. Early studies of the relationships between assimilation and the characteristics of the plant attempted to correlate CO_2 fixation with composition of leaves, for example pigment content, nitrogen or protein content, cell surface area, and stomatal frequency (see Heath 1969). However the failure of this approach to explain and predict assimilation rates under different conditions, combined with greater understanding of the fundamental processes, has led to the concept that the rate of photosynthesis by the whole system, be it algal cell or intact leaf, is the result of the integrated cellular biophysics and biochemistry. Therefore a more mechanistic understanding of the system is rapidly developing. Mathematical and statistical models on the relationships between photosynthesis and environmental factors, such as curves fitted to measured relationships between photosynthesis and light or leaf protein content (see Thornley 1976) are valuable for practical purposes but are not considered further here. Mechanistic models provide understanding of the underlying processes and their interactions which control photosynthetic rate and the response to environment. Such models are of value in understanding complex interactions between environment and productivity of plants, and of the constraints on the plant. Models will be invaluable in the improvement of crops and agricultural practice such as in the use of nitrogenous and other fertilizers to improve yield of crops *via* increased photosynthetic rate, and to assess the role of genetic engineering techniques in modifying photosynthetic capacity.

In this chapter the general responses of C3 and C4 plant photosynthesis to light, CO_2, O_2 and temperature are considered and the underlying metabolic causes of their different responses are analysed qualitatively; more detailed and rigorous mathematical treatments are to be found in the references.

Response of photosynthesis to environment

Photosynthesis by leaves depends on (1) rates of synthesis of NADPH and ATP (which are functions of the light reactions and electron transport), (2) rate of synthesis of RuBP which is controlled by the PCR cycle, (3) rate of carboxylation of RuBP which is a function of the activity of the RuBP carboxylase enzyme and of the ratio of RuBP oxygenase to carboxylase activity (i.e. of photorespiration to gross photosynthesis), and (4) rate of supply of CO_2 to the enzyme active sites.

Rate of gross photosynthesis (P_g) which is the rate of CO_2 assimilation before respiratory losses, is determined by irradiance, CO_2 and O_2 concentrations and by temperature, and is very species dependent. It may be measured by the assimilation of $^{14}CO_2$ by a leaf during the first few seconds of exposure to the radiotracer during steady-state photosynthesis in $^{12}CO_2$, when the rate of assimilation is constant, before $^{14}CO_2$ loss by respiration. Net photosynthesis, P_n, is the difference between P_g and the rate of CO_2 loss by photorespiration, R_l, and dark respiration, R_d:

$$P_n = P_g - (R_l + R_d) \quad [11.1]$$

The ratio of P_g to (R_l + R_d) is not constant but depends on species and environment. Dark respiration is the only form of respiration in darkness and dominates at very low irradiance but R_l is greater than R_d during assimilation with high O_2/CO_2 ratios. At present it is difficult to distinguish between the two types of respiration and to measure their rate during photosynthesis (Ch. 8). C4 plants have very small (R_l + R_d) so P_n is almost as large as P_g under most conditions.

Light response curves for C3 and C4 plants

Representative relationships are shown in Fig. 11.1. In darkness P_g is zero and both C3 and C4 plants release CO_2 from dark respiration. The rate of R_d increases with warmer temperatures and may be greater when leaves contain large amounts of carbohydrates; such 'burning off' may help to regulate cell metabolism by removing unwanted carbohydrates. This form of respiration is by an 'alternative pathway' oxidase in mitochondria, which produces only one, rather than the three ATP per molecule of NADH oxidized of normal respiration. With increasing irradiance P_g increases until, at the light compensation point, P_g = (R_l + R_d) and P_n is zero. The quantum flux at light compensation varies with species and environment; shade plants generally have lower values than sun plants (e.g. 20 compared with 80 μmol quanta $m^{-2} s^{-1}$). With increased irradiance, to 200 μmol quanta $m^{-2} s^{-1}$, P_g increases faster than respiration and P_n increases almost linearly with photon flux, as electron transport is directly proportional to the number of photons captured. In dim light and with small P_n, leaves in air are not limited by CO_2 supply or small stomatal conductance (g'_{st}), because increasing the CO_2 in the atmosphere does not increase assimilation. However, as P_n increases and particularly at small conductance, CO_2 concentration inside the leaf decreases until it becomes limiting and further increase in light does not increase P_n (see p. 213 for the dependence of internal CO_2 partial pressure on external CO_2 and stomatal and boundary layer conductances). With g for CO_2 of 0.6–0.3 mol $m^{-2} s^{-1}$ and 30 Pa CO_2 and 21 kPa O_2

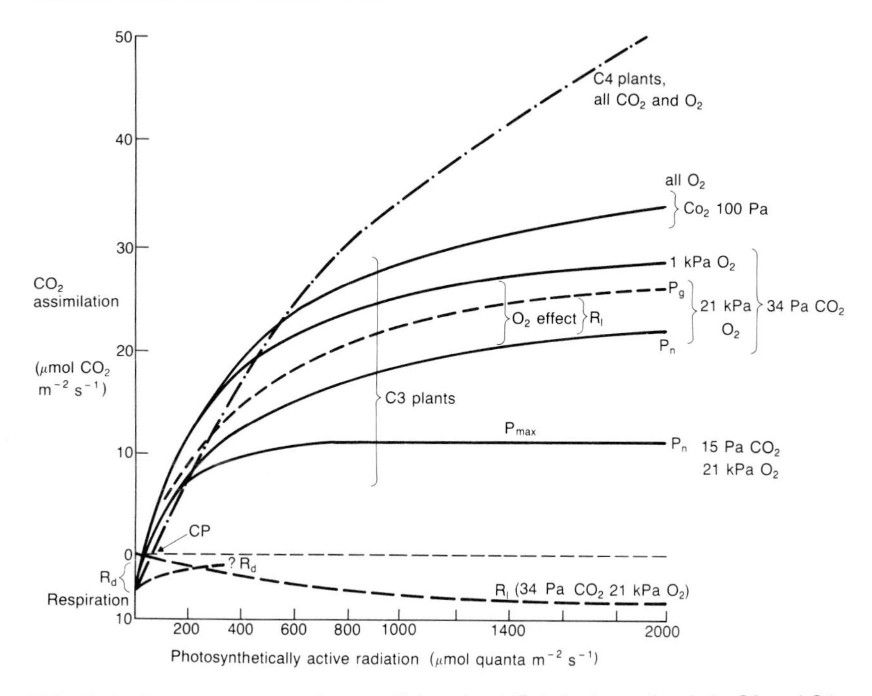

FIG. 11.1 Response curve of gross (P_g) and net (P_n) photosynthesis in C3 and C4 plant leaves with irradiance at different CO_2 supply; in bright light (> 1500 μmol quanta m^{-2} s^{-1}) assimilation may decrease due to photoinhibition. Dark respiration (R_d) occurs in both types but photorespiration (R_l) is much greater in C3 plants particularly in high O_2. Light compensation (cp) and maximum rate of photosynthesis (P_{max}) are shown. The initial linear slope gives the photochemical efficiency. The smaller P_g in C3 species in air (estimated from $^{14}CO_2$ uptake) than in low O_2 is caused by consumption of RuBP by the oxygenase reaction in high O_2. Schematic after information from references.

in the atmosphere the internal CO_2 concentration in the leaf becomes limiting at P_n of about 30 μmol m^{-2} s^{-1} in C3 leaves when irradiance is about 1000 μmol quanta m^{-2} s^{-1}; P_n does not increase and may even decrease at greater irradiance due, in part, to excessive energy load on the light-harvesting systems, particularly PSII, which inhibits energy transduction and slows assimilation. The light-saturated rate of photosynthesis is called P_{max}. C3 leaves in high CO_2/O_2 conditions have large P_{max}, comparable to C4 plants (although species dependent and requiring that the plants are grown at high irradiance for full expression of capacity). C3 leaves may have P_{max} of 40 μmol CO_2 m^{-2} s^{-1} in low O_2. However, with increased O_2/CO_2 ratio, P_g and P_n decrease until at zero ambient CO_2 there is net evolution of respired CO_2. The decrease in photosynthesis is caused by both the greater R_l, which evolves CO_2 so offsetting P_g, and also because photorespiration consumes part of the RuBP produced which is therefore unavailable to the carboxylation reaction. This is discussed shortly.

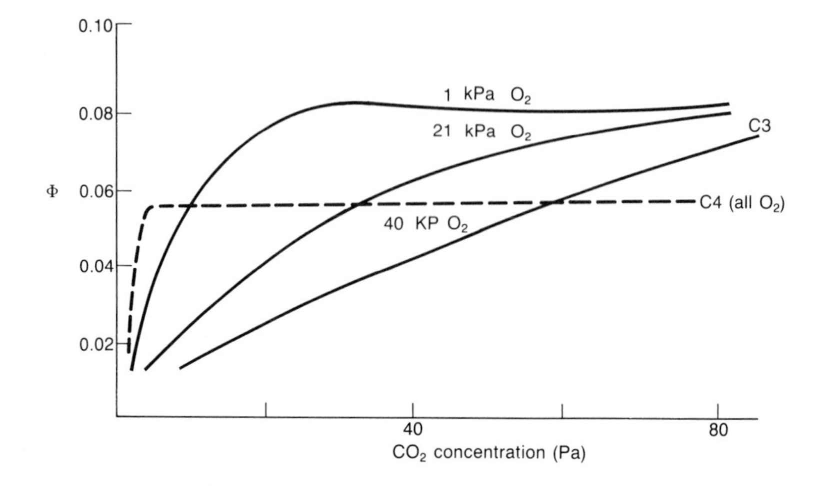

FIG. 11.2 Photochemical efficiency of C4 and C3 leaves as a function of O_2 and CO_2 concentration. (Modified from Osmond, Björkman and Anderson 1980, Figure 9.16).

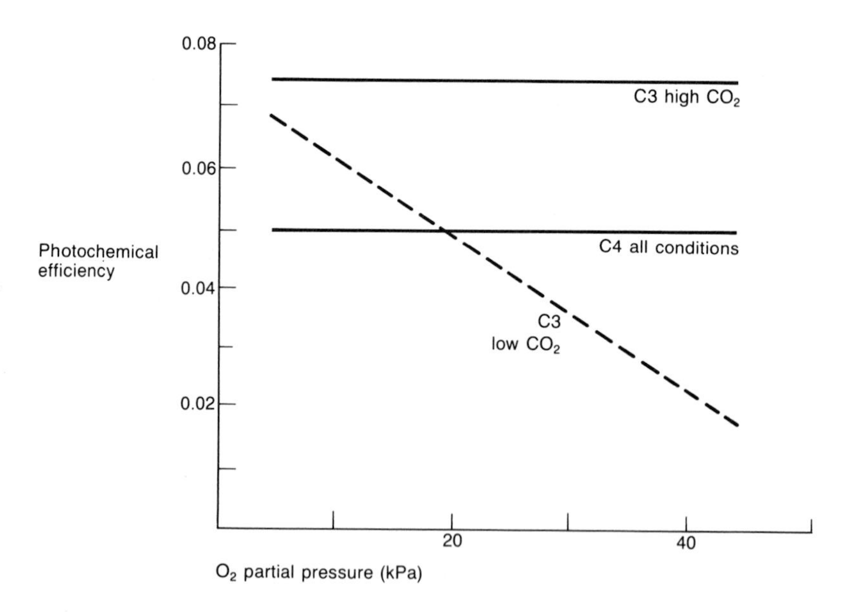

FIG. 11.3 Photochemical efficiency of C4 and C3 leaves with oxygen and CO_2 concentration. (Modified from Osmond, Björkman and Anderson 1980, Figure 9.16).

A large ratio of O_2 to CO_2 partial pressure inhibits assimilation even in dim light.

Photochemical efficiency, ϕ, or quantum yield (mol CO_2 mol photons^{-1}), is given by the initial slope of the light response curve, when CO_2 is not limiting. C3 leaves (Fig. 11.2) have ϕ of 0.05, that is 1 mol CO_2 per 20 photons absorbed when in air. Eliminating R_l with low O_2 concentration (1 kPa) increases ϕ to 0.07 or about 1 CO_2 per 14 photons. In contrast C4 leaves have ϕ of 0.05 mol CO_2 mol quanta^{-1} in all O_2 conditions (Fig. 11.2). Photochemical efficiency of C3 leaves in low oxygen (Fig. 11.3) is greater than that of C4 plants at high CO_2 but as O_2 increases so more CO_2 is required to overcome the effects of photorespiration; the photochemical efficiency of C4 plants is independent of the gas phase partial pressure of O_2 or of CO_2 above 5–10 Pa CO_2. Thus the absence of significant photorespiration in C4 species due to the efficient PEP carboxylase reaction in accumulating CO_2 at the RuBP carboxylation site eliminates the oxygenase reaction and makes them largely independent of CO_2 supply. The lower ϕ of C4 leaves is due to the greater requirement for ATP in the biochemical pathways of CO_2 assimilation and their relative inefficiency in dim light also stems from this.

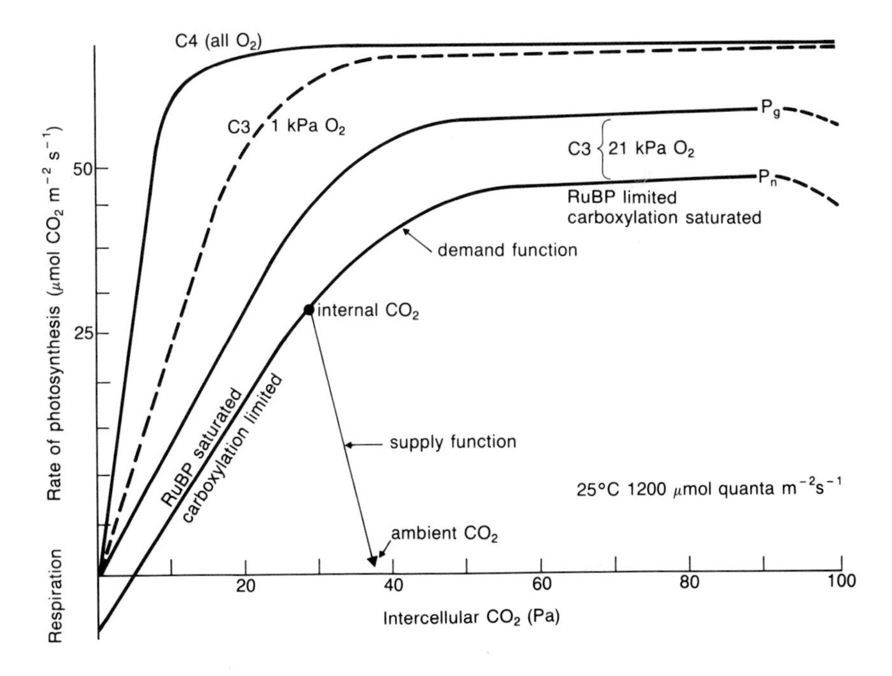

FIG. 11.4 Response of gross (P_g) and net (P_n) photosynthesis to CO_2 concentration and oxygen for C3 leaves, which respond to O_2, and C4, which do not. The 'demand function' is the requirement for CO_2 and depends on the characteristics of RuBP carboxylase/oxygenase; RuBP non-limited and limited parts of the response curve are indicated.

CO_2 response curves for C3 and C4 plants

Response of net photosynthesis of C3 leaves to CO_2 in the atmosphere (Fig. 11.4) at saturating irradiance, depends on the O_2 concentration. In low oxygen and zero CO_2, CO_2 release is small and predominantly from 'dark' respiration. Assimilation rises as CO_2 increases in both C3 and C4 plants, with similar carboxylation efficiency (CO_2 assimilated/CO_2 partial pressure) shown by the

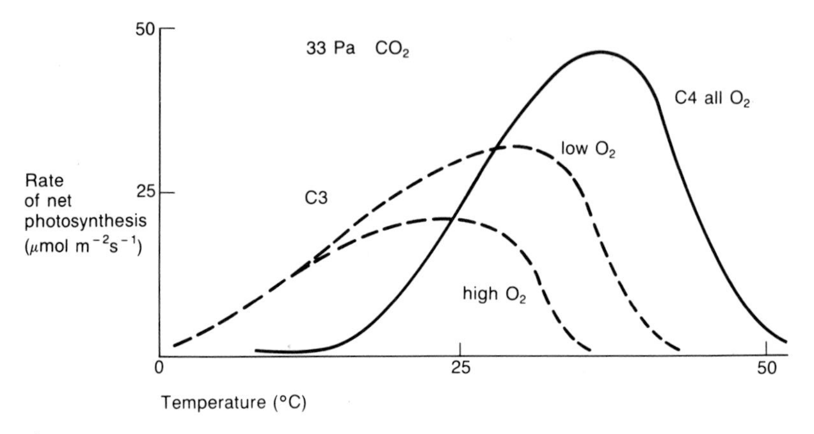

FIG. 11.5 Relative rates of net photosynthesis of an 'average' C3 and C4 plant in relation to temperature and the effects of O_2.

initial slope of the 'demand function'. With increased O_2, assimilation of C4 plants is unaffected but C3 leaves assimilate less. In air P_n of C3 leaves is substantially less than P_g, due to photorespiration which also causes the large respiration rate at low CO_2. With greater CO_2 concentration P_g increases until $P_g = (R_l + R_d)$ and P_n is zero, which is the CO_2 compensation concentration or partial pressure, with values of about 0–1 and 4–5 Pa CO_2 in 1 and 21 kPa O_2 respectively in C3 leaves. In C4 plants it is below 1 Pa CO_2. As CO_2 increases, P_n of C3 plants in low O_2 rises linearly and saturates at about 20 Pa CO_2 partial pressure. However, in 21 kPa O_2, P_n of C3 leaves increases less for a given increase in CO_2 and saturates only at 40–60 Pa CO_2. With low CO_2 the carboxylase is RuBP saturated and CO_2 limited but at high CO_2 the control is reversed. The decreased carboxylation efficiency of C3 plants with increased O_2 is due to photorespiration. By relating the measured assimilation rate to internal CO_2 the relative importance of RuBP and CO_2 limitation to assimilation may be assessed.

Temperature and photosynthesis

At low temperatures (< 10 °C) C3 photosynthesis is more efficient than C4 (Fig. 11.5), due in part to factors such as membrane stability which are outside the realm of the present discussion of photosynthesis. Some temperate C4 species (e.g. *Spartina townsendii*) are adapted to low temperature but the adaptations to the photosynthetic mechanism are not understood. At low temperature and in bright light photochemical damage to the light-harvesting systems may occur in the non-adapted species as the energy load on the carotenoid 'safety valve', superoxide dismutase (SOD) and catalase systems becomes excessive. With increased temperature C3 photosynthesis rises (assuming good water supply) to a broad optimum between 15 and 25 °C; above 30 to 35 °C assimilation decreases rapidly and stops at about 40 °C. C4 plants in contrast have an optimum at 25 to 35 °C and may not be damaged until 45 to 50 °C. The inhibition in C3 species is caused by increased photorespiration, with sensitivity to heat increasing as the CO_2/O_2 ratio decreases. C4 plants again benefit from greatly reduced photorespiration. Both C3 and C4 plants are, however, subject to high-temperature effects on enzymes and membranes. C4 plants adapted to cool growth conditions may be sensitive to high temperatures, whereas C3 species may adapt to the heat, *Larrea* for example, being able to grow in the extreme heat of Death Valley, California.

Cool conditions decrease the ϕ of C4 plants more than of C3 (Fig. 11.6) but with warm temperatures ϕ of C3 leaves decreases compared with C4. At low O_2 the responses of C3 and C4 plants are similar. Combination of hot conditions and low CO_2/O_2 ratio in the leaf, caused for example, by decreasing the stomatal conductance, greatly favours the RuBP oxygenase and loss of carbon so that C3 photosynthesis becomes much less efficient as temperature rises. Also the solubility ratio O_2/CO_2 favours oxygenation. C4 photosynthesis is

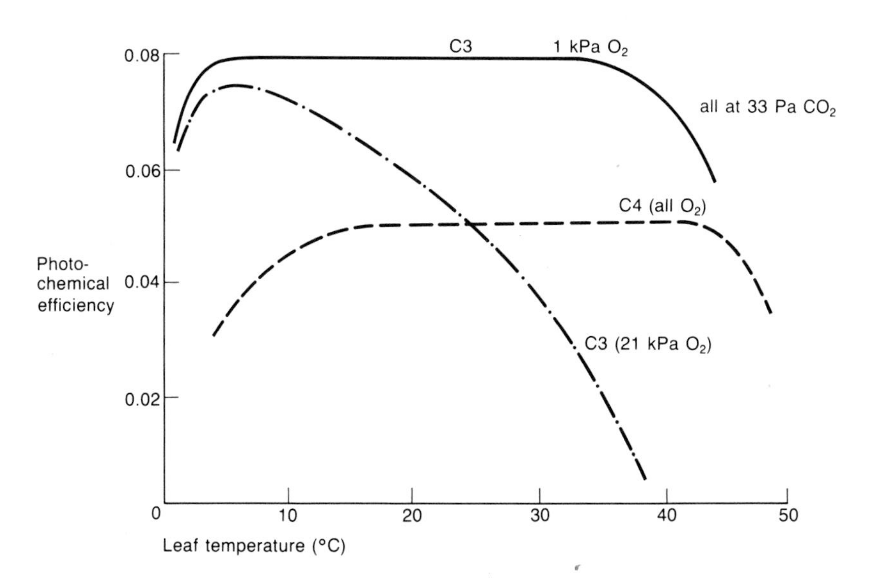

FIG. 11.6 Photochemical efficiency of C4 and C3 species in relation to temperature in air or low oxygen and dilute CO_2; generalized responses. (Based on data from Osmond, Björkman and Anderson 1980).

independent of O_2, CO_2 and temperature (to about 40 °C), because of the PEP carboxylase reaction, whereas photorespiration offsets P_g of C3 plants in air and decreases P_n. It should be noted that the rates of assimilation quoted and the values of temperature and gas partial pressures etc., are generalized, and individual C3 and C4 species differ considerably, reflecting their ecological adaptations.

Carbon dioxide and light compensation

Carbon dioxide and light response curves of C3 species in air show P_n to be zero at positive ambient CO_2 partial pressure and at low irradiance. These are the CO_2 compensation concentration and light compensation values. Carbon dioxide compensation may also be measured by enclosing a leaf in a gas-tight chamber under illumination and allowing it to remove or add CO_2 to the atmosphere until equilibrium (measured by infra-red gas analysis) is attained. Carbon dioxide compensation of C3 leaves is low in cool temperatures with low O_2, but increases as temperature and oxygen rise, particularly when O_2 exceeds the point where the oxygenation/carboxylation ratio, α, exceeds 2 and temperatures are greater than 35 °C. In contrast C4 plants have small compensation concentrations independent of O_2 and temperature up to higher values. Light compensation of C3 leaves is achieved in very dim light in low O_2, high CO_2 and cool temperatures but increases as they increase. C4 plants have higher light compensation in those conditions but it remains small up to much higher O_2 and temperature conditions than C3 plants. Below light compensation, respiration exceeds assimilation and the CO_2 partial pressure increases for both C3 and C4 leaves to very large values in closed systems. When assimilation is inhibited, for example by water stress, the compensation values also rise, even at bright light, in C3 and C4 species. As CO_2 compensation concentration and light compensation are not fixed values but change with conditions the frequently used terms CO_2 and light compensation points are not justified. Compensation values are relatively easy to measure and reflect the photorespiratory characteristics of the plant; they have therefore been used to select for variations in R_l. Although the majority of plants examined fall into the high (C3) or low (C4) groups, some intermediate forms have been found.

Photorespiration

Photorespiratory CO_2 release is estimated most accurately as the difference between gross and net assimilation, the former measured as $^{14}CO_2$ uptake and the latter as $^{12}CO_2$ assimilation under steady conditions where the rate of CO_2 fixation is constant and the internal carbon fluxes also. A leaf in a suitable chamber (see Fig. 10.2), under constant conditions, is allowed to assimilate $^{12}CO_2$. A gas of identical composition but with ^{14}C-labelled CO_2 of known specific activity is substituted for the $^{12}CO_2$ without changing conditions and, using a suitable detector (e.g. ionization chamber) the depletion of ^{14}C during the first 30–60 s is measured. As the $^{14}CO_2$ requires many minutes to saturate the pools of assimilates and to be evolved as photorespiration, this technique measures gross assimilation under constant conditions. However, respiratory CO_2 is probably partially reassimilated in the tissue before it escapes to the atmosphere; this 'short circuit' causes an underestimation of the rate of R_l. Other methods used to measure efflux of carbon in R_l include measuring the rate of CO_2 release from a leaf immediately after darkening following a period of photosynthesis. There is rapid CO_2 evolution, in the 'post-illumination burst', which lasts for up to 5 minutes before decreasing to the rate associated with dark respiration. It arises from the continued flux of carbon through the glycolate pathway after the light reactions and P_g have stopped and decreases with cessation of the photosynthetically produced glycolate flux. R_l is also measured by feeding intermediates of the glycolate pathway and measuring CO_2 or $^{14}CO_2$ efflux. Inhibitors of the pathway have also been used to stop R_l; the accumulation of intermediates or changes in CO_2 production are then measured to estimate R_l. However, these methods suffer to some extent by altering the steady state conditions, and often decrease photosynthesis and therefore R_l (Zelitch 1979). Photorespiration is about 25–30 per cent of P_n in air and the proportion increases with larger O_2 and smaller CO_2 partial pressures, high temperatures and water stress. The increase is due to the characteristics of RuBPc/o and to the combined photorespiratory and PCR cycles (Figs 11.7 and 11.8). As glycine synthesis is proportional to the ratio of

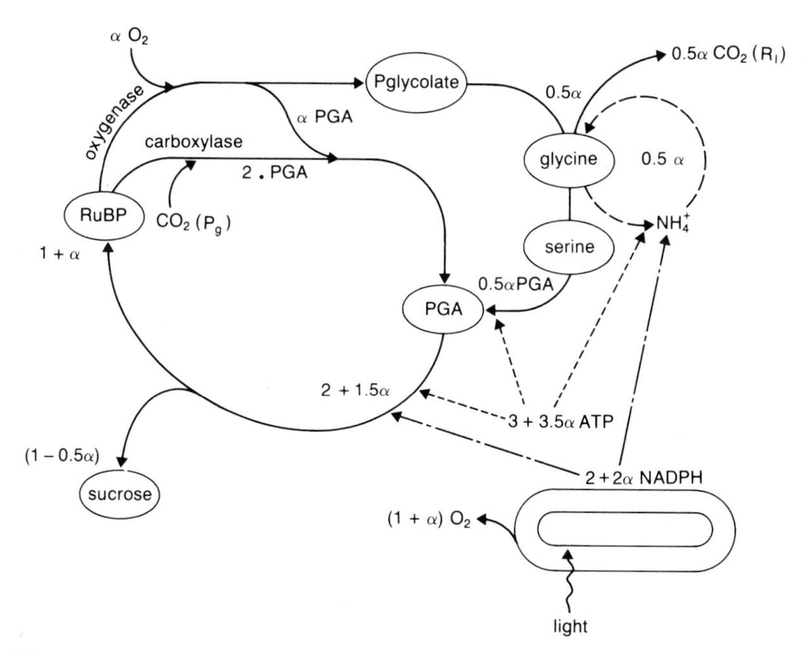

FIG. 11.7 Scheme of carbon fluxes in the combined photosynthetic carbon reduction and photorespiratory cycles in C3 leaves. P_g is gross photosynthesis, R_l the rate of photorespiration, α the RuBP oxygenase/carboxylase ratio which depends on the O_2/CO_2 ratio inside the leaf. This figure should be compared with Fig. 8.4 which shows the glycolate pathway in greater detail.

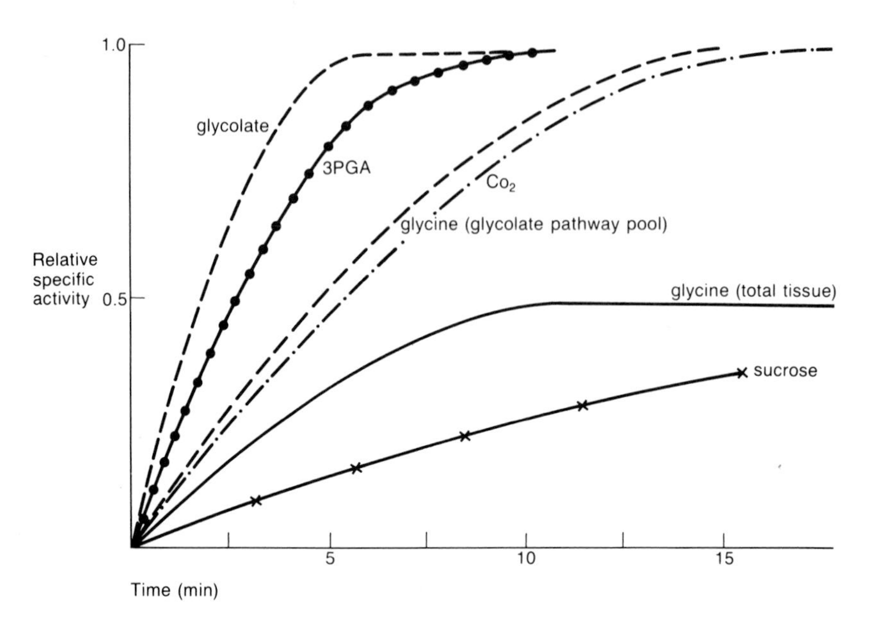

FIG. 11.8 Change in the ^{14}C specific activity of photosynthetic and photorespiratory metabolites, and photorespired CO_2, relative to the specific activity of the $^{14}CO_2$ supplied at time 0. The rate of change depends on the flux of carbon and the size of pools of intermediates (schematic).

oxygenase to carboxylase, α(p. 136), so with internal CO_2 concentration in the leaf of about 30 Pa and 21 kPa O_2, α is about 0.4 and $R_l = \alpha/2 \times P_g$. At high assimilation rate (40 μmol m^{-2} s^{-1}) R_l is 8 μmol m^{-2} s^{-1} and P_n is 32 μmol m^{-2} s^{-1}, that is R_l is 25 per cent of P_n. With very small α, R_l tends to zero and P_g is close to P_n. However as the carbon not used to form glycine is recycled to regenerate RuBP so carboxylation increases. Thus the stimulation of photosynthesis by low O_2 has two causes; inhibition of CO_2 loss and increased CO_2 assimilation. Conditions causing smaller stomatal conductance, such as water stress, may decrease P_g and decrease the CO_2/O_2 ratio in the tissue and increase α, so the proportion of R_l to P_n rises. C4 plants have similar rates of P_g and P_n and are little affected by conditions. Increasing temperature increases α, partly because the solubility of CO_2 is less than that of O_2. R_d also increases, so dark respiration offsets photosynthesis.

Limiting factors of photosynthesis

Assimilation rate is governed by the slowest, or rate-limiting step in the complete process. Blackman in 1906, introduced the concept of limiting factors – the rate of a complex process is determined by the rate of the slowest part, be it supply of material or of an individual chemical reaction or process. Increasing the rate of the slowest process increases the overall rate until another factor becomes limiting, as shown (Fig. 11.3) by increased assimilation when CO_2 concentration is increased in bright light. However, in very bright light and concentrated CO_2 the rate is not increased because internal physiological and biochemical processes are limiting. Complex metabolic systems adjust to conditions, optimizing the supply of materials and energy to the demands for the products, both within and between the different parts ('sub-systems') of the whole system. Light-harvesting and electron transport and the enzymatic reactions of the PCR cycle are subject to complex controls, both feedback and feed-forward, that is, the product of a reaction modifies the processes leading to further reactions, or stimulates later ones, so increasing the overall rate. The modification may be both stimulation or inhibition. Such processes are non-linear functions of particular inputs to the system and show complex responses, which appear to provide long term metabolic stability over a wide range of conditions in relation to growth, reproduction etc. In complex systems several factors may interact, each being close to the 'control point' at which the system is 'designed' to work or, in the case of plants, selected in evolution to function optimally. Interaction between several processes provides subtle control of the rate of the overall process. As conditions within the system or external to it alter so one or other of the control processes regulates until the system returns to a long term equilibrium, with perhaps somewhat different factors controlling the overall rate, yet each still close to the optimal value.

In the very short term (from fractions of seconds to hours) assimilation is controlled by the supply of light, CO_2 etc. as well as by enzyme amount and kinetics. The overall rate is governed in the long term by the amount of components of the system, for example of electron transport carriers in thylakoid membranes and the amount of membrane or RuBP carboxylase. These are determined by conditions during the growth and development of the leaf which change its structure and composition and capacity for photosynthesis. As an example, inadequate nitrate supply impairs synthesis of enzymes (e.g. RuBP carboxylase) and therefore CO_2 assimilation. It also inhibits formation of structural proteins and pigments, and therefore decreases the formation of thylakoids, slowing electron transport and light-harvesting. Deficient phosphorus supply inhibits assimilation by decreasing intermediary metabolism and the formation of assimilates. Iron deficiency inhibits processes in which iron is an electron carrier, for example in the Rieske Fe–S centres in thylakoids.

Photosynthesis is not generally a linear function of nutrient supply; deficient plants may grow much less but maintain those leaves which are formed with sufficient and balanced nutrient to allow a viable system to develop and therefore permit assimilation to take place at quite high capacity. However, such leaves may senesce rapidly as materials are remobilized to younger organs. This shortens the assimilation life of the leaf and, with severe deficiencies, impairs assimilation. Long term effects of nutrition and other conditions on development of the photosynthetic apparatus and its function are not well understood, despite the agronomic importance.

Metabolic models of the regulation of photosynthesis

Light and CO_2 response curves may be analysed using models (Fig. 11.7) which combine the processes of electron transport and synthesis of NADPH and ATP with models of enzyme activity, to calculate RuBP synthesis. This may be related to the characteristics of RuBPc/o and the equations governing CO_2 supply to show what controls the dynamics of assimilation with changing environmental conditions. Farquhar and co-workers have abstracted biochemical processes into mathematical equations, describing the relationships as a function of conditions. They have simulated the photochemical and carboxylation efficiencies of leaves and have concluded that the assimilation of C3 leaves is controlled by carboxylation at low CO_2 in air but by the rate of RuBP synthesis at high CO_2 partial pressures. The equations, which are greatly simplified here to outline the principles, describe assimilation in relation to light (photon flux), I, CO_2 and O_2 partial pressure, C and O respectively, and temperature, which will not be further discussed here. Rate of electron transport, J, from water to $NADP^+$ is related to the saturated rate of electron transport, J_{max} (determined in isolated thylakoids) and to I by:

$$J = \frac{J_{max} \times I}{I + 2.1 J_{max}} \qquad [11.2]$$

where 2.1 expresses the proportionality of the response. As I increases so J saturates, as observed experimentally. J_{max} is principally a function of the rate of e^- and H^+ transport through the plastoquinone pool, so shade grown plants, with less PQ than those grown in bright light, have smaller J_{max}. Assimilation by plants grown in intermediate light may be limited by electron transport in very bright light.

NADPH production is related to $0.5 J$ (2 e^- reduce $NADP^+$) and depends on the concentration of $NADP^+$ and of the $NADP^+$-ferredoxin reductase enzyme. With insufficient of the oxidized nucleotide acceptor or insufficient enzyme, synthesis of NADPH will be small even if electron transport is large:

$$\text{rate of NADPH synthesis} = 0.5 \times J \min\{[NADP^+]/[\text{enzyme sites}]\} \qquad [11.3]$$

where min { } means 'minimum of' and takes the value of 1 or the value of the $[NADP^+]/[\text{enzyme site concentration}]$ ratio if less than 1. Thus the rate varies according to the ratio of acceptor to enzyme but cannot exceed 1, that is synthesis of NADPH cannot exceed $0.5 J$. If there is no $NADP^+$ or enzyme then the ratio is 0 and synthesis stops. To achieve maximum rate of synthesis, the $NADP^+$ should be slightly in excess of enzyme capacity, so that enzymes sites are always filled. Measurements suggest that the amount of enzyme activity is unlikely to limit the supply of NADPH to assimilation.

ATP production is more complex, because three modes of electron flow coupled to photophosphorylation provide flexibility. In linear electron flow, J and ATP synthesis are related, as 3 H^+ are consumed per ATP and 2 H^+ are transported into the thylakoid per electron. In cyclic photophosphorylation one electron gives 1 H^+ and synthesis of 1 ATP is related to $\frac{1}{3}$ of cyclic electron flow, J_c, as 3 e^- gives 3 H^+ and 1 ATP. ATP synthesis depends on the concentrations of ADP and CF_1 enzyme active sites:

$$\text{rate of ATP synthesis} \approx \tfrac{2}{3}J + \tfrac{1}{3}J_c \times \min\{1, [\text{ATP}]/[\text{CF}_1 \text{ enzyme sites}]\} \quad [\mathbf{11.4}]$$

where min { } again denotes that the rate will depend on the ratio should it decrease below 1 and that otherwise ATP formation is proportional to electron transport. As electron transport, and ADP or active enzyme site concentration decrease so does ATP production. Saturation is achieved when electron transport reaches a maximum provided there are sufficient enzyme sites saturated with ADP and P_i. Evidence suggests that there is adequate enzyme capacity, but a rather large concentration of ATP + ADP is needed to maintain RuBP synthesis by the enzyme ribulose-5-phosphate kinase. Thus RuBP regeneration is controlled by NADPH and ATP synthesis and by the activity of PCR cycle enzymes (although enzyme activity and NADPH synthesis are usually in excess of ATP synthesis) and the availability of 3PGA as substrate:

$$\text{rate of RuBP synthesis} = \frac{[\text{PGA}]}{2R_p} \times \frac{[\text{ATP}]}{[\text{ATP}]\,[\text{ADP}]} \times M \quad [\mathbf{11.5a}]$$

where R_p is the maximum potential pool size of RuBP without any demand by carboxylation or oxygenation. M is the maximum rate of reduction of 3PGA to RuBP and depends on the capacity of the PCR cycle enzymes, which is thought to be sufficient to maintain RuBP production at the rate determined by electron transport. If NADPH synthesis is slower than ATP formation then:

$$\text{rate of RuBP synthesis} = \frac{[\text{PGA}]}{R_p} \times \frac{[\text{NADPH}]}{[\text{NADPH}] + [\text{NADP}^+]} \times M \quad [\mathbf{11.5b}]$$

and eqn 11.5b applies.

Rate of RuBP carboxylation, A, when the enzyme is saturated with RuBP, depends on CO_2 and O_2 and the enzyme characteristics only, given by Farquhar and von Caemmerer (1982) as:

$$A = V_{c_{max}} \times \frac{C - \Gamma^\star}{C + K_c(1 + O/K_o)} - R_d \quad [\mathbf{11.6}]$$

where $\Gamma^\star$ is the CO_2-light compensation value and is equal to $(0.5\, V_{o_{max}} K_c O)/(V_{c_{max}} K_o)$. It is related to α, the oxygenase to carboxylase ratio by $2\Gamma^\star/C$. C and O are partial pressures of, and K_c and K_o are Michaelis-Menten constants for, CO_2 and O_2 respectively; R_d is 'dark' respiration in the light and $V_{c_{max}}$ and $V_{o_{max}}$ are velocities of RuBP carboxylase and oxygenase respectively. When CO_2 pressure is low the enzyme is RuBP saturated; removing O_2 stops competition with CO_2 and increases assimilation.

In high CO_2 *in vivo*, particularly if the photon flux is low, the concentration of free RuBP may be insufficient to saturate all the enzyme sites and assimilation is expressed by:

$$A = J \times \frac{C - \Gamma^\star}{4.5\,C + 10.5\,\Gamma^\star} - R_d \qquad [11.7]$$

so A depends on the rate of electron transport and therefore on light, as it clearly does, not only on the partial pressures of CO_2 and O_2, as eqn 11.6 suggests.

By combining two equations, one giving the RuBP saturated carboxylation rate and one describing the unsaturated rate, it is possible to calculate the CO_2 at which the transition from saturated to unsaturated assimilation occurs:

$$CO_2 \text{ transition} = \frac{K_c\,(1 + O/K_o)\,J/(4.5\,V_{c_{max}}) - 7/3\,\Gamma^\star}{1 - J/(4.5\,V_{c_{max}})}$$

$$\text{Recall that } \Gamma^\star = \frac{0.5\,V_{o_{max}}\,K_c O}{V_{c_{max}}\,K_o} \qquad [11.8]$$

It can be seen that the transition will depend on the photorespiratory process as determined by oxygen concentration on the rate of carboxylation, and on the electron transport processes, J, which are determined by photon flux. Bright light ensures more rapid RuBP synthesis so the transition occurs at a higher CO_2 partial pressure than in dim light. In bean, von Caemmerer and Farquhar found that the transition occurred at about 22 Pa in dim light (950 μmol quanta $m^{-2}\,s^{-1}$) and at 30 Pa in 1400 μmol quanta $m^{-2}\,s^{-1}$. Removing O_2 increases the CO_2 pressure at which RuBP limitation occurs, as oxygenation is suppressed and RuBP is made available for carboxylation. The rate of change from RuBP saturating (CO_2 limiting) to RuBP limiting (CO_2 saturating) conditions occurs more rapidly than if it depended only on the RuBP carboxylase saturation kinetics.

Leaves in air have an internal CO_2 partial pressure which is determined by the rate of assimilation and stomatal conductance. Measurement of the assimilation rate with internal CO_2 partial pressure suggests that the transition from RuBP saturating to limiting conditions occurs close to the normal internal CO_2 partial pressure, slightly below ambient, when RuBP synthesis and carboxylation are in approximate balance at average photon flux. If a leaf photosynthesizing in intermediate light is suddenly brightly illuminated, 3PGA is consumed as RuBP is synthesized and carboxylation increases, until a new steady state is achieved at a smaller CO_2 partial pressure and higher assimilation rate. A decrease in photon flux slows formation of RuBP, decreases assimilation and the pools of assimilates and internal CO_2 increase. Possibly there is a balance, determined genetically and by environmental conditions during growth, between the capacity of the leaf to synthesize RuBP and its capacity for carboxylation. If either were much larger than the other, the leaf would not assimilate more CO_2. However, producing the enzyme and membranes requires nitrogen and energy so that the balance between processes would be inefficient use of resources. Approximate balance (co-limitation) between RuBP regenerating and consuming systems would be the most efficient ecologically.

Leaf composition and assimilation

To determine what steps in the photosynthetic process limit the overall rate, the maximum rates of partial processes may be calculated from their characteristics, as far as they are known. The approach is only semi-quantitative and uses values determined *in vitro* which may not be applicable *in vivo*. However, it is a useful illustration of how the system functions. The maximum rates of processes are calculated from the amount of pigments or enzyme per m^2 of leaf (this basis is adopted because it determines the interception of light in most plants), from the number of enzyme active sites, and the number of moles of substrate converted in unit time. This is a function of the binding characteristics of the enzyme sites, that is, rates of binding substrate, of reaction and of release of products (the turnover time (measured in seconds; the reciprocal is the turnover number, s^{-1})). For 1 m^2 of leaf:

$$\frac{\text{mol substrate}}{\text{converted } s^{-1}} = \frac{\text{mol}}{\text{enzyme}} \times \frac{\text{active enzyme}}{\text{sites (mol enzyme)}^{-1}} \times \frac{\text{mol substrate}}{\text{(mol active sites)}^{-1}\ s^{-1}} \qquad [11.9]$$

Complex parts of the system, as well as single enzymes, may be treated in this way by substituting the measured amounts of the components and the rate at which reactions occur.

Light absorption and transduction

Chlorophyll captures photons and energy is transferred at rates much greater than 10 events s^{-1}. A leaf containing 560 μmol chlorophyll m^{-2} (see Table 4.1 (Ch. 4)) could trap 5600 μmol photons $m^{-2}\ s^{-1}$ or twice the number of PAR photons incident upon a leaf at noon in the brightest environments. With a photon flux of 1500 μmol quanta $m^{-2}\ s^{-1}$ and with efficiency of 3 photons per e^- (allowing for inefficiency in capture of some wavelengths) the potential rate of e^- transport would be 500 μmol $m^{-2}\ s^{-1}$. With 4 e^- per CO_2 (this efficiency is about 12 photons per CO_2 fixed) the rate of CO_2 fixation would be 100 μmol CO_2 $m^{-2}\ s^{-1}$ or twice the rate observed for C3 plants under ideal conditions (50 μmol CO_2 $m^{-2}\ s^{-1}$). Thus light capture is not the limiting factor in assimilation at high photon flux. Even leaves 50 per cent deficient in chlorophyll because of Mg^{2+} or nitrogen shortage, can capture sufficient photons in normal light for photosynthesis, but more severe chlorosis limits assimilation. Under natural conditions leaves of shade plants have less chlorophyll per unit area of leaf but more chlorophyll in light-harvesting antennae per reaction centre than sun plants. This increases the chances of capturing a photon in dim light but in bright light they capture photons inefficiently. Sun plants have less chlorophyll per reaction centre and absorb and utilize photons more efficiently in bright light than shade plants as the electron transport capacity and enzyme capacity is increased. Increasing the chlorophyll content of leaves, by excessive nitrogen fertilization for example, may not increase photon capture or photosynthesis, as it is not a limiting factor. Leaves grown in dim light conditions may have 20 per cent less chlorophyll and a slightly smaller ratio of reaction centres to chlorophyll than those in high light so the light gathering efficiency is similar. If the number of reaction centres of PSII and PSI is each about 1 per 300 chlorophylls, then with 560 μmol chlorophyll m^{-2} leaf this is approximately 2 μmol reaction centres m^{-2}. If turnover time is between 100 and 1000 s^{-1}, 200–2000 μmol m^{-2} s^{-1} of electrons could enter the transport chain; the lowest rate is probably an underestimate as turnover time is much faster. With 4 e^- needed per CO_2 the potential rate is between 50 and 500 μmol CO_2 m^{-2} s^{-1}, up to 10 times greater than the maximum rates of photosynthesis so the concentration of reaction centres probably does not limit photosynthesis. Reaction centre concentration is greater in bright-light adapted plants, allowing faster photochemistry; photosynthesis of shade plants may be limited in bright light by the concentration of reaction centres.

Electron transport is rate limited by the transfer of electrons *via* plastoquinone (p. 92). Leaves contain about 10 mol of PQ per mol of PSII and PSII reaction centres, that is 20 μmol m^{-2} leaf. An electron passes from PQ to PSI in about 20 ms, a turnover number of 50 s^{-1}, giving a potential rate of 20 μmol PQ m^{-2} × 50 s^{-1} or 1000 μmol e^- transferred m^{-2} s^{-1}. With 4 e^- required per CO_2, assimilation is potentially of the order of 250 μmol CO_2 m^{-2} s^{-1}, five times greater than the fastest observed rates in good conditions and unlikely to be limiting. Plants grown in bright light have more PQ and other electron transport components per unit area than those from shade; this is probably an adaptive mechanism to increase the rates of electron transport. In bright-light grown *Atriplex triangularis* the capacity of photosystem-driven electron transport was 3 to 4 times that of dim-light grown plants, but the amount of PQ and photosystem increase by only 1.3 times. It is not understood why leaves do not form excess PQ and electron transport components under all conditions. Perhaps the energetic and material costs are too great in dim light, whereas in bright light the extra energy capture and photosynthesis are greater than the investment in photosynthetic machinery.

Rate of NADPH synthesis

The rate of e^- transport controls $NADP^+$ reduction if $NADP^+$ supply is adequate. The concentration of NADPH is about 0.1 mM in leaves and if all is in the chloroplast stroma (volume 1.4×10^{-5} m^3 m^{-2} leaf) the amount per unit area of leaf is 1.4 μmol m^{-2}. With a maximum rate of assimilation of 50 μmol CO_2 m^{-2} s^{-1} and with 2 NADPH per CO_2 assimilated the amount of NADPH required is 100 μmol m^{-2} s^{-1} and the turnover time will be about 16 ms. Reduction of $NADP^+$ requires 2 e^- and the potential rate of electron transport is 20 mmol m^{-2} s^{-1} so the potential is 10 mmol of $NADP^+$ reduced m^{-2} s^{-1}. With 1500 μmol quanta m^{-2} s^{-1} or 500 μmol e^- m^{-2} s^{-1} this rate of NADPH synthesis is $2\frac{1}{2}$ times greater than that required for observed CO_2 fixation.

However the rate of $NADP^+$ reduction is possibly much slower as it is an enzymatic reaction. The concentration of NADP reductase in the stroma is 80 μM or about 1 μmol m^{-2}; with a turnover time of 1 ms (1000 s^{-1}, rather fast for many complex enzymes) and 2 e^- per NADPH, the amount of $NADP^+$ reduced would be about 500 μmol m^{-2} s^{-1} or about 250 μmol CO_2 m^{-2} s^{-1}, greatly in excess of the observed values. A less realistic turnover rate of 10 ms (i.e. turnover number 100 s^{-1}) would give a rate of 25 μmol CO_2 m^{-2} s^{-1}, considerably less than the highest rates of photosynthesis measured. However, with rapid turnover and a high concentration of the enzyme, NADPH synthesis is unlikely to be a major limitation in assimilation.

Photophosphorylation

The potential rate of ATP synthesis may be calculated from the enzyme complement and proton transport. Estimates of CF_1 are 0.42 mg CF_1 mg chl^{-1} or about 1 mol CF_1 per 840 mol chl in bright light; there may be less in dim light. With 0.56 mmol of chlorophyll m^{-2} of leaf there is 0.67 μmol CF_1 m^{-2} of leaf. With a calculated turnover time of 4 ms (turnover number 250 s^{-1}) synthesis of ATP will be $0.67 \times 250 = 168$ μmol ATP m^{-2} s^{-1}. Three ATP are required for assimilation of 1 CO_2 so the rate of CO_2 assimilation would be $168/3 = 56$ μmol m^{-2} s^{-1}, close to the maximum rates of CO_2 assimilation in C3 leaves. This is an upper limit to the rate, for fewer CF_1 per chlorophyll would decrease the rate of assimilation. Also ATP is required for other processes. Allowing one extra ATP the rate of synthesis is decreased to 42 μmol m^{-2} s^{-1}, inadequate for rapid photosynthesis. Some estimates give more CF_1 per unit of chlorophyll, increasing the potential rate of ATP synthesis. However estimates of the *in vitro* rate of photophosphorylation (cyclic or non-cyclic) are small (250 μmol ATP (mg chl)$^{-1}$ h^{-1} equivalent to 35 μmol ATP m^{-2} s^{-1}) and cannot match the required rates of 150 μmol ATP m^{-2} s^{-1} for high rates of photosynthesis. Neither electron flow nor H^+ transfer limit ATP synthesis. With 2 mmol e^- m^{-2} s^{-1} and 2 H^+ per e^-, the potential H^+ flux is 4 mmol H^+ m^{-2} s^{-1} and with 3 H^+ per ATP gives 1.3 mmol ATP m^{-2} s^{-1}. A requirement of 3 ATP per CO_2 would give about 400 μmol CO_2 m^{-2} s^{-1}, greatly in excess of the observed rate. It is the enzyme reaction rate which limits synthesis of ATP. Shade plants have a smaller CF_1/chlorophyll ratio than sun plants and their photosynthesis saturates at smaller rates; with 1 CF_1/1800 chlorophylls the maximum rate of ATP synthesis would be more than halved and also the maximum rate of photosynthesis, as observed. Therefore, given the amount and turnover time of CF_1 in leaves, the rate of ATP synthesis is a greater potential limitation on the overall rate of assimilation than light capture, electron transport or NADPH synthesis. If electron and H^+ transport is increased by bright light, CF_1 would have to turn over faster or more CF_1 would have to be synthesized to avoid limitation. If P_i or ADP were limiting, as they may be in rapidly photosynthesizing tissues which have large pools of phosphorylated intermediates, then the potential rate of ATP synthesis would be smaller. Concentration of ATP in chloroplasts must be kept large to ensure a high energy charge for conversion of ribulose-5-phosphate to ribulose bisphosphate.

Rate of RuBP synthesis

Calculating the rate of activity of PCR cycle enzymes is complex as turnover times and amounts are not well defined, but several control points have been identified. Fructose bisphosphatase and sedoheptulose bisphosphatase are two main control points regulated by light. Rates of these enzyme reactions are sufficient for fast photosynthesis. Activity of phosphoribulokinase and other PCR cycle enzymes probably does not limit assimilation as rates are in excess of the measured rates of photosynthesis.

Carboxylation and oxygenation by RuBPc/o

RuBPc/o has a low affinity for CO_2 (12–20 μM for C3 plants), is competitively inhibited by O_2 (p. 135) and the amount of enzyme is important; concentration is 0.5 mM (giving the very considerable concentration of 275 g protein litre^{-1}) or 7 μmol RuBPc/o per m^{-2} leaf. Each mole of RuBPc/o has eight potentially active sites giving 56 μmol active sites m^{-2} leaf or 4 mM concentration. The enzyme has a turnover time of 0.5 s (turnover number 2 s^{-1}) so the potential rate of carboxylation is 112 μmol CO_2 m^{-2} s^{-1}, greater than the maximum rates of CO_2 assimilation under the best conditions, but potentially limiting. However, not all sites are active and this decreases the potential activity under normal atmospheric conditions. More importantly the CO_2 concentration in leaves is about 10 μM so the maximum rate of carboxylation will be smaller than the potential, similar to measured rate of P_n (net rate of photosynthesis). Decreasing O_2 decreases R_l and the competition for RuBP and increases assimilation by *c.* 40% so the limitation to photosynthesis is in the enzyme characteristics. The high rate of carboxylation is achieved at the expense of very large concentrations of protein and consequently demands much energy and nitrogen.

These simple calculations suggest that rates of ATP synthesis and RuBPc/o activity are potentially limiting under different conditions. At large photon flux and with high CO_2 and without photorespiration ATP may be limiting; some C4 chloroplast modifications (e.g. agranal chloroplasts) may increase ATP synthesis. Even with photorespiration ATP may limit, as it is consumed for synthesis of RuBP which is not used for CO_2 fixation.

Analysis of carbon dioxide and light response curves

It is useful to reconsider the response curves in the light of the previous discussions. Without CO_2, C3 plants in 21 kPa O_2 and bright light photorespire carbon from storage. Electron transport to $NADP^+$, and ATP and RuBP synthesis are limited by the rate of turnover of the $NADP^+$ and ATP pools; the electron transport chain and NADP are probably highly reduced and energy charge high. Long periods under these conditions may damage leaves if excessive energy is not dissipated, particularly if R_l (photorespiration) is prevented. At the CO_2 compensation concentration α is 2 and the flux of CO_2 from the glycolate pathway equals P_g. With further increase in CO_2 concentration P_g increases and also rates of NADPH, ATP and RuBP synthesis, but these are still in excess of demand and the carboxylation reaction is RuBP saturated but limited by CO_2. As CO_2 concentration increases further, the enzyme becomes more CO_2 but less RuBP saturated, and P_n increases slowly as RuBP synthesis limits assimilation under CO_2 saturation conditions. Carbon dioxide supply is limiting when the carboxylation rate ('demand function') exceeds the rate of CO_2 supply ('supply function') as shown in Fig. 11.3, and internal CO_2 concentration decreases. Even with large stomatal conductance internal CO_2 concentration is smaller than ambient CO_2 in air, and P_n is limited by CO_2 supply in bright light. Possibly this enables the photosynthetic system to function at optimum efficiency without large fluctuations in rate.

Net photosynthesis is affected by temperature through RuBP synthesis and the carboxylation reaction. At low temperatures (below 5 °C in many C3 and 10 °C in C4 plants) RuBP synthesis is limiting. It increases with warmer temperatures. However, P_g of C3 plants in air is offset by increasing R_l. Comparison of P_n *versus* internal CO_2 response curves for leaves under different conditions or between species enables the causes of decreasing P_n to be analysed. Decrease in P_{max} indicates deficient synthesis of RuBP, and change in the initial slope of the relationship shows altered carboxylation efficiency, because of inadequate amount of RuBP carboxylase for example. Measurement of conductance in relation to external CO_2 supply shows the limitation imposed by diffusion processes. The CO_2 response curves for C4 leaves may be analysed in the same way but understanding of the relationships is not as advanced as for C3 leaves.

Compartmentation in photosynthesis

Photosynthetic metabolites in the cell occur in particular compartments or 'pools' which may be in solution in the stroma or other organelles or bound to enzymes. Influx and efflux of metabolites determines the amount of material in the pools. Information about the size of pools and fluxes is essential for the analysis of assimilation in relation to environment, genotype etc. Radiotracers (e.g. ^{14}C) may be used to analyse the size of pools, and the direction and magnitude of the fluxes, if the total radioactivity and specific radioactivity (SA = (^{14}C in pool/total C in pool)) of materials in the pools is known. SA is determined experimentally by feeding $^{14}CO_2$ for a given time, rapidly killing the tissue, extracting and separating the metabolites and measuring the amount of ^{14}C and ^{12}C and their ratio. Analysis of photosynthetic carbon fluxes may be made with leaves photosynthesizing with $^{12}CO_2$ at the steady rate (i.e.

constant P_n with time) and then rapidly supplying $^{14}CO_2$ of known SA without changing P_n. In the first seconds of labelling the ^{14}C in 3PGA increases and, after a period which depends on the rates of flux into and out of the pools and on the pool size, saturates at, or close to, the SA of the feeding gas (i.e. 3PGA has an SA relative to the feeding gas of 1). With time, metabolites and storage compounds derived from 3PGA acquire label. Changes in the SA of compounds with time show the metabolic sequences. Rates of flux from one compound to another may be calculated from:

$$\text{flux per unit time} = \frac{1}{\text{SA of precursor}} \times \frac{1}{\text{time}} \times {}^{14}\text{C in product} \qquad [\mathbf{11.10}]$$

It is possible, for example, to show that the flux of carbon into glycine of the glycolate pathway in leaves is much greater in air than in 1 kPa O_2, that glycine exists in two pools (at saturation the SA of glycine is less than that of the feeding gas) and that serine in leaves is not derived exclusively from the glycolate pathway. Mathematical analysis of compartmental systems is sophisticated and provides a powerful tool for understanding photosynthetic metabolism and factors controlling it.

Photosynthesis and fluorescence in leaves

The interaction between light harvesting, e^- transport, NADPH and ATP synthesis and CO_2 fixation in intact tissues may be analysed by measuring transient changes in chlorophyll *a* fluorescence concomitant with O_2 evolution and CO_2 assimilation during changes in light and gas concentrations or with application of inhibitors. Fluorescence measurement is illustrated in Fig. 11.9.

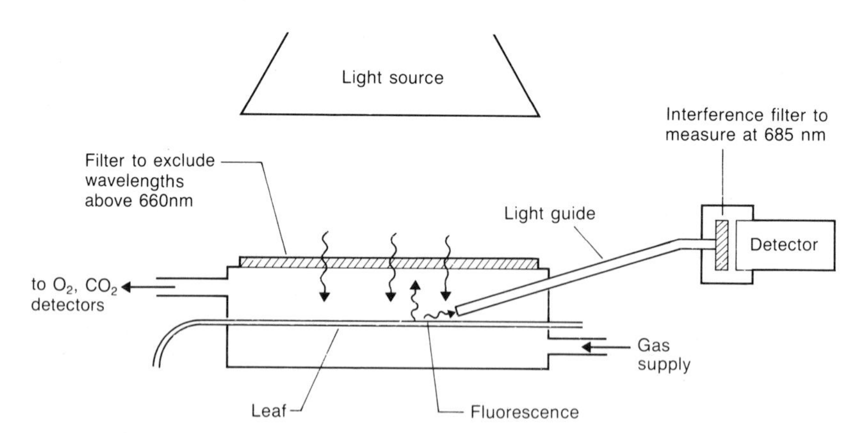

FIG. 11.9 System for measuring fluorescence emission from leaves together with O_2 and CO_2 exchange under controlled gas phase conditions in a leaf chamber. The leaf is illuminated by light from combinations of sources and filters giving light of narrow (e.g. laser) or wide band (incandescent lights). Fluorescence of particular wavelength (selected by filters) is measured by a photomultiplier or photodiode detector.

Light of particular wavelengths (e.g. 450 or 650 nm) produced by optical filters or lasers, or white light, excites chlorophyll and the fluorescence emitted (from chl *a* of PSII only at room temperature) between 680 and about 720 nm, or a particular wavelength (frequently 685 nm) is measured with a fast responding detector. Much information has been obtained during photosynthetic induction following illumination after darkness. Figure 11.10 summarizes and schematizes the responses. Fluorescence increases very quickly (ms) to a value F_0, due to emission from the chlorophyll matrix during excitation transfer. All the photosystem reaction centres and acceptors are able to take e^- (i.e. the 'traps' are open). At F_0 the reaction centres of PSII eject electrons to the primary acceptors which reduce plastoquinone (Fig. 11.10). From F_0 to I, PSII is oxidized, water is split, producing O_2, but the reduction of PSII electron acceptors is slow so that fluorescence rises. As PSII is oxidized there is some excitation transfer and fluorescence drops (the I to D drop). However PQ is reduced faster than e^- is transferred to cytochrome *f* and PSI, the acceptors of e^- from PSII and pool of PQ are reduced, e^- cannot be ejected from the PSII reaction centre, energy accumulates in the antenna and fluorescence increases to the peak P. Some water splitting takes place during the D to P stage and O_2 is evolved. The $\Delta\psi$ component of the proton motive force increases quickly but the ΔpH increases slowly so that ATP synthesis is probably minimal. However NADPH synthesis is more rapid. Enzymes of the PCR cycle are activating during this phase and therefore there is limited CO_2 assimilation. Oxidation of PQ, as e^- passes to PSI and reduces $NADP^+$ is accompanied by a flux of H^+ into the thylakoid increasing ΔpH, which drives phosphorylation. More rapid ejection of electrons uses excitation and decreases fluorescence (the P to S phase). Fluctuations in fluorescence (S to M) are caused by, and characteristic of, interactions between processes with very different time constants as the rates adjust to an equilibrium. DCMU infiltrated into the leaf blocks the reduction of PQ and the small pools of intermediate acceptors are reduced very rapidly so fluorescence rises faster than for the normal tissue and to a larger value which remains constant. PQ is not reduced nor O_2 released. The difference between P in the DCMU-treated and untreated tissues shows that PQ is never fully reduced *in vivo* because e^- flux to PSI keeps the pool partly oxidized.

The relative contributions of electron transport in the thylakoid, $\Delta\psi$ and ΔpH build up, and the movement of Mg^{2+} and other ions to fluorescence changes are not yet established, particularly in steady state assimilation. In the D to P stage electron transport dominates, the P to M transitions probably reflect phosphorylation. With time, fluorescence decreases to a low, relatively constant value (T) and O_2 evolution and CO_2 fixation increase, often with damped oscillations, to constant rates with a partially reduced PQ pool. However during the period from P to T the transfer of excitation energy which is predominantly from LHC to PSII in the O to M phase, redistributes or spills over (p. 78) to PSI as the protein kinase in the thylakoid regulates the association of photosystems in the membranes. Excitation energy in PSI increases,

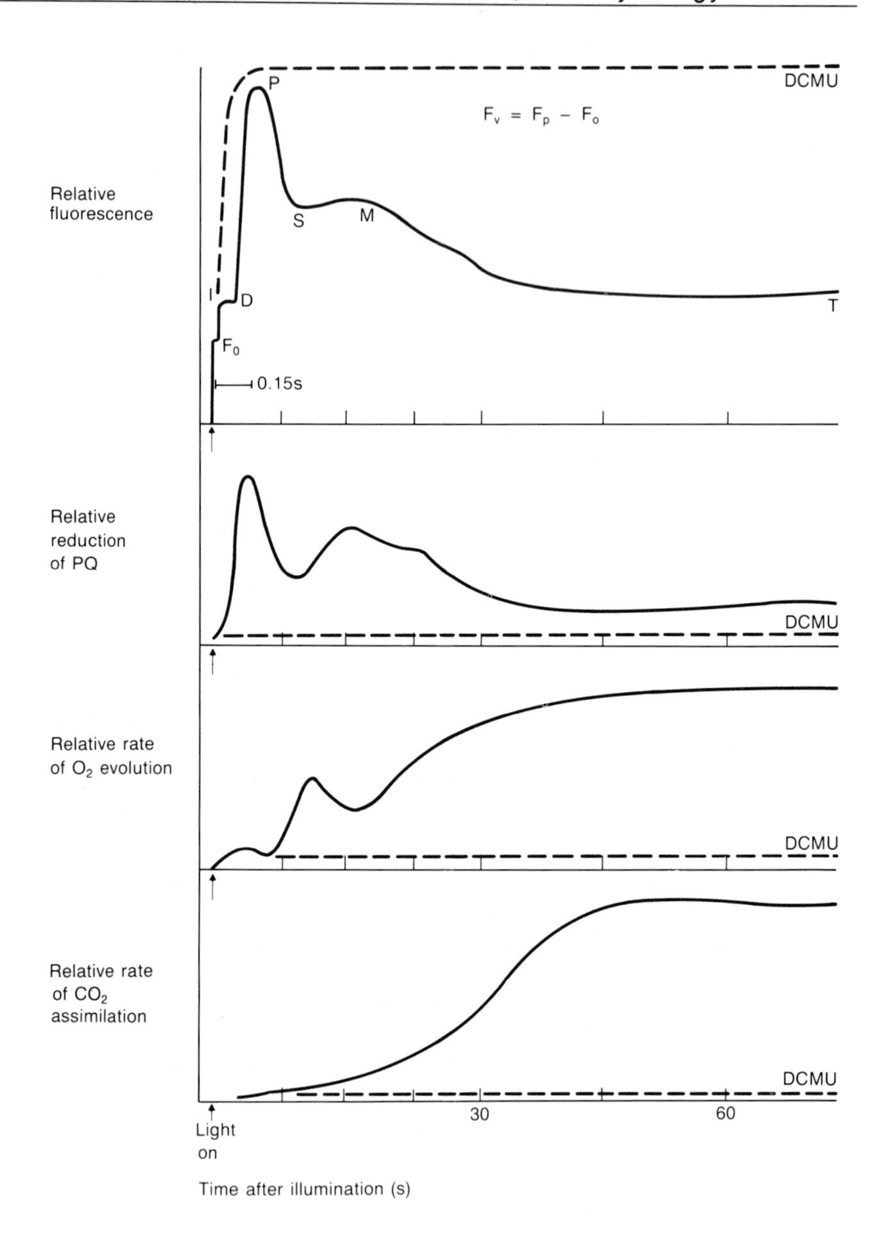

FIG. 11.10 Fluorescence transients following illumination (Kautsky curve) in leaves without or with DCMU, which blocks electron transport. Oxygen evolution and CO_2 assimilation are related to the redox state of plastoquinone, which reflects the rates of electron transport and of formation and consumption of the components of the thylakoid proton motive force; see text for details.

and with it the efficiency of energy use; fluorescence decreases as photosynthesis increases.

Inverse relation between assimilation and fluorescence is often observed as the biochemical processes use excitation energy. If e^- flow is blocked after PSII, fluorescence increases as energy 'backs up'. However e^- movement may be stopped by other processes which prevent use of photons in the lumen, for example deficient phosphate supply. Then ΔpH increases, e^- transport slows and the fluorescence increases whilst CO_2 assimilation falls. If ΔpH is destroyed by uncouplers then e^- transport increases (if $NADP^+$ is reduced or there is active pseudocyclic or cyclic electron transport) and fluorescence decreases together with CO_2 assimilation. If water splitting is blocked then PQ is not reduced but fluorescence increases.

Changing O_2 and CO_2 concentration around leaves alters fluorescence and assimilation and provides information on the energetics and internal fluxes in photosynthesis. Thus if a leaf in constant illumination in CO_2-free air is provided with 100 Pa CO_2, assimilation increases rapidly to a large rate, presumably because there is excess RuBP in the 'pool' for immediate carboxylation. As H^+ and e^- are consumed to regenerate RuBP, fluorescence decreases. However within a few minutes assimilation drops and fluorescence rises rapidly, probably because ATP synthesis is too slow to regenerate RuBP fast enough to maintain the previous rate of assimilation and the electron transport backs up. There may be several damped oscillations as the system adjusts to conditions. However, changes in assimilation and fluorescence are not always reciprocal. With inhibitors of photophosphorylation, the relationship depends on the mode of action of the substance. In leaves, for example, phlorizin increases fluorescence because it blocks the H^+ flow through CF_0-CF_1 and therefore slows e^- transport. Gramicidin, which makes thylakoids leaky to H^+, increases e^- flow and decreases fluorescence. Both slow or stop assimilation. Also, increasing O_2 at 30 Pa CO_2 decreases both P_n and fluorescence, as energy is probably used by photorespiration or e^- transport to O_2. Stress (e.g. water deficit and photoinhibition) increases fluorescence as consumption of excitation energy is impaired.

Fluorescence induction kinetics of chlorophyll *a* at room temperature may be interpreted as the product of several processes consuming the excitation energy absorbed by the pigments. For a given radient energy, I, fluorescence, F, is related to the fraction, β, of the energy reaching chlorophyll *a* of PSII of concentration [chl *a* II], and the quantum yield, ϕ_f, by:

$$F = I \times \beta \times [\text{chl } a \text{ II}] \times \phi_f \qquad [11.11]$$

where quantum yield is the fluorescence emission relative to the rate of excitation used by radiationless dissipation (heating, d) transfer to other molecules (t) and photochemistry (p) (see Ch. 3). The rate constant K (with units of s^{-1}) for each process of energy loss expresses the rate of decay; the greater the constant the shorter the intrinsic lifetime, τ_0, of the state, that is $K = 1/\tau_0$. Given that K_f, K_d, K_t and K_p are the rate constants for fluorescence, radiationless decay, transfer to other molecules and photochemistry then the lifetime of an excited pigment molecule is:

$$\tau = \frac{1}{K_f + K_d + K_p + K_t} \qquad [11.12]$$

The fluorescence given out in relation to the quanta of light absorbed is, if β and [chl *a* II] are constant throughout the fluorescence changes, the quantum yield of fluorescence:

$$\phi_f = \frac{\text{intensity of fluorescence}}{\text{intensity of absorbed light}} = \frac{K_f}{K_f + K_d + K_t + K_p \cdot P} \qquad [11.13]$$

where P is the fraction of PSII reaction centres which can accept energy from chlorophyll.

At F_0 on the induction curve (Fig. 11.9) the PSII reaction centres are fully oxidized, P = 1 and quantum yield of fluorescence, ϕ_{fo}, is:

$$\phi_{fo} = \frac{K_f}{K_f + K_d + K_t + K_p} \qquad [11.14]$$

At maximum fluorescence, F_m, in the presence of DCMU which prevents e^- moving to PQ, P = 0 and

$$\phi_{fm} = \frac{K_f}{K_f + K_d + K_t} \qquad [11.15]$$

Variable fluorescence, F_v, is:

$$F_v = F_m - F_o = \phi_{f_m} - \phi_{f_o} \qquad [11.16]$$

and shows the quantum yield of photochemistry. The maximum yield of photochemistry (i.e. reduction of PQ) is given by:

$$\phi_p = \frac{K_p}{K_f + K_d + K_t + K_p} \qquad [11.17]$$

and $\phi_p = F_v/F_m$ if I, β and [chl *a* II] and the amount of PQ are constant. F_p, the maximum fluorescence at P on the induction curve, without DCMU approaches F_m with DCMU and is a measure of F_m for the physiological, unpoisoned system. Thus the ratio of F_v to F_m or F_p, provides a measure of the photochemical processes in the thylakoid. If PQ is reduced then F_v/F_p is small and the photochemical quantum yield is small, and there is little or no decrease in fluorescence with time. This may occur if electron transport is blocked by lack of NADPH or other acceptors, or by slow ATP synthesis which causes H^+ to accumulation in the thylakoid and exert 'back pressure' on electron transport. Conversely if there is active consumption of NADPH etc., then e^- transport and ΔpH decrease, fluorescence falls and CO_2 assimilation increases.

However, uncouplers which dissipate ΔpH and allow e^- transport and thus PQ reduction decrease fluorescence even though CO_2 assimilation is small. Thus fluorescence is a most important probe to estimate photochemical efficiency and show how light absorption and chemical processes are linked. Fluorescence indicates the energy state of the chlorophyll matrix and is influenced by electron and proton transport and therefore the 'down-stream' events which control them. It is a sensitive probe of events in the chloroplast and in conjunction with measurements of gas exchange and metabolites provides information on the rapid changes in photosynthesis and the effects of environment and physiological conditions.

References and Further Reading

Atkins, G. L. (1969) *Multicompartment Models in Biological Systems*, Methuen and Co., London, Halsted Press, Division of John Wiley and Sons, New York.

Baker, N. R. and **Bradbury, M.** (1981) Possible applications of chlorophyll fluorescence techniques for studying photosynthesis *in vivo*, pp. 355–73 in Smith, H. (ed.), *Plants and the Daylight Spectrum*, Academic Press, London.

Berry, J. and **Björkman, J.** (1980) Photosynthetic response and adaptation to temperature in higher plants, *A. Rev. Plant Physiol.*, **31**, 491–543.

Björkman, O. (1981) Ecological adaptation of the photosynthetic apparatus, pp. 191–202 in Akoyunoglou, G. (ed.), *Photosynthesis*, vol. VI, *Photosynthesis and Productivity, Photosynthesis and Environment*, Balaban International Science Services, Philadelphia.

Boardman, N. K. (1977) Comparative photosynthesis of sun and shade plants, *A. Rev. Plant Physiol.*, **28**, 355–77.

Cowan, I. R. (1982) Regulation of water use in relation to carbon gain in higher plants, pp. 589–613 in Lange, O., Nobel, P. S., Osmond, C. B. and Ziegler, H. (eds), *Encyclopedia of Plant Physiology*, (N.S.), vol. 12B, *Physiological Plant Ecology II*, Springer-Verlag, Berlin.

Davies, D. D. (1980, ed.) *The Biochemistry of Plants*, vol. 2, *Metabolism and Respiration*, Academic Press, New York.

Farquhar, G. D. (1979) Models describing the kinetics of ribulose bisphosphate carboxylase-oxygenase, *Arch. Biochem. Biophys.*, **193**, 456–68.

Farquhar, G. D. and **Sharkey, T. D.** (1982) Stomatal conductance and photosynthesis, *A. Rev. Plant Physiol.*, **33**, 317–45.

Farquhar, G. D. and **von Caemmerer, S.** (1982) Modelling of photosynthetic response to environmental conditions, pp. 549–87 in Lange, O., Nobel, P. S., Osmond, C. B. and Ziegler, H. (eds), *Encyclopedia of Plant Physiology* (N.S.), vol. 12B, *Physiological Plant Ecology II*, Springer-Verlag, Berlin.

Graham, D. (1980) Effects of light on 'dark' respiration, pp. 526–80 in Davies, D. D. (ed.), *The Biochemistry of Plants*, vol. 2, *Metabolism and Respiration*, Academic Press, New York.

Graham, D. and **Chapman, E. A.** (1979) Interactions between photosynthesis and respiration in higher plants, pp. 150–62 in Gibbs, M. and Latzko, E. (eds), *Encyclopedia of Plant Physiology* (N.S.), vol. 6, *Photosynthesis II*, Springer-Verlag, Berlin.

Guinn, G. and **Mauney, J. R.** (1980) Analysis of CO_2 exchange assumptions:

feedback control, pp. 1–16 in Hesketh, J. D. and Jones, J. W. (eds), *Predicting Photosynthesis for Ecosystem Models*, vol. II, CRC Press Inc., Boca Raton, Florida.

Lichtenthaler, H. (1981) Adaptation of leaves and chloroplasts to high quanta fluence rates, pp. 273–87 in Akoyunoglou, G. (ed.), *Photosynthesis*, vol. VI, *Photosynthesis and Productivity, Photosynthesis and Environment*, Balaban International Science Services, Philadelphia.

Lorimer, G. H. and **Andrews, T. J.** (1981) The C_2 chemo- and photorespiratory carbon oxidation cycle, pp. 330–74 in Hatch, M. D. and Boardman, N. K. (eds), *The Biochemistry of Plants*, vol. 8, *Photosynthesis*, Academic Press, New York.

Nobel, P. S. and **Longstrett, D. J.** (1981) Effects of environmental factors on leaf anatomy, mesophyll cell conductance and photosynthesis, pp. 245–54 in Akoyunoglou, G. (ed.), *Photosynthesis*, vol. VI, *Photosynthesis and Productivity, Photosynthesis and Environment*, Balaban International Science Services, Philadelphia.

Osmond, C. B., Björkman, O. and **Anderson, D. J.** (1980) *Physiological Processes in Plant Ecology. Towards a Synthesis with Atriplex*, Springer-Verlag, Berlin.

Papageorgiou, G. (1975) Chlorophyll fluorescence: an intrinsic probe of photosynthesis, pp. 320–71 in Govindjee (ed.), *Bioenergetics of Photosynthesis*, Academic Press, New York.

Pearlman, J. G. and **Lawlor, D. W.** (1981) Tracer experiments and compartmental modelling in analysis of plant metabolism, pp. 91–108 in Rose, D. A. and Charles-Edwards, D. A. (eds), *Mathematics and Plant Physiology*, Academic Press, London.

Sesták, Z, Catsky, J. and **Jarvis, P. G.** (1971, eds) *Plant Photosynthetic Production, Manual of Methods*, Dr W. Junk, The Hague.

Thornley, J. H. M. (1976) *Mathematical Models in Plant Physiology*, Academic Press, London.

Willmer, C. M. (1983) *Stomata*, Longman, London.

Zelitch, I. (1979) Photorespiration: Studies with whole tissues, pp. 353–62 in Gibbs, M. and Latzko, E. (eds), *Encyclopedia of Plant Physiology* (N.S.), vol. 6, *Photosynthesis II*, Springer-Verlag, Berlin.

CHAPTER *12*

Photosynthesis, plant production and yield

Mass of standing vegetation in different habitats varies widely, from almost nothing in some deserts to hundreds of tonnes of dry matter per hectare in tropical forests, and the rates of dry matter production also differ greatly. Crops of well watered and fertilized C3 cereals in temperate zones have the potential to produce more than 25 t dry matter ha^{-1} in an 8-month growing season and in the tropics sugar cane (C4) forms over 80 t dry matter ha^{-1} $year^{-1}$. All organic matter is derived from photosynthesis and accumulation of inorganic matter in vegetation (usually < 10% of dry matter) requires photosynthetic energy. Total net photosynthesis per unit ground area (m^2) is a function of light energy absorbed per unit leaf area, the response of net photosynthesis to light, total leaf area (m^2) and the number of days of assimilation:

$$\text{Total net photosynthesis m}^{-2}\text{ season}^{-1} = \text{photosynthesis m}^{-2}\text{ leaf day}^{-1} \times \text{m}^2\text{ total leaf m}^{-2}\text{ ground} \times \text{day season}^{-1} \quad [\mathbf{12.1}]$$

Assimilation may be calculated (although difficult to achieve in practice) for an entire growing season by integration of the photosynthetic rate of individual leaves over light, temperature, CO_2 supply, water stress etc., and then over the entire leaf area for each day. The daily assimilation may then be summed over the season to give total assimilation. Net photosynthesis is a function of light intensity (photon flux), CO_2 concentration in the atmosphere within the vegetation, temperature, water supply, nutrition and the physiological state of the plant, for example leaf age and reproductive state. There is considerable interaction between environment and plant growth which controls P_n (net rate of photosynthesis) and total net photosynthesis; poor nutrition, for example, slows leaf growth and causes early leaf senescence. A smaller leaf area over the life of the crop, even if P_n is little affected, decreases assimilation and growth. Water stress for short periods may decrease stomatal conductance and inhibit P_n more than area, and thereby inhibit growth. Vegetation is subject to many fluctuations in environment which change or decrease growth (generally called 'stresses') and each combination may influence different plant processes, with different quantitative and qualitative effects, although modifying the same plant mechanisms. For each plant species and community, net photosynthesis per unit leaf area is optimal under a particular range of conditions. Differences caused by environment and between species are explicable in terms of the interactions between environment, biochemistry and physiology. Understanding of environmental control of net photosynthesis is developing rapidly, particularly at the level of organization of the leaf, as discussed in Chapter 11. In crops the interaction between processes is semi-quantitatively understood.

Light is the driving force for production, and temperature, nutrition and water regulate production; they are also responsible for seasonal variations, with particular combinations of conditions important in different years. Conditions during development of the photosynthetic system modify the structure of light-harvesting, energy-transducing and enzyme systems and thereby alter the efficiency of assimilation. However, although understanding of the mechanisms is developing, quantitative models of the changes in biochemistry and physiology in relation to environment are not well developed, despite the potential importance in understanding the control of productivity in vegetation.

Maximum production of photosynthate depends on the photon flux incident upon the canopy and its absorption by leaves in different layers, and on the efficiency of conversion to assimilate. About 10–15 per cent of incident PAR (photosynthetically active radiation) is reflected and transmitted by leaves and only part of the energy not captured by upper leaves in the canopy is absorbed by other foliage, thus limiting overall efficiency of light capture to *c.* 85–90 per cent of incident PAR. As P_n is saturated at photon fluxes of half or less of full sunlight, particularly in C3 and shade plants, efficiency decreases markedly at intensity above saturation of P_n. Leaves adapted to bright light are more effective at using radiation at the top of vegetation canopies and shade leaves use dim light in the lower canopy. Absorption of blue and red wavelength by upper leaves, enriches light in the lower canopy in the green and infra-red (which has photomorphogenic effects on plants) and requires changes in light absorbing characteristics of leaves. In dim light assimilation is a linear function of intensity and is at its most efficient. Plant with larger photochemical efficiency in dim light will have greater growth rates for a given light absorption and may be more successful in competition, for example in dense vegetation or other shady habitats. Natural vegetation canopies are stratified, with the success of different species in occupying niches mainly dependent on their ability to capture light. Plants of very dim light may be unable to adjust to bright light without photochemical damage and some have developed structural and other characteristics, such as the ability to change leaf orientation to light, for protection. Stratification of leaves of different efficiency within a canopy allows very effective light absorption and high productivity. In crop monocultures, leaves are of similar efficiency throughout the canopy and may be less efficient at extremes of radiation than those of adapted species. However, leaves in the lower parts of canopies are usually older and have lower efficiency and less total capacity for assimilation than young leaves in the top of the canopy. C3 plants are more efficient in dim to intermediate light intensities, whereas C4 are more efficient in bright than in dim light. Canopy architecture influences the efficiency of light utilization; thus C3 crops suffering nutrient or water stress produce less leaf than unstressed crops and so intercept less light overall.

However the light intercepted is of higher intensity where assimilation may be higher but the conversion is less efficient. In contrast C4 crops would absorb less total light with less leaf area but would absorb it with higher efficiency. Vegetation does not receive bright radiation at all times. Many habitats are dimly lit for long periods, with clouds and twilight, when the sun's elevation is low. Efficiency in dim light is then of great importance and C3 species would have an advantage.

Oxygen concentration in the canopies of actively photosynthesizing crops is effectively constant. However CO_2, which is at much lower partial pressure, may decrease substantially, for example by 20 per cent from 34 to about 28 Pa within dense, vigorously assimilating vegetation, where turbulent transfer of gases is restricted. Internal CO_2 partial pressure will then decrease further, depending on stomatal conductance. C3 crops are more affected than C4 because of the inefficient RuBP carboxylase reaction. Stomatal conductance may also decrease due to water shortage, high temperature etc., and further restrict CO_2 supply and decrease photosynthesis; again the effect is relatively greater in C3 plants.

Response of P_n to temperature is also very important in determining assimilation (see Ch. 11). Warm conditions favour vegetation with a high temperature optimum for assimilation, for example C4 over C3 species. Often relatively small differences in temperature have large effects on productivity, particularly when plants approach the limits of their adaptation. Differences between seasons in crop production may be related to interactions between temperature and light, thus C3 crops may produce more in a cool, relatively dimly lit but long growing season than in a warm, bright but short season. The effects of conditions, both singly and in combination, on growth become more pronounced as they depart from the 'broad optima' to which particular plant species are adjusted. Thus the productivity and yield stability of maize decreases more the further into temperate zones that the plant is grown, because of the cooler and shorter growing seasons. However, the complex responses of growth and yield to environment involve many biochemical and physiological responses in addition to those of photosynthesis.

Respiration is the main process other than assimilation which determines dry matter production. Total net photosynthesis (eqn 12.1) if calculated from net photosynthesis during illumination, allows for both photo- and dark respiration of leaves in the light, but neglects respiration during the light period for other organs, and for the whole plant in darkness. Plant productivity is, therefore, dependent on respiratory losses throughout the period of crop growth (eqn 12.2):

$$\begin{aligned}\text{dry matter production} = {} & \text{total net photosynthesis m}^{-2}\text{ ground day}^{-1} \\ & \times \text{ number of days season}^{-1} \\ & - \text{ total respiration m}^{-2}\text{ ground day}^{-1} \\ & \times \text{ number of days season}^{-1}\end{aligned} \qquad [12.2]$$

$$(\text{g m}^{-2}\text{ season}^{-1} = \text{g m}^{-2}\text{ day}^{-1} \times \text{day season}^{-1} - \text{g m}^{-2}\text{ day}^{-1} \times \text{day season}^{-1})$$

As with photosynthesis the rate of respiration per unit dry mass of different organs must be correctly integrated over total mass, age, and conditions throughout the life of the crop. Respiration is required to provide energy and substrates for all biochemical processes including turnover of cell structure and also formation of new growth, although part of the requirement may come directly from photosynthesis. Respiration is often divided into 'maintenance' and 'growth' components corresponding to the demands of the existing system and to the requirements for synthesis of new cell components. However, although this distinction is a useful concept to analyse the interaction between respiration and plant functions, the type of respiration is identical, only the use to which the products are put differs. It is difficult to distinguish between the two functions of respiration.

Two mechanisms of dark respiration consuming carbohydrates have been identified, coupled in two ways with the electron transport chain in the mitochondria and ATP production. In one, the normal electron transport is coupled to 3 ATP synthesis and in the other, electrons are consumed by an alternative oxidase pathway. Electron flow in this gives only 1 ATP, and 'short circuits' the normal process and is less energetically efficient. It may provide a spill-over mechanism to regulate carbohydrate and reductant energy in plants. Respiration is greatly dependent on temperature, but other factors influence it, rapid P_n in the previous light period stimulating it, for example, probably by increasing carbohydrates in the plant; it is possible that carbohydrates are consumed by the alternative pathway when in excess of the needs of normal respiration. Respiration is important as it proceeds throughout each day and uses much assimilate, perhaps 50 per cent of total production. Small rates of respiration, as in shade adapted plants, minimize loss of energy and carbon and thus increase the efficiency of energy use under limited light. Plants adapted to bright light have relatively large respiration; presumably energy and carbon are not limiting factors in their growth. Plants appear to require rapid respiration for rapid growth, and in species adapted to adverse conditions metabolic activity is much lower. Perhaps large respiration is an inevitable consequence of maintaining a large assimilatory capacity. There is evidence for rye grass, that cultivars differ widely in the ratio of respiration to assimilation and have correspondingly different growth rates. Combination of low respiration with high capacity for photosynthesis gives the greatest dry matter production in a range of environments. Both processes may be optimized to different conditions so their role in determining total crop production may vary with environment.

Photosynthetic rate is the major determinant of dry matter production; C4 and C3 crops, with average P_n in bright light of 30 and 13 μmol CO_2 m^{-2} leaf s^{-1} respectively, have growth rates of 22 and 12 g m^{-2} ground day^{-1}. Yield of C4 crops is correspondingly greater than C3; average United States yields of the C3 plants wheat and soybean are 2.5 t ha^{-1} and of rice 5.8 t ha^{-1}, whilst those of the C4 species maize and sorghum are 7.8 and 4.5 t ha^{-1} respectively. However, in experiments on many crops, P_n measured on individual leaves over short periods (minutes to hours) at high irradiance, has not correlated closely with dry matter production or yield. This apparent anomaly is readily understood as a consequence of the neglect of the factors which contribute to production. Measurements of CO_2 exchange over longer periods and of larger areas of crop have generally agreed better with production. Short-term estimates of P_n, particularly in bright light, give only the contribution of younger, active leaves to total assimilation and neglect the contribution of older, shaded leaves. Leaf area changes in response to conditions rather more than assimilation per unit area, and variation in leaf area may dominate production. Also the amount of respiration per unit of dry matter and of the total crop is not usually measured. As both assimilation and respiration may change to different extent depending on conditions and are not included in short term measurements the lack of correlation is perhaps not surprising. Whole crop measurements over longer periods are therefore expected to correlate better with production.

Crop production is closely related to light interception, as expected if PAR determines P_n, and therefore to leaf area. Efficiency of conversion of light to dry matter in cereals and sugar beet is close to 2 g dry matter MJ^{-1} of PAR absorbed (Fig. 12.1). Conversion of CO_2 to crop dry mass depends on the type

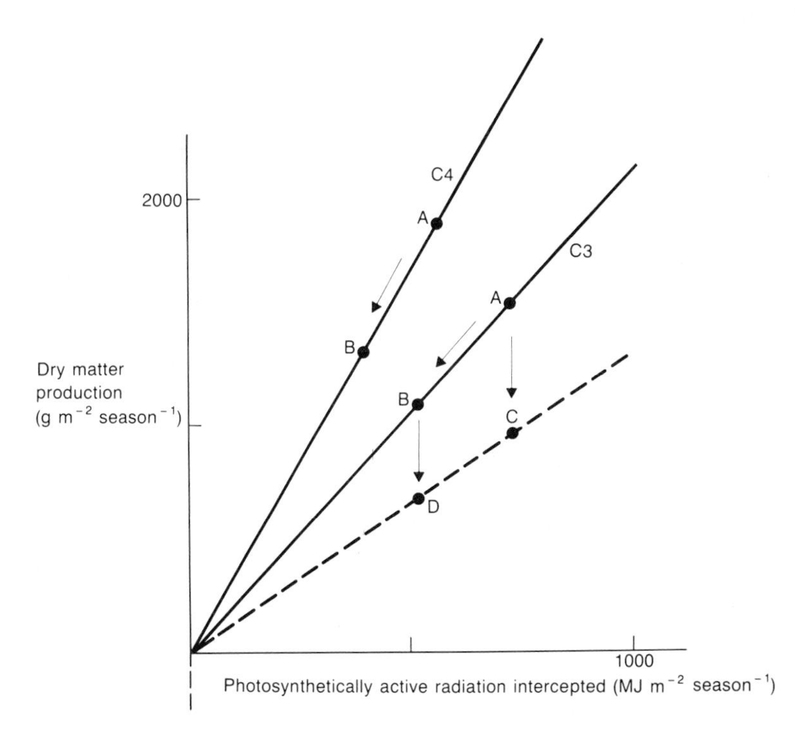

FIG. 12.1 Dry matter production as a function of photosynthetically active radiation for C4 and C3 crops with adequate water, nutrients and temperature. Changes in leaf area decrease light interception and move production from A to B. Impaired metabolism caused by stress decreases production at a given light interception (A → C, illustrated for C3 only). Stresses which decrease both area and efficiency decrease production *via* A → B → D.

of organic molecules synthesized and their relative proportion. One gram of CO_2 gives 0.4 g fats, 0.62 g starch and 0.5 g protein, an average value (allowing for the proportions of fats, proteins etc., in dry matter) is 0.58 g dry matter. Plants producing much oil or protein produce less dry matter than those forming carbohydrates, but the energy yield may be similar. Each gram of CO_2 assimilated is equivalent to an energy content of 38 kJ g^{-1} in fats, 12.6 kJ g^{-1} in proteins and 17 kJ g^{-1} in starch, with an approximate value of 15–20 kJ g^{-1} in dry matter from the leaves of a range of species. From the known total energy requirement for a given dry matter production and the energy in the crop, the efficiency of energy conversion of crops and natural vegetation can be calculated. For a C3 crop requiring 0.5 MJ g^{-1} dry matter and producing 25 tonnes dry matter per hectare the energy input is 1250 MJ m^{-2} and the energy content (taking the upper value of 20 kJ g^{-1}) is 50 MJ m^{-2}, an efficiency of about 4 per cent. Alternatively, assuming that the average energy of a mole of PAR photons is 0.2 MJ and that 1 g dry matter is equivalent to 1.7 g CO_2 (0.039 mol CO_2), then the quantum yield is less than 0.02 (mol CO_2/mol photons) an efficiency much smaller than the theoretical conversion efficiencies or those obtained under ideal conditions. The conversion efficiency of C4 crops is 50 per cent greater than C3 but still very inefficient in use of light energy.

Light interception increases almost linearly as the leaf area per unit ground surface (called the leaf area index, LAI) increases to 2–3, and at LAI > 4 most of the light is absorbed, often approaching 90 per cent. Additional leaves absorb little extra light with increasing LAI, and lower leaves suffer mutual shading so total production may not increase. However, total respiration increases in proportion to dry matter and consumes an increasingly large part of assimilation as a crop grows; therefore dry matter production may increase little with larger LAI. In young crops the ratio of assimilation to respiration is relatively constant so the efficiency of production is constant. Differences in total production between crops may reflect differences in basic efficiency, but these are small, as far as is known, within the C3 and C4 groups. Differences between species with season and environment may largely reflect variation in photosynthetic area which determines the position on curve A in Fig. 12.1. Temperature usually determines the leaf area development of temperate crops and has little effect on efficiency so that productivity moves along the curves according to radiation absorbed. However, under nutrient or water deficiency the efficiency of conversion of light to dry matter decreases at a given light absorption (A→C). As metabolic processes seem to vary less than plant morphology, much variation comes from change in photosynthetic area.

Water use of natural vegetation and crops depends on the atmospheric conditions, for example humidity and wind speed, radiation load on the leaves and stomatal control. C3 plants have water use efficiencies of very approximately 1–3 mg dry matter g^{-1} water transpired, and C4 crops 3–8. These differences are related to the larger stomatal conductances and smaller assimilation rates, particularly under light saturating conditions, of C3 compared to C4 plants. In temperate climates the advantages of C4 plants are minimized but in bright but dry environments, their advantages increase.

Nitrate reduction is also essential for growth and efficient P_n; if rates of NO_3^- conversion to amino acids are slower than the potential rate of protein synthesis (which is genetically controlled but regulated largely by temperature), expansion of the photosynthetic surface is slowed and senescence increased. If more severe, deficiency causes metabolic changes which lead to decreased P_n. Changes in plants are not linearly related to nitrate reduction; with very high rates of nitrate assimilation small changes in supply may have little effect on efficiency or growth of leaf area. However, when nitrate is deficient, there may be large effects on leaf area and somewhat smaller effects on P_n. Rate of nitrate reduction depends on the supply of reductant and carbon skeletons from photosynthesis and on the amount and turnover of enzymes as well as on NO_3^- supply. Thus productivity is a complex function of environment and plant characteristics and is not directly related to maximum rates of P_n.

Photosynthesis and yield (i.e. a particular part of a crop required for human use) are indirectly related. Dry matter is distributed, 'partitioned', between harvestable organs (e.g. cereal grains and root tubers) and organs (e.g. cereal straw) which are not consumed. Distribution depends on the number of potential storage sites, their capacity for assimilate and supply of assimilate. In cereals, for example, yield is determined by the number of ears produced per unit ground area, the number of grains per ear and the mass per grain. Conditions which prevent grain or ear formation (nutrient or water stress) decrease yield irrespective of photosynthesis although there are generally some related effects of stress as the parts of the system are very closely integrated. Crop yield is then limited by the storage capacity for assimilate (often called the 'sink' capacity) under some conditions and the rate of production of assimilate ('source' of supply) in others. The relation of assimilation to yield and dry matter production depends on environment and plant characteristics and many processes, which occur at different times and respond in different ways to conditions, are involved. Photosynthesis is the main driving force for plant growth, productivity and yield but it is not directly proportional to them.

Present crop yields, worldwide, are much smaller than potential, because of poor nutrition, drought and poor varieties. Better husbandry would increase the food supply for a rapidly increasing world population without increasing intrinsic, or biochemically determined, photosynthetic efficiency. Plant breeding and selection have increased yields largely by improving harvest index (= yield of harvestable material/total dry matter produced). Also, selection of greater yielding crops appears to have decreased photosynthetic efficiency per unit leaf area. Thus modern cereals have greater leaf area but smaller P_n than older varieties so that total productivity is similar under comparable nutrition, but harvest index is larger. Further improvements in yield may come from selection for larger harvest index but there are limits to the reductions that can be made in support and assimilatory organs. In the longer term, greater photosynthetic efficiency is needed. Genetic manipulation of enzymes (e.g. RuBP carboxylase) or of processes (e.g. photophosphorylation) which may limit assimilation, may lead to higher efficiency and therefore to increased yield. Increased efficiency of production is, however, unlikely to result from changes in a single metabolic process in P_n and alterations to other, distantly related processes will be necessary. Thus it is not yet established if inhibition of glycolate pathway metabolism would increase production of assimilates but make C3 crop plants more susceptible to photoinhibition under stress conditions. Sink capacity must also increase in proportion to the increased production if increased photosynthetic efficiency is to be fully utilized.

References and Further Reading

Christy, A. L. and **Porter, C. A.** (1982) Canopy photosynthesis and yield in soybean, pp. 499–511 in Govindjee (ed.), *Photosynthesis*, vol. II, *Development, Carbon Metabolism and Plant Productivity*, Academic Press, New York.

Elmore, C. D. (1980) The paradox of no correlation between leaf photosynthetic rate and crop yields, pp. 155–67 in Hesketh, J. D. and Jones, J. W. (eds), *Predicting Photosynthesis for Ecosystem Models*, CRC Press Inc., Boca Raton, Florida.

Gifford, R. M. and **Evans, L. T.** (1981) Photosynthesis, carbon partitioning and yield, *A. Rev. Plant Physiol.*, **32**, 485–509.

Gifford, R. M. and **Jenkins, C. L. D.** (1982) Prospects of applying knowledge of photosynthesis toward improving crop production, pp. 419–57 in Govindjee (ed.), *Photosynthesis*, vol. II, *Development, Carbon Metabolism and Plant Productivity*, Academic Press, New York.

Gifford, R. M., Thorne, J. H., Hitz, W. D. and **Giaquinta, R. T.** (1984) Crop productivity and photoassimilate partitioning, *Science*, **225**, 801–8.

Legg, B. J., Day, W., Lawlor, D. W. and **Parkinson, K. J.** (1979) The effects of drought on barley growth, models and measurements showing the relative importance of leaf area and photosynthetic rate, *J. Agric. Sci., Camb.*, **92**, 703–16.

Monteith, J. L. (1977) Climate and the efficiency of crop production in Britain, *Phil. Trans. R. Soc., Lond. B*, **281**, 277–94.

Nasyrov, Y. S. (1978) Genetic control of photosynthesis and improving of crop productivity, *A. Rev. Plant Physiol.*, **29**, 215–37.

Portis, A. R. Jr. (1982) Introduction to photosynthesis: Carbon assimilation and plant productivity, pp. 1–12 in Govindjee (ed.), *Photosynthesis*, vol. II, *Development, Carbon Metabolism and Plant Productivity*, Academic Press, New York.

Thornley, J. H. M. (1976) *Mathematical Models in Plant Physiology*, Academic Press, London.

Zelitch, I. (1982) The close relationship between net photosynthesis and crop yield, *Bioscience*, **32**, 796–802.

Index